AGRONOMIC RICE PRACTICES AND POSTHARVEST PROCESSING

Production and Quality Improvement

AGRONOMIC RICE PRACTICES AND POSTHARVEST PROCESSING

Production and Quality Improvement

Edited by

Deepak Kumar Verma
Prem Prakash Srivastav
Altafhusain B. Nadaf

Apple Academic Press Inc. | Apple Academic Press Inc.
3333 Mistwell Crescent | 9 Spinnaker Way
Oakville, ON L6L 0A2 Canada | Waretown, NJ 08758 USA

First issued in paperback 2021

Exclusive worldwide distribution by CRC Press, a member of Taylor & Francis Group

No claim to original U.S. Government works

ISBN 13: 978-1-77463-401-1 (pbk)
ISBN 13: 978-1-77188-712-0 (hbk)

CIP data on file with Canada Library and Archives

Library of Congress Cataloging-in-Publication Data

Names: Verma, Deepak Kumar, 1986- editor. | Srivastav, Prem Prakash, editor. | Nadaf, Altafhusain B., editor.

Title: Agronomic rice practices and postharvest processing : production and quality improvement / editors: Deepak Kumar Verma, Prem Prakash Srivastav, Altafhusain B. Nadaf.

Description: Waretown, NJ : Apple Academic Press, 2019. | Includes bibliographical references and index.

Identifiers: LCCN 2018047039 (print) | LCCN 2018050431 (ebook) | ISBN 9780429488580 (ebook) | ISBN 9781771887120 (hardcover : alk. paper)

Subjects: LCSH: Rice.

Classification: LCC SB191.R5 (ebook) | LCC SB191.R5 A655 2019 (print) | DDC 633.1/8--dc23

LC record available at https://lccn.loc.gov/2018047039

Apple Academic Press also publishes its books in a variety of electronic formats. Some content that appears in print may not be available in electronic format. For information about Apple Academic Press products, visit our website at **www.appleacademicpress.com** and the CRC Press website at **www.crcpress.com**

CONTENTS

ABOUT THE EDITOR

Deepak Kumar Verma is an agricultural science professional and is currently PhD Research Scholar with specialization in Food Processing Engineering at the Agricultural and Food Engineering Department, Indian Institute of Technology, Kharagpur, West Bengal, India. In 2012, he received a *DST-INSPIRE Fellowship* for PhD study by the Department of Science & Technology (DST), Ministry of Science and Technology, Government of India. Mr. Verma is currently assigned for research on "Isolation and Characterization of Aroma Volatile and Flavoring Compounds from Aromatic and Non-aromatic Rice Cultivars of India," whereas during master's degree, his research was assigned entitled "Physico-chemical and Cooking Characteristics of Azad Basmati (CSAR 839-3): A Newly Evolved Variety of Basmati Rice (*Oryza sativa* L.)." He earned his BSc degree in Agricultural Science in 2009 from the Faculty of Agriculture, Gorakhpur University, Gorakhpur, and MSc (Agriculture) in Agricultural Biochemistry in 2011 with first rank and also received a department topper award from the Department of Agricultural Biochemistry, Chandra Shekhar Azad University of Agricultural and Technology, Kanpur, India. Apart from his area of specialization as plant biochemistry, he has also built up a sound background in *plant physiology, microbiology, plant pathology, genetics and plant breeding, plant biotechnology and genetic engineering, seed science and technology, food science and technology,* etc. In addition, he is member of different professional bodies, and his activities and accomplishments include conferences, seminar, workshop, training, and also the publication of research articles, books, and book chapters.

**Prem Prakash Srivastav, PhD,** is Associate Professor of Food Science and Technology in the Agricultural and Food Engineering Department, Indian Institute of Technology, Kharagpur, West Bengal, India. He has graduated from Gorakhpur University, Gorakhpur, and received his MSc degree with a major in Food Technology and a minor in Process Engineering from G. B. Pant University of Agriculture and Technology, Pantnagar, India. He was awarded a PhD from the Indian Institute of Technology, Kharagpur. He teaches various undergraduate-, postgraduate-, and PhD-level courses and has guided many research projects at the PhD, master's, and undergraduate levels. His research interests include development of specially designed convenience, functional, and therapeutic foods; extraction of nutraceuticals; and development of various low-cost food-processing machineries. He has organized many sponsored short-term courses and completed sponsored research projects and consultancies. He has published various research papers in peer-reviewed international and national journals, and proceedings, and many technical bulletins and monographs as well. Other publications include books and book chapters along with many patents. He has attended, chaired, and presented various papers at international and national conferences and delivered many invited lectures at various summer/winter schools. He has received best poster paper awards from ISAE, 2009; ICTF, 2010; FOQSAT, 2011; and IFT (USA), 2014. He is life member of various professional bodies, namely, ISTE, AFST(I), IDA, and AMI, and is a member of the American Society of Agricultural and Biological Engineers and the Institute of Food Technologists (USA).

Altafhusain Nadaf, PhD, is working as an Associate Professor at the Department of Botany, Savitribai Phule Pune University, Pune, in the area of biochemistry and molecular genetics of scented rice for past 13 years. He was awarded an Erasmus Mundus Action 2 India4EU II scholarship to visit the University of Bologna, Italy, as a Visiting Professor (2014) and a DST-BOYSCAST Fellowship to work as a Visiting Fellow at the Centre for Plant Conservation Genetics, Southern Cross University, Lismore, Australia, for one year in 2010–2011. He has received research grants from several funding agencies, including the DST-Fast Track Scheme for Young Scientists and the DBT-Rapid Grant for Young Investigators (RGYI). He has successfully guided several PhD and MPhil students and women scientists (under DST-Women Scientists Scheme: A & B). He is working as a reviewer for many international journals. He has in his credit two books and more than 50 research papers published in peer-reviewed national and international journals of repute. He has presented his research work at national and international level. His research interests are biochemistry and molecular genetics of rice aroma volatiles. He has characterized scented rice varieties of India and other plant species containing rice aroma volatiles following HS-SPME-GC-MS/FID approach. At the molecular level, he has also characterized the betaine aldehyde dehydrogenase2 (*badh2*) gene in the plant species having 2-acetyl-1-pyroline (2AP) expression. In addition, he has revised Indian Pandanaceae thoroughly with the contribution of three new species and assessed phylogenetic relationship among the taxa using cpDNA regions. Some of the species are now being characterized for salt-tolerant genes and bioactive compounds.

CONTRIBUTORS

Umair Ashraf
Research Scholar, Department of Crop Science and Technology, College of Agriculture,
South China Agricultural University, Guangzhou 510642, PR China
Scientific Observing and Experimental Station of Crop cultivation in South China,
Ministry of Agriculture PR China, Guangzhou 510642, PR China.
E-mail: umairashraf2056@gmail.com

Arnab Banerjee
Assistant Professor, University Teaching Department, Department of Environmental Science,
Sarguja Vishwavidyalaya, Ambikapur 497001, Chhattisgarh, India. E-mail: arnabenvsc@yahoo.co.in

Zahoor A. Bhat
Assistant Professor/Scientist, Division of Genetics and Plant Breeding, Sher-e-Kashmir University of
Agricultural Sciences and Technology of Kashmir, Shalimar, Srinagar 191121, Jammu and Kashmir,
India. E-mail: zahoor.bhat@gmail.com

Sudhanshi Billoria
Research Scholar, Department of Agricultural and Food Engineering, Indian Institute of Technology,
Kharagpur 721302, West Bengal, India. E-mail: sudharihant@gmail.com

Ashaq Hussain
Associate Professor (Agronomy), Faculty of Agriculture, Sher-e-Kashmir University of
Agricultural Sciences and Technology, Wadura, Sopore 193201, Jammu and Kashmir, India.
E-mail: ahshah71@gmail.com

Manoj Kumar Jhariya
Assistant Professor, University Teaching Department, Department of Farm Forestry,
Sarguja Vishwavidyalaya, Ambikapur 497001, Chhattisgarh, India. E-mail: manu9589@gmail.com

Gazala H. Khan
Senior Research Fellow, Centre for Plant Biotechnology, Sher-e-Kashmir University of Agricultural
Sciences and Technology of Kashmir, Shalimar, Srinagar 191121, Jammu and Kashmir, India.
E-mail: moazin_khan@yahoo.co.in

Aabid Hussain Lone
Senior Research Fellow, Faculty of Agriculture, Sher-e-Kashmir University of Agricultural Sciences
and Technology, Wadura, Sopore 193201, Jammu and Kashmir, India. E-mail:aabidlone08@gmail.com

Dipendra Kumar Mahato
Senior Research Fellow, Indian Agricultural Research Institute, Pusa Campus, New Delhi 110012,
India. E-mail: kumar.dipendra2@gmail.com

Zhaowen Mo
Associate Professor, Department of Crop Science and Technology, College of Agriculture,
South China Agricultural University, Guangzhou 510642, PR China
Scientific Observing and Experimental Station of Crop cultivation in South China,
Ministry of Agriculture PR China, Guangzhou 510642, PR China.
E-mail: scaumozhw@126.com, zwmo@scau.edu.cn

Mukesh Mohan
Associate Professor, Department of Agricultural Biochemistry, College of Agriculture,
Chandra Shekhar Azad University of Agriculture and Technology, Kanpur 208002, Uttar Pradesh,
India. E-mail: drmukeshmohan@rediffmail.com

S. Najeeb
Associate Professor/Senior Scientist, Mountain Research Centre for Field Crops,
Sher-e-Kashmir University of Agricultural Sciences and Technology of Kashmir, Shalimar,
Srinagar 191121, Jammu and Kashmir, India. E-mail: najeeb_sofi@rediffmail.com

G. A. Parray
Associate Director Research, Mountain Research Centre for Field Crops, Sher-e-Kashmir University
of Agricultural Sciences and Technology of Kashmir, Shalimar, Srinagar 191121, Jammu and Kashmir,
India. E-mail: parray_2005@rediffmail.com

Parmeshwar Kumar Sahu
Research Scholar, Department of Genetics and Plant Breeding, Indira Gandhi Krishi Vishwavidyalaya,
Raipur 492012, Chhattisgarh, India. E-mail: parmeshwarsahu1210@gmail.com

Manish Kumar Sharma
Research Scholar and RAEO, Department of Agronomy, Indira Gandhi Krishi Vishwavidyalaya,
Raipur 492012, Chhattisgarh, India. E-mail: mksharma003@gmail.com

Asif B. Shikari
Associate Professor/Senior Scientist, Centre for Plant Biotechnology, Sher-e-Kashmir University of
Agricultural Sciences and Technology of Kashmir, Shalimar, Srinagar 191121, Jammu and Kashmir,
India. E-mail: asifshikari@gmail.com

Parmeet Singh
Assistant Professor (Agronomy), Faculty of Agriculture, Sher-e-Kashmir University of
Agricultural Sciences and Technology, Wadura, Sopore 193201, Jammu and Kashmir, India.
E-mail: parmeetagron@gmail.com

Ram Kumar Singh
Professor, Department of Agronomy, Institute of Agricultural Sciences, Banaras Hindu University,
Varanasi 221005, Uttar Pradesh, India. E-mail: rksingh_agro@rediffmail.com

Prem Prakash Srivastav
Associate Professor, Department of Agricultural and Food Engineering, Indian Institute of Technology,
Kharagpur 721302, West Bengal, India. E-mail: pps@agfe.iitkgp.ernet.in

Xiangru Tang
Professor, Department of Crop Science and Technology, College of Agriculture,
South China Agricultural University, Guangzhou 510642, PR China
Scientific Observing and Experimental Station of Crop Cultivation in South China,
Ministry of Agriculture PR China, Guangzhou 510642, PR China. E-mail: tangxr@scau.edu.cn

Mamta Thakur
Research Scholar, Department of Food Engineering and Technology, SLIET Longowal 148106 Punjab,
India. E-mail: thakurmamtafoodtech@gmail.com

Deepak Kumar Verma
Research Scholar, Department of Agricultural and Food Engineering, Indian Institute of Technology,
Kharagpur 721302, West Bengal, India.
E-mail: deepak.verma@agfe.iitkgp.ernet.in; rajadkv@rediffmail.com

Gaurav S. K. Verma
Research Scholar, Department of Agronomy, Institute of Agricultural Sciences,
Banaras Hindu University, Varanasi 221005, Uttar Pradesh, India. E-mail: gauraviasbhu@gmail.com

V. K. Verma
Research Scholar, Department of Agronomy, Institute of Agricultural Sciences,
Banaras Hindu University, Varanasi 221005, Uttar Pradesh, India. E-mail: vermaagribhu@gmail.com

Shabir H. Wani
Assistant Professor/Scientist, Division of Genetics and Plant Breeding, Sher-e-Kashmir University of
Agricultural Sciences and Technology of Kashmir, Shalimar, Srinagar 191121, Jammu and Kashmir,
India. E-mail: shabirhussainwani@gmail.com

Shafiq A. Wani
Professor, Centre for Plant Biotechnology, Sher-e-Kashmir University of Agricultural Sciences and
Technology of Kashmir, Shalimar, Srinagar 191121, Jammu and Kashmir, India.
E-mail: shafiqawani@gmail.com

Dhiraj Kumar Yadav
University Teaching Department, Department of Farm Forestry, Sarguja Vishwavidyalaya,
Ambikapur 497001, Chhattisgarh, India. E-mail: dheeraj_forestry@yahoo.com

ABBREVIATIONS

AAs	amino acids
AAS	atomic absorption spectrophotometer
AC	amylose content
AE	agronomic efficiency
AMF	arbuscular mycorrhizal fungi
ANFs	antinutritional factors
ANN	artificial neural network
ASEAN	Association of South East Asian Nations
ASV	alkali spreading value
ATS	Association of Tefey Saina
AWD	alternate wetting and drying
B:C ratio	benefit to cost ratio
BD	bulk density
BGA	blue green algae
BILs	backcross inbred lines
BLB	bacterial leaf blight
BMP	best management practices
BOD	biological oxygen demand
BRRI	Bangladesh Rice Research Institute
BV	biological value
CA	conservation agriculture
CAR	Cordillera Administrative Region
CF	continuous flooding
CF	crude fiber
CGR	crop growth rate
CHO	carbohydrate
CMS	cytoplasmic male sterility
CNRMBN	China National Rice Molecular Breeding Network
COD	chemical oxygen demand
CP	crude protein
CRA	climate resilient agriculture
CRRI	Central Rice Research Institute
CSA	climate smart agriculture
CT	conventional tillage
CT-I	conventional transplanted rice with flooded irrigation systems
CT-II	conventional transplanted rice with intermittent irrigation systems
CTR	conventional transplanted rice

CTW	conventional tilled wheat
DADPW	days after disappearance of ponded water
DAP	diammonium phosphate
DAS	days after sowing
DAT	days after treatment
DAT	day after transplanting
Db	dry weight basis
DDS	dry-direct-seeded
DEFRA	Department for Environment, Food and Rural Affairs
DGP	donor gene pool
DMP	dry matter production
DOM	degree of milling
DS	dry seeded
DS	direct seeding
DSR	direct-seeded rice
D-SR	dry-seeded rice
EAAs	essential amino acids
EC	electrical conductivity
EGP	elite gene pool
Eh	redox potential
EPoE	early post emergence
eQTL	expression quantitative trait loci
ES	embryo-sac
ET	evapotranspiration
FAME	fatty acid methyl ester
FAO	Food and Agriculture Organization
FAOUN	Food and Agriculture Organization of the United Nations
FBD	fluidized bed drying
FE	food energy
FID	flame ionization detector
FSS	ferrous sulfate spray
FSSAI	Food Safety and Standards Authority of India
FYM	farmyard manure
GBS	genotyping by sequencing
GC	gas chromatography
GCMs	general circulation models
GFDL	General Fluid Dynamics Laboratory
GHG	greenhouse gases
GISS	Goddard Institute of Space Studies
GLC	gas liquid chromatography
GM	green manure
Gn1a	grain number 1a
GPE	grain production efficiency

GSMs	general circulation models
GWAS	genome-wide association studies
HA	hot air
HDT	high day temperature
HI	harvest index
HNT	high night temperature
HRR	head rice recovery
HRY	head rice yield
HYV	high-yielding varieties
IAA	indole acetic acid
ICM	integrated crop and resource management
ICRISAT	International Crops Research Institute for the Semi-Arid Tropics
IE	internal efficiency
IF-V	intermittent flooding through the vegetative stage
IGP	Indo-Gangetic plains
INM	integrated nutrient management
IPCC	Intergovernmental Panel on Climate Change
IPM	integrated pest management
IPNS	Integrated Plant Nutrition System
IR	infrared
IRRC	Irrigated Rice Research Consortium
IRRI	International Rice Research Institute
ISU	irrigation service units
IWD	intermittent wet and dry
IWM	integrated weed management
IYR	International Year of Rice
KLaC	kernel length after cooking
LAI	leaf area index
LDL	low-density lipids
LNP	Leb Nok Pattani
MAGIC	multiparent advanced generation intercross
MAS	marker-assisted selection
MC	moisture content
MOP	muriate of potash
MOST	Ministry of Science and Technology
mQTL	metabolomic quantitative trait loci
MS	monosaccharides
MSRI	modified system of rice intensification
MT	mechanical transplanting
MW	microwave
NAM	nested association mapping
NARES	National Agricultural Research and Extension Systems
NARS	National Agricultural Research Systems

NIL	near-isogenic line
NPR	net photosynthetic rate
NPS	nonpoint source
NUE	nitrogen use efficiency
Os CDPK	Oryza sativa calcium-dependent protein-kinase
Os MAP	Oryza sativa mitogen-activated protein
OsDREB	Oryza sativa dehydration-responsive element binding
PBDSR	permanent bed direct-seeded rice
PBW	permanent bed wheat
PE	physiological efficiency
PER	protein efficiency ratio
PM	poultry manure
PN	photosynthetic produce
PSB	phosphate-solubilizing bacteria
PU	prilled urea
PUSFAs	polyunsaturated fatty acids
QTL	quantitative trait loci
RBD	randomized block design
RD	recommended dose
RDF	recommended dose of fertilizer
RDN	recommended dose of nitrogen
RDW	respiratory dry weight
REY	rice-equivalent yield
RH	relative humidity
RIL	recombinant inbred line
RKN	root knot nematodes
RLD	root length density
RPR	root pulling resistance
RPT	row paddy transplanter
RSR	recommended seed rate
SalTol	salinity tolerance
SC	seedling casting
SCI	system of crop intensification
SFAs	saturated fatty acids
SIMRIW	simulation model for rice and weather relationship
SLC	surface lipid content
SOC	soil organic carbon
SP 1	suphanburi 1
SRES	special report on emission scenario
SRI	system of rice intensification
SSNM	site-specific nutrient management
SSP	single superphosphate
STR	stress tolerant rice cultivars

TGFC	temperature gradient field chamber
TNAU	Tamil Nadu Agriculture University
TP	transplanting
TP	total phosphorus
TPR	transplanted rice
TR	transplanted rice
UKMO	United Kingdom Meteorological Office
UNGA	United Nations General Assembly
USFAs	unsaturated fatty acids
VAR	vector autoregression
VER	volume expansion ratio
WDS	wet direct-seeded
WHO	World Health Organization
WPET	evapotranspiration from rice area
WPET+E	evapotranspiration from rice area + evaporation from nonrice area
WPI	water productivity under irrigation
WPI+R	total water input
WS	wet seeding
WSR	wet-seeded rice
WU	water uptake
WUE	water use efficiency
ZEU	Zn-enriched urea
ZnHI	zinc harvest index
ZTDSR	zero-tilled direct-seeded rice
ZTW	zero-tilled wheat

PREFACE

Rice is the staple food of around three billion people, most of them in Asia, which accounts for 90% of global rice consumption. Rice constitutes a major source of nutrition and contributes a significant share of dietary energy in a number of Asian countries. Among 23 species of the genus *Oryza*, *Oryza sativa* L. is cultivated in Asia and *O. glaberrima* Steud. in West Africa. *O. sativa* L. is further differentiated into *indica* and *japonica*. Rice has immense diversity, and it is estimated that more than 100,000 varieties of rice exist in the world. India has an ancient heritage of rice cultivation and has over 70,000 cultivars of rice germplasm.

Agronomic practices encompass many areas of conservation—from practicing reduced-tillage methods, which lessen the need to till the soil before each crop, to managing planting populations, which ensures crops are not over- or undercrowded, and therefore are in optimal growing conditions. These small changes in farmers' routines can yield major dividends at harvest. The agronomic practices majorly include using appropriate seeding rates and fertilizer, keeping narrow rows, using hybrid maturities, and applying starter fertilizer doses.

After using effective agronomic practices, proper postharvest processing and handling is an important part of modern agricultural production. Postharvest processes include the integrated functions of harvesting, cleaning, grading, cooling, storing, packing, and transport. Postharvest technologies occur between the producer and the consumer—processes protect produce to preserve quality, reduce damage, travel distances, grade and categorize, document sources, and label. Postharvest handling involves the practical application of engineering principles and knowledge of fruit and vegetable physiology to solve problems. Therefore, it needs a closer coordination between all segments of the industry from the grower to the consumer, with great emphasis on proper postharvest handling, a multidisciplinary and systems approach to problem-solving, a greater use of computer control and communications technology, and a greatly renewed emphasis on mechanization.

The increased grain yield and improved rice quality are absolutely necessary to feed the world's galloping population and to maintain its health and nutrition. Thus, the recent approach for rice production includes the improvement of both yield and grain quality to cater for consumer demand and also to increase the nutritional level of the general public.

Presently, the India is producing more surplus than is needed nationally. Improvement in quality provides assurance that the surplus will find a rewarding market. The genetic makeup of grain is the major factor influencing the quality of rice. Modern programers continually strive to refine and improve the genetic characteristics that influence quality in the most desirable product. Another factor is the environment under which the plant is grown, such as the light rainfall and temperature. Similarly, soil and the management practices affect the grain quality. Careful harvesting and postharvest handling may maintain or even improve the rice quality. Quality characteristics in rice may be categorized into three broad areas. (1) Physical characteristics include moisture content, shape, size, whiteness, translucency, chalkiness, head rice, broken rice, brewers, green kernels, and yellow kernels. (2) The analysis of physicochemical characteristics of rice include amylose content, protein content, gel consistency, volume of expansion of cooked rice, water absorption, and cooking time. (3) The organoleptic properties of cooked rice include color, aroma, hardness, stickiness, and consistency.

This book, *Agronomic Rice Practices and Postharvest Processing: Production and Quality Improvement,* addresses these three important aspects of rice. For convenience of the readers, the book has been divided into four parts. *Part 1* takes into account *Recent Trends and Advances for Higher Production and Quality Improvement* and consists of four chapters covering climate-resilient agriculture practices in rice through an Indian perspective; climatic effects, relative performance, and constraints in direct-seeded versus transplanted rice of Asia; constraints in temperate rice culture and interventions to mitigate the challenges; and recent advances and trends in system of rice intensification. *Part 2* covers *Nutrient Management for Rice Production and Quality Improvement,* which consists of three chapters. Effect of zinc on growth, yield, and quality attributes of rice for improved rice production; integrated nutrient management in transplanted rice by pelleting technique; and effect of different organic sources of nutrition on growth, yield, and quality of rice are discussed in detail. *Part 3* addresses *Weed Management for Improved Rice Production* through a chapter describing precautions and future implications in weed management and integrated weed management in zero-till direct-seeded rice. *Part 4* deals with *Postharvest Processing for Rice Quality Improvement.* It includes two chapters discussing, effect of parboiling on different physicochemical and cooking properties of rice and nutritional quality evolution in rice.

The book has taken the present shape due to the contributions of renowned scientists, researchers, and professors through their tireless research. We are sure that this book will be a useful guide for the rice researchers working in the area of agronomic practices, postharvest processing, and quality improvement in rice.

Deepak Kumar Verma
Prem Prakash Srivastav
Altafhusain B. Nadaf
Editors

PART 1

Recent Trends and Advances for Higher Production and Quality Improvement

CHAPTER 1

CLIMATE-RESILIENT AGRICULTURE PRACTICES IN RICE: AN INDIAN PERSPECTIVE

ARNAB BANERJEE[1*] and MANOJ KUMAR JHARIYA[2]

[1]University Teaching Department, Department of Environmental Science, Sarguja Vishwavidyalaya, Ambikapur 497001, Chhattisgarh, India

[2]University Teaching Department, Department of Farm Forestry, Sarguja Vishwavidyalaya, Ambikapur 497001, Chhattisgarh, India

*Corresponding author. E-mail: arnabenvsc@yahoo.co.in

ABSTRACT

Agriculture is the major occupation of people residing in the developing countries like India where climatic changes are a serious concern. Increase in temperature, changes in the rainfall patterns, and increased frequency of extreme climatic conditions such as floods and severe droughts are being common in different agroecological zones of India, thus posing a threat to the India's food security. To deal this, the climate-resilient farming is the best option which aims to lower the impact of agriculture on climate change and to increase the resiliency of farms on account of a changing climate. The production of rice in India is concentrated to the northeastern part where the increased precipitation and submergences of rice plots in addition to water stress condition cause significant damage in every cropping period, whereas the crop production is being affected by the drought in the upland areas. Therefore, there is a strong need of appropriate strategies and recent advanced agricultural technology for the betterment of rice farmers of the country. Climate-resilient farming is one such technique for improving the economic and environmental security of poor people.

1.1 INTRODUCTION

The world is blessed with diverse resources for efficient utilization and benefit for mankind. Due to overexploitive nature of human being, they are now facing severe problems due to changing environmental/climatic events. Agriculture is the backbone for most of the Third World nations including India. Interestingly, agriculture supports 60% of Indian population as livelihood. Climatic change is a global phenomenon having its severe impact on developing countries due to lack of adequate technology and expertise along with lesser resources to cope up with climate change phenomenon. Climate change is itself a big challenge from food security perspective imposing tremendous pressure on boosting up agricultural productivity. Therefore, climate resilience is a dwindling phenomenon under high seasonal variability of climatic events. Vulnerability refers to the problems associated with human wellbeing and development (Kasperson et al., 2003). Therefore, assessing the vulnerability toward climate-sensitive perspectives requires considerable importance.

1.2   RICE PRODUCTION TRENDS IN INDIA

Among different cereal crops, Rice is one of predominant food crops of India, representing one-fourth of gross cropped area of the country under rice cultivation along with nearly half of total food grain and cereal production in Indian agriculture (Verma et al., 2012, 2013, 2015; Verma and Srivastav, 2017). Rice producing potential seems to have increased considerably roughly four times during last 60 years. At present, India is reported as lower level of productivity of rice than the global average (2.7 t/ha). In comparison to other countries, rice production in India is considerably higher than other developing nations.

Rice farming happens to be the basic livelihood for most of the people in developing countries such as India, Nepal, Bhutan, China, and Bangladesh under Asian subcontinent. From world's perspectives, Latin American countries were not very much upgraded in terms of rice production. In rice production, considerable variation was found in terms of dryland and wetland farming as well as agroecological climatic variation (Nguyen, 2002). According to Varma (2017), the rice cultivated area has been increased to 43 million ha in the year 2014 from 40 million ha in 1980.

As per average trend in rice area, largest area coverage was attributed toward eastern and central east part of India. It was observed that a dramatic decline in the area coverage was recorded from various 5-year plans, especially, in southern states of India. Among the states in India, Assam and Odisha reflected positive trend in terms of productivity. Gradual improvement in agricultural sectors has promoted all-round development in rice farming systems throughout Indian subcontinent. Higher productivity under agriculture sector is strongly associated with cultivation of high-yielding varieties (HYV) coupled with improved scientific technology.

1.3 RICE PRODUCTION AND ITS PROBLEMS

Different types and nature of problems are associated with rice production in different states and area of India. The maximum rice-growing area happens to be the eastern part of India. This zone is accompanied by increased precipitation and submergences of agricultural plots under rice cultivation along with water stress condition in every cropping period resulting significant damage to agriculture. Drought causes considerable setback to crop production in upland areas.

The soil substratum shows inertness toward nutrient application in some cases. Lack of HYV along with quality acts as constraint factors toward rice production in certain areas. The problem is further aggravated by pest outbreak along with gap in extension programs from lab to land conditions.

1.4 CLIMATE CHANGE AND AGRICULTURAL OUTPUT

Agriculture promotes economic as well as social development in developing countries; vulnerability of agriculture from climate change perspective has drawn attention of scientific community throughout the world (Fischer et al., 2002; Kurukulasuriya and Rosenthal, 2003; IISD, 2003). Reported literature reflects the ill effect of changing climates upon tropics although causing minor increases in crop yields on short-term basis in a few areas (Maddison et al., 2007). In agriculture system, the various components including soil substratum and its associated processes along with outbreak of harmful biological agents happen to be the significant factor affecting different agricultural processes. These effects are reflected through various abiotic stresses with considerable variations.

1.5 CHALLENGES TOWARD ADAPTATION FOR CLIMATE CHANGE

Biotic communities by their own have specific defense mechanism to resist changes occurring in today's climate by inherent scientific and traditional knowledge. From today's perspectives, adaptation due to rapid changes in various climatic elements is incapable of self-regulation. The recent processes need the scientific tactics having governmental support so that the variations and probable shifting of climatic elements can be minimized thus fulfilling basic needs of farmer.

Moreover, the immediate implementation of such tactics is also a concern which must be easy to adapt without considerable diversification and build on existing knowledge and practices. Managing tactics followed by farmers rely toward farming under changing scenario of climate which includes agricultural entrepreneurship toward secured earnings, use of existing varieties with less divergence, broadening farm business to diminish threat, additional financial support, pledging, marketing, and promotions of agricultural commodities (Chambers, 1989). Under vulnerability with surprising deviations of climatic elements on farming, appropriate strategies are necessary to combat ill effect of alteration in the climate.

Resiliency in climate change means integration toward sustainable agricultural practices against various climate extremes. Such perturbations and instabilities may comprise events such as drought, flooding, heat/cold wave, inconsistent precipitation, extensive dry spells, and biological and climatic threats. Climate resilient agriculture (CRA) executes cautious and upgraded management regimes of natural resources by implementation of suitable strategies. For developing nations like India, now major challenge is to feed 17.5% of the global population with meager proportion of agricultural resources. Depleting nature of agricultural resource base accompanied through global perturbations results in unsuitable cropping environment. A road toward orienting sustainable agricultural practices under climatic extremes can be supported by appropriate technologies, laboratory to land programs, and extension activities among farming communities.

1.6 CLIMATIC VARIABILITY AND INDIAN FARMING SYSTEMS

Variability of climate phenomenon happens to be influencing the agriculture sectors both directly and indirectly. The nature and extent of consequence may fluctuate as per degree and level of climate change, geographical location, and nature of production. Quantification of climatic variability trend and its impacts required precise investigation and agroecological modeling. As per various reports, focus was given to the local situations in terms of predicting climatic circumstances. The focal points include:

1. Variability in crop output.
2. Use of alternative production systems such as integrated nutrient management.
3. Environmental influences mainly associated with various edaphic factors as well as homogenization of crop architecture.

1.7 IMPACTS ASSOCIATED WITH FARM OUTPUT

Reduction of duration of crops associated with changes in temperature regimes. Thermal increment boosts up early ripeness of various agricultural crops. In yearly cropping system, duration of crop maturity reflects significant reductions up to 15–21 days, hampering farm outputs. Thermal variability alters the reproductive biology of various crops (wheat, paddy, sunflower, etc.). Additionally, abnormal climatic influences such as flood and drought lead to disease and pathogen outbreaks among crops under agriculture system. Agroecological modeling suggests decline in agricultural productivity at its various spheres especially on paddy, wheat, and maize. In other crops (legumes and oilseeds), the effect seems to be little bit altered.

1.7.1 RICE

Indian monsoon has now become progressively surprising and unpredictable. The number of concentrated rainfall organized with considerable reduction in monsoonal period since last half a century. As a result, combination of climatic extremes has promoted the cultivation of paddy under wet condition. Rainfed farming is severely affected due to such alteration in the climate. The event of dry spell has pronounced influence over extra precipitation on rice farming (Auffhammer et al., 2012). As per future predictions, rice production is expected to decrease to a considerable level in the upcoming days in developing nations such as India (Soora et al., 2013).

1.7.1.1 ELEVATED CO_2 AND RICE GROWTH

Amplified atmospheric CO_2 concentrations can be considered as significant factors promoting a global perturbation which adversely affect the yield attributes of different crop species in agriculture as CO_2 and photosynthesis are very much interrelated (Watanabe and Kume, 2009). In case of paddy,

C3 metabolism leads to incorporation of CO_2 as plant biomass. C4 mode of biochemical process is much more effective in comparison to C3 due to presence of *Rubisco* enzyme and its affection toward O_2 and CO_2. High level of CO_2 promotes carboxylation level resulting into acceleration of photosynthetic rate of C3 plants. Elevated CO_2 usually raise number of yield and productivity parameters in rice (Kim et al., 2001; Cheng et al., 2001). Issues accredited to affect the susceptibility of development of paddy include several agroecological factors. Nevertheless, the scientific exploration has yet to reach its ultimate points of acceptability (Zhong et al., 2004).

High level of CO_2 enhanced paddy growth, biomass yield (15–30%), and photosynthetic rate (30–40%), which is subjected to genotype and environmental attributes. Elevated CO_2 may have negligible impact on nutrient cycling in agroecosystem. It is supposed to be linked with leaf attributes area growth toward CO_2 level. The estimated deviations in temperature and CO_2 have shown contrasting influence on production. Rise in temperatures cut the growing season which in turn declined yields, whereas higher CO_2 level improved rice production (Erda et al., 2005). As per Long et al. (2004), when the C3 species such as paddy are grown under elevated CO_2 level, the proportion of photosynthetic produce (PN) is increased by improvement of CO_2 concentration and total hindrance in plant respiration in presence of light. Still, the stimulatory impact of high CO_2 concentration decline progressively due to continuity between CO_2 concentration and paddy crop (Chen et al., 2005). Prolonged duration of higher level of CO_2 results into lesser PN in photosynthetic area in a plant which in turn is comparatively lesser than foliage under ambient conditions of CO_2 concentration level (Xu et al., 1994).

The rice experiments reflect an adjustment toward higher CO_2 level in plants under C3 pathway. Control over photosynthetic process of plants in response to higher CO_2 level is mediated by factors such as proportional reduction of foliage nitrogen followed by other plant physiological process which would inhibit the functionality of stomata which causes minute amount of nitrogen mobilization (Kanemoto et al., 2009). Adjustment of photosynthetic potential, carbon capture, and storage are distinctly more under elevated CO_2 (Leakey et al., 2009). Nevertheless, numerous studies revealed that elevated rate of photosynthesis leads to transformation into bulk volume of paddy yield (Jablonski et al., 2002). Increased rice production indicated significant deviation, because of various factors such as genotypic differences and interfaces among CO_2 and various climatic features (De Costa et al., 2006). Higher CO_2 caused positive impact in physiological processes of rice regulating yield throughout the cropping period. The total photosynthetic rate and its enrichment were found to be significantly higher during the initiation period of crop growth and subsequently declined. Improvement of carbon capture and storage declined with successive stages of paddy (Sakai et al., 2001).

1.7.1.2 ELEVATED CO_2 AND METABOLIC ACTIVITY IN PADDY

During growth and developmental phase of rice crop, the metabolic activity is yet to be understood properly. Research findings reveal respiratory loss of carbon to a significant amount during growth and developmental phase of crop species in tropics (Pritchard and Amthor, 2005). Furthermore, biochemical interactions between environmental factors, water, and nutrients promoting growth and development of crop species is yet to be revealed properly. Thermal regimes of short duration may accelerate crop metabolic activity, while few evidences are available in terms of impact of long-duration metabolic rate on crop species (Pritchard and Amthor, 2005). The metabolic activity in plant system significantly varied with enhancement of CO_2 level. Mixed response of metabolism in plant system has been reported by different workers in earlier time (Griffin et al., 1996; Cheng et al., 2000; Baker et al., 2000; Gielen et al., 2003; Wassmann and Dobermann, 2007; Leakey et al., 2009). In the absence of light, the respiratory activity based on high level of ambient CO_2 has been reported by some workers (Sakai et al., 2001; Uprety et al., 2003). Furthermore, rice respiratory dry weight (RDW, specific respiration coefficient—expressed as biomass of standing crop on dry weight basis) was checked by higher CO_2 level, while coefficient value of aerial coverage of vegetation in absence of light (represented as ground area) were promoted by high level of CO_2 (Baker et al., 1992; Sakai et al., 2001). According to Baker et al. (2000), the negative impact of increased CO_2 concentration on the ground area in short spell was recorded along with the promotive output (in terms of apparent dark respiration rate on temporal basis) during the absence of light. Same impacts of higher level of CO_2 on RDW were reported by other workers (Uprety et al., 2003). Increase in temperature would promote growth and development of crop species and also would be associated with maintaining respiratory function. Higher level of ambient CO_2 were found to be insignificant on RDW perspectives in the foliage of the vegetation which is represented by standing crop biomass on dry weight basis during the growth period, thus reflecting zero influence of nitrogen in foliage part of paddy (Xu et al., 2006). At some specific plant parts, accelerated rate of respiration with enhance level of CO_2 causes significant reduction in carbon assimilation in plants, but higher respiratory activity might promote productivity and yield attributes, therefore, providing more capacity to do work for proper translocation of photosynthate in different plant parts (Leakey et al., 2009).

1.7.1.3 ELEVATED CO_2 AND WATER USE

The variability of climate would likely exert pressures on natural water bodies in addition to rapid population rise, land-use change, and socioeconomic development along with urbanization. Research work projects that there would be a huge reduction in freshwater content in most part of Asia with lesser water availability in river in the upcoming two or three decades. Global warming expedites the distortion of global hydrological cycle, resulting into scarcity of water among the tropical climatic countries. The intensity of rainfall may rise globally in spite of lowering of average level of rainfall as well as interval between precipitation events would promote climatic extreme events (IPCC, 2007).

Changes in the climatic pattern significantly impact the availability and utility of water to crop plants. However, increase in thermal regime may promote evaporative process, but higher CO_2 content inhibits the utilization of water by crop by crop alteration in the crop physiology. Rice production greatly depends on availability of water, and irrigated lowlands comprise more than half area of paddy harvesting which promotes two to three times more crop yield as compared to nonirrigated regions (IRRI, 2002). Pospisilova and Catsky (1999) reported several case studies of utilization of water by crop plants due to enhancement of ambient CO_2 level. According to their observations, rate of photosynthetic process increases more in most cases while reduction in the value of stomatal conductance (gs) and water removal process from vegetation is quite high in most of the cases. As a result, ambient CO_2 increment promotes the capacity of water use in plant system along with lowering of development of water stress condition revealed through plant water potential parameters. gs value reduced significantly due to higher CO_2 level (De Costa et al., 2003; Yoshimoto et al., 2005; Ainsworth, 2008). Various condition of environment imposes significant influence over gs in higher CO_2 level (Ainsworth and Rogers, 2007). As per Shimono et al. (2010), there is a temporal variation of gs value in paddy under higher exposure to CO_2. From the research results, important findings regarding impact of higher level of CO_2 on gs and other parameters under varied condition may give us an idea of possibilities of requirement of water under rainfed farming system of paddy.

Reported research work reveals higher CO_2 level has a negative impact on rate of transpiration and therefore, promotes the plant to be efficient in water use. The gs reduces considerably due to closing of guard cells on leaves. According to the reports of Dafeng et al. (2001), major foliage cover of plants may uptake more gs value as well-nutrient level of vegetative part is reduced under exposure to higher level of CO_2. The growth performance under water stress condition increases manifold under higher level of CO_2 exposure.

1.7.1.4   RICE AND HEAT STRESS

Rice crop is very much susceptible during reproductive growth phase. Water stress and higher level of temperature may be disastrous for paddy farmers. Low temperature imposes more critical effect on rice production in comparison to high temperature. Hence, area for cultivation might increase with higher temperature level. Thus, in tropics, higher temperature leads to lesser reproductive output (Swaminathan, 1984). Gradual increases in temperature have a profound influence on rice spikelet making it infertile. High temperature stimulates infertility in spikelets (Yoshida and Parao, 1976; Kim et al., 1996).

Nearly doubling of CO_2 level reduces the temperature limit for spikelet distortion under high temperature in comparison to the flowering at normal temperature and low level of pollen formation which were key vulnerable phases during the successive period of paddy (Farrell et al., 2006). Furthermore, high thermal stress conditions during anthesis resulted in the development of infertility and give zero seed output (Satake and Yoshida, 1978). This is due to deprived pollen production and dryness in anther, hence least figures of sprouting (Prasad et al., 2006). The distinction in genetic material toward spikelet infertility may also alter by thermal regimes which can be interpreted by wide temperature limits, exemplified by several species (Prasad et al., 2006).

Some workers (Mathauda et al., 2000) reported that increase of growth and yield attributes of paddy during dry spell of a time a correlation were found between productivity and lowest recorded temperature. It was recorded that grain yield goes down to a considerable amount with change in scale of thermal regimes during dry cropping period, while influence of highest temperature on production rate of paddy was found to be insignificant (Peng et al., 2004). Sheehy et al. (2006) validate reduction in yield at specific temperature.

High night temperature (HNT) has no effect on photosynthesis but significantly impacts chlorophyll biosynthesis, nutrient content in foliage, percentage germination, and fertility of paddy floral inflorescence. Additionally, higher temperature during night hinders plant growth by delaying rice panicle emergence. It also affects the yield negatively to a significant level in comparison to normal temperature. The main reason behind changes in productivity and yield of rice is due to improper translocation of biomass in the grains without least interference by photosynthesis (Mohammed and Tarpley, 2009).

HNT in comparison with day temperature hinders the growth and development at various phases of grain filling along with cell size reduction in paddy (Morita et al., 2005). Higher temperature increases the metabolic rate of plants (Amthor, 2000), more maintenance respiration cuts the total assimilates offered to growth and yield (Monteith, 1981). HNT at the stage of reproduction lowers the stimulatory effect of higher CO_2 level on yield of brown rice (Cheng et al., 2009). This was not due to lesser enhancement of dry weight, as a consequence improper allocation of biomass to grain due to lower spikelet's fertility level. The increasing levels of foliage growth with probable lesser temperature influence were the reasons for increasing total biomass by HNT at the time of reproduction. Besides these, secondary factors that give their profound influence in terms of low production include incompatibility between the thermal regime of day and night on vegetative and reproductive growth of crop plants. Therefore, more precise basic exploration is required for assessing the impact of thermal regime on plant physiology leading to higher productivity of paddy during night (Peng et al., 2004).

1.8 POLICIES TOWARD CRA

1.8.1 DEVELOPMENT OF RICE VARIETIES CAPABLE OF COMBATING CLIMATIC EXTREMES

Development of newer varieties of paddy capable of adapting themselves to climate variability reflecting higher level of adjustment to various global climatic phenomena as well as biological processes hampering crop yield and productivity are essential to reduce vulnerability and boost up productivity in rice systems under stress conditions. Research and developmental organizations of international repute has provided several reports on rice varieties adapted to climate change. Strategies to promote these varieties at large scale have been taken; scientific and systematic approach constraints the proper implementation at field level. Absence of agro-industry-based support as well as no proper dimension toward cultivation of rice varieties adapted to climatic variability is a significant problem. Under such circumstances, collaborative efforts are required as follows: (1) economical subsidy toward promoting cultivation of climate adaptive varieties of rice at field level, (2) proper training and development of technical expertise from paddy cultivation perspectives, (3) distribution of scientific knowledge, and (4) development of exchange program between farming community and scientific experts from rice cultivation perspectives.

1.8.2 CLIMATIC VARIABILITY AND AGRICULTURAL PROTECTION

Small-scale farmers in India and other developing countries face severe problem to combat the impact of climatic variability on crop production due to their lower economic conditions and thus, are incapable of implementing advanced technologies at field level. Providing economic support in terms of subsidies to promote advanced technologies at the field level could be an adequate step to combat the present scenario of climate extremes. Such step can boost up agricultural productivity on one hand and on the other may boost up the socioeconomic conditions of the farming community for maintaining their daily livelihood. For developing nations, orientation of paddy farming can be done with due consideration of the climate. Assessing damage cost of the farmer is often misleading due to shortage of time and money. Alternative strategies for damage cost assessment focus more on conditions at the field level. As per the opinion of agro-based countries, different types of insurance scheme providing safety and security for the farmer may give a positive output. Further, implementing such types of schemes required scientific support as well as adequate infrastructure.

1.8.3 SEQUENTIAL MAINTENANCE OF MOISTURE REGIME IN RICE

It is an indigenous technology for rice cultivation practice which involves alternate flooding and subsequent release of water from rice fields during crop production for rice-producing countries in Southeast Asia. Alternate wetting and drying promotes water conservation without hampering the yield

potential of rice. Such strategies have applied impacts on farming community as on one hand lesser water use is taking place and on other, they lead to
energy conservation. Research works reveals that such strategies efficiently
reduces water use up to one-third portion along with reduction of methane
generation up to almost half.

1.8.4   RICE CROPPING CALENDAR FOR SUSTAINABLE PRODUCTION

In Southeast Asia, climate-oriented cultivation practices are a sustainable
approach in all developing countries having agricultural base. Orientation of
sowing time can be regulated as per occurrence of precipitation pattern under
scenario of climatic variability. The criteria for selecting optimal cropping
calendar include (1) rainfall trend analysis, (2) predicting yield potential of
crops, and (3) harmonizing precipitation pattern and crop productivity.

1.8.5   UTILIZATION OF ROW PADDY TRANSPLANTER

Implementation of row paddy transplanter (RPT) was done for timely
planting of paddy as well as cost reduction in paddy cultivation. Kharif
paddy was grown as rainfed crop in the village which faced erratic rainfall.
Farmers faced the problem of managing sufficient labor to transplant the
paddy in short time that leads to late transplanting of Kharif paddy causing
below-optimum yield of the crop. RPT technology revealed that the cost of
production per hectare was less by around 3500 rupees with 10–15% more
yield in comparison to conventional manual transplanting.

1.8.6   WATER SAVING SYSTEM OF RICE INTENSIFICATION PADDY CULTIVATION WITH SHORT DURATION VARIETY

Under conventional system, most of the farmers prefer to grow Kharif paddy
variety Ranjit, having crop duration of about 150 days under rainfed condition. It has become a problem for the farmers to transplant this variety of
paddy within the month of July. Demonstration on system of rice intensification (SRI) paddy cultivation using high-yielding short-duration variety
Gomoti having 130 days duration were implemented at the field level to

overcome the problem of water scarcity due to less water requirement in SRI practices as well as shorter duration of the variety.

1.9 ADAPTATIONS FOR SMALL-SCALE FARMERS TOWARD AGRICULTURAL PRODUCTIVITY

Small holder farmers of Third World nations can effectively maintain the productivity by increasing gross cropped area along with scientific support. According to Thorlakson (2011) following strategies must be carried out for sustainable agricultural production:

1. Exploration for wider genetic diversity of crops;
2. Enhance additional source of income from secondary agricultural outputs;
3. Proper marketing facilities;
4. Improvement and dissemination of existing agricultural knowledge through technical expertise development;
5. Development of efficient communication systems to meet up the gap between production and marketing sectors;
6. Wholesale involvement of local stakeholders toward adoption of newer climate-smart technology; and
7. Appropriate prediction of dry spell and climatic events along with water conservation and effective plantation scheme keeping in mind the maintenance of soil health.

Adaptive policies often required considerable financial support to implement in the field level. Farmers of developing nation such as India are under severe threat of climatic extremes and need to reorient themselves under the condition of long dry spell, discontinuous rainfall pattern, and higher temperature regimes. Indigenous technologies can cope up the climatic extremes to a considerable amount. Such indigenous practices need to be encouraged at the farming community to reduce the burden of economy in agriculture.

1.9.1 COMMON SEEDLING STOCK DEVELOPMENT FOR LONG-DURATION PLANTATION

Rice has got its indigenous advantage being cultivated in low land due to retention of higher water level under higher precipitation circumstances.

Cultivation of long-duration paddy varieties provides benefit in terms of maintenance of productivity of both *Rabi* and *Kharif* crops. Sowing of paddy's seedling with subsequent delay in terms of time of sowing may hamper the productive potential as well as on productivity of other crops. Long-duration rice varieties can be efficiently used under such circumstances. However, this has got its own negative consequences in terms of unavailability of land for rice transplantation in rainfed areas. This may further aggravate the problem of acute shortage of fodder content for livestock.

1.10   CRA PRACTICE FOR RICE PRODUCTION

Rice cultivation in areas with insufficient rainfall is a big challenge for farmers of developing countries. Therefore, establishment of a common seed stock is an adaptive strategy to combat such problems. The technology implies germination and maintenance of a stock of rice seedlings with adequate irrigation with 15 days interval. In case of delay in occurrence of monsoon by 15 days, common seedling stock can be established with rice variety having long crop duration to promote 1 month older seedlings to be transplanted. Further delay up to 1 month provokes establishment of secondary seedling stock with rice variety having medium crop duration to apply 1-month-old seedlings. Stock of rice crop varieties having short crop duration can be implemented for transplanting 1 month older seedlings in case of further delay in occurrence of monsoon. This system of transplantation of rice seedlings in a climate-friendly way can be a suitable strategy to combat vulnerability of climatic extremes.

1.11   ROLE OF BIOTECHNOLOGY FOR IMPROVING RICE PRODUCTIVITY

The objectives of improving rice productivity comprise different dimensions in itself. At one hand, care should be taken to develop stress-tolerant varieties under climatic extreme as well as maintaining the qualitative and quantitative attributes. Climatic variability has its significant impact in various spheres of paddy yield and plant pathogenic consequences at cellular, molecular, and physiological level (Sreenivasulu et al., 2015).

Current scenario in the field of biotechnology has widened the scope of genetic variability to be efficiently used with due consideration of improving yield. Such approaches can be mediated through appropriate gene sequencing and further incorporation into host plant genome in a planned manner. From future perspective both at in vivo and in vitro conditions, biotechnology imparts higher productivity through (1) genetic recombination to promote quality attributes in crop plants to combat against climate change, (2) hereditary change to combat agricultural pollution in various dimensions aiming toward sustainable agriculture, and (3) improvement of nutrient utilization efficiency of paddy plants through incorporation of genetic materials from microbes.

1.11.1   RICE APOMIXIS

Molecular approaches incorporating various genetic recombination and alterations in diverse nature of genetic material include suitable selection process which may be the future research area under paddy cultivation.

1.11.2   HYBRID RICE RESEARCH

Recombinant transgenic verities of paddy may boost up the agricultural productivity in our country in near future. Such type of genetic mosaic developed in various agricultural sectors have shown positivity on regional basis, although maintenance of productivity should be kept in mind with due consideration of stress-tolerant quality in crop plant. Variability in precipitation pattern is a larger problem under the era of climate change. Appropriate genetic inbreeding for producing HYV needs to be generated to combat such climatic variability. This would lead to boost up the agroeconomy of our country.

1.11.3   SHORTENING OF RICE BREEDING SEASON

Changes in genetic architecture are suitable alternatives to reduce the crop duration period which aids in monetary, time, and labor-intensive farming system. Proper investigation to identify the various genetic phenomena could lead to understanding of high-yielding expressions in rice. The technology may also incorporate genetic material from nonpaddy species.

1.12   DEVELOPMENT OF ECO-FRIENDLY AGROTECHNOLOGY TOWARD IMPROVING RICE PRODUCTION IN INDIA UNDER THE SCENARIO OF CLIMATE CHANGE

1.12.1   RESOURCE UTILIZATION AND MANAGEMENT IN AGRICULTURE

From future perspective, paddy farming requires efficient use of natural resource to boost up the productivity. A strategy includes advance agrotechniques being compatible with agroecosystem and having an eco-friendly approach. Such approaches require policy formulation and subsequent support from the government sectors aiming toward improving the paddy production scenario in Indian subcontinent. Reutilization of residual crop material may be a suitable strategy which would promote sustainability toward soil physicochemical status.

1.12.2   CROP RESIDUE MANAGEMENT

Recycling practices in agro-industry sector promote conservation of our natural assets as well as would help our country to develop its natural resource base. Scientific exploration in various fields of energy output based on agroproducts needs to be properly developed to sustain our natural resources. Manual transplanter developed by CRRI, Cuttack promotes efficient utilization of paddy produce in agricultural sector.

1.12.3   MECHANIZATION OF FARM

Strategies need to be formulated in our country for efficient utilization of agroresources. In this context, farm mechanization can be a suitable option to achieve the desired target. Such policies not only improve productivity but also saves times, money, and human resources. Mechanization strategy includes: (1) modification in tillage activities for optimum availability of nutrients to crop plants, (2) use of seed drill equipment aiming toward improving the seed output of crop species, and (3) use of energy-efficient transplanting system and mechanized weeding system to promote maximum crop growth.

1.12.4 POSTHARVEST TECHNOLOGY AND VALUE ADDITION

From global perspective, lack of appropriate processing technology causes a significant loss in agricultural productivity in developing countries and therefore, poses a severe threat in terms of food production. Utilization of secondary produce from paddy cultivation can be done to promote value in paddy farm produce. Newer technologies need to be developed to promote newer options to agricultural sectors.

Food habit of an area is an essential prerequisite of yield and productivity in agricultural sector. Therefore, changes in terms of developing newer products to sustain the agricultural production of rice are the need of the hour. Brown rice is rich in vitamin B as well as other micronutrients in comparison to processed-rice varieties. Some varieties can be effectively maintained under psychrophilic conditions. Researches on several policed varieties were known to promote other food products associated with paddy. Such processes also generate opportunities under agro-industry to convert paddy produce into several important useful commodities.

1.12.5 DEVELOPMENT OF ZINC- AND IRON-RICH RICE VARIETIES

Micronutrients have differential effect on human health. Some micronutrients promote growth and development as well as help to maintain the physiological processes of the body. Deficiency of some micronutrients often leads to health-related problem of various age groups. Newer solutions to overcome the health-related issues have revealed development of iron- and zinc-rich genetic material by DNA recombination technology under plant breeding is a suitable step that can be incorporated in rice varieties. Such strategies not only improve quantity but also the quality of the food grain. Transgenic incorporation of specific genetic material through genetic recombination could promote and enrich nutrient content in rice plants.

1.12.6 C4 RICE DEVELOPMENT

Biochemical process under C3 mechanism helps paddy and other crop species to inculcate ambient CO_2 and as a result, lose net carbon gain and productivity to a considerable amount via metabolic activity. Therefore, species under C3 lose their competitive attitude with respect to other crops who have generated processes such as "CO_2 pump"—the C4 mode of biochemical operations helps to fix more ambient CO_2 in foliage, biomass and therefore, to combat metabolic loss of carbon.

This mechanism for construction of C4 mode of operation in paddy is a major step to improve agricultural production of rice. Such strategy includes genetic conversion of C3 pathway to C4 mode. This may promote C4 rice to enhance production and productivity at considerable level, irrespective of different factors limiting growth, development, and production of rice.

During strategy formulation, prior consideration should be given to incorporate the internal configuration of C4 plants into C3 rice plant. The carbon photosynthetic system provides insight of experimental feasibility toward technological interventions of genetic material of C4 plants to upgrade the productivity. Such techniques help to increase amount of CO_2 in a single cell and thus favoring C4 mode of operation.

1.13   SUMMARY AND CONCLUSION

Indian farming is bestowed with indigenous process which leads to harvest food in a sustainable manner. At its various dimensions of activities, eco-friendly technologies should be the main focus that would conserve the natural resources. Change in climate is likely to impose additional problems toward productivity, demand, and supply of food in near future. In the Indian context, agricultural dependency of Indians promotes resource depletion and lack of proper management. Various works reported that Asian countries farming communities is enriched with traditional knowledge and wisdom.

CRA practices have promoted various dimensions such as development of stress tolerance rice varieties, crop insurance, cropping calendar designing for rice, integrating the output of climate resilient technology with natural resource management policies, and capacity building among the farmer community which would help them to combat climatic changes and its deleterious effect on world agriculture.

Appropriate planning should be done to promote combating power in our agricultural systems under climatic variability. Several agricultural techniques should be designed for various agroecological conditions of India significantly hamper to combat climatic variability if properly tested and executed at the field level. Some strategies that help adapt to climate change in Indian agriculture are soil organic carbon build up, other eco-friendly technologies, water harvesting with recycling for supplemental irrigation, growing drought- and flood-tolerant varieties, water-saving technologies, location-specific agronomic and nutrient management, improved livestock feed, and feeding methods. Intuitional interventions promote collective action and build resilience among communities.

There are no modest solutions toward complexity of problems from global point of view. Scientific as well as technological approaches are beneficial but insufficient to deal with environmental challenges. Climate-resilient farming can offer prospects for upgrading the livelihood of poor people through endowment of economic and environmental security. Under the present scenario of climate change, CRA practice is an inevitable truth toward sustainable rice production in agro-based country such as India. Different strategies accompanied by technological advancement need to be properly implemented at the field level keeping in mind the socioeconomic upliftment of poor farmer community of India.

KEYWORDS

- agricultural productivity
- climate resilient
- cropping systems
- mechanization
- organic farming
- simulation modeling
- sustainable production

REFERENCES

Ainsworth, A. E. Rice Production in a Changing Climate: A Meta-analysis of Responses to Elevated Carbon Dioxide and Elevated Ozone Concentration. *Glob. Change Biol.* **2008,** *14,* 1642–1650.

Ainsworth, E. A.; Rogers, A. The Response of Photosynthesis and Stomatal Conductance to Rising (CO_2): Mechanisms and Environmental Interactions. *Plant Cell Environ.* **2007,** *30,* 258–270.

Amthor, J. S. The McCree–de Wit–Penning de Vries–Thornley Respiration Paradigms: 30 Years Later. *Ann. Bot.* **2000,** *86,* 1–20.

Auffhammer, M.; Ramanathan, V.; Vincent, J. R. Climate Change, the Monsoon, and Rice Yield in India. *Clim. Change* **2012,** *111* (2), 411–424.

Baker, J. T.; Allen Jr., L. H.; Boote, K. J.; Pickering, N. B. Direct Effects of Atmospheric Carbon Dioxide Concentration on Whole Canopy Dark Respiration of Rice. *Glob. Change Biol.* **2000,** *6,* 275–286.

Baker, J. T.; Laugel, F.; Boote, K. J.; Allen Jr., L. H. Effects of Daytime Carbon Dioxide Concentration on Dark Respiration in Rice. *Plant Cell Environ.* **1992**, *15*, 231–239.

Chambers, R. Editorial Introduction: Vulnerability, Coping and Policy. *IDS Bull.* **1989**, *20*, 1–7.

Chen, G. Y.; Yong, Z. H.; Liao, Y.; Zhang, D. Y.; Chen, Y.; Zhang, H. B.; Chen, J.; Zhu, J. G.; Xu, D. Q. Photosynthetic Acclimation in Rice Leaves to Free-air CO_2 Enrichment Related to Both Ribulose-1,5-bisphosphate Carboxylation Limitation and Ribulose-1,5-bisphosphate Regeneration Limitation. *Plant Cell Physiol.* **2005**, *46*, 1036–1045.

Cheng, W.; Sims, D. A.; Luo, Y.; Colemen, J. S.; Johnson, D. W. Photosynthesis, Respiration, and Net Primary Production of Sunflower Stands in Ambient and Elevated Atmospheric CO_2 Concentrations: An Invariant NPP:GPP Ratio? *Glob. Change Biol.* **2000**, *6*, 931–941.

Cheng, W. G.; Inubushi, K.; Yagi, K.; Sakai, H.; Kobayashi, K. Effects of Elevated Carbon Dioxide Concentration on Biological Nitrogen Fixation, Nitrogen Mineralization and Carbon Decomposition in Submerged Rice Soil. *Biol. Fertil. Soils* **2001**, *34*, 7–13.

Cheng, W. G.; Sakai, H.; Hasegawa, T. Interaction of Elevated CO_2 and Night Temperature on Rice Growth and Yield. *Agric. For. Meteorol.* **2009**, *149*, 51–58.

Dafeng, H.; Yiqi, L.; Weixin, C.; Coleman, J. S.; Dale W. J.; Daniel A. S. Canopy Radiation- and Water-use Efficiencies as Affected by Elevated CO_2. *Glob. Change Biol.* **2001**, *7*, 75–91.

De Costa, W. A. J. M.; Weerakoon, W. M. W.; Abeywardena, R. M. I.; Herath H. M. L. K. Response of Photosynthesis and Water Relations of Rice (*Oryza sativa*) to Elevated Atmospheric Carbon Dioxide in the Subhumid Zone of Sri Lanka. *J. Agron. Crop Sci.* **2003**, *189*, 71–82.

De Costa, W. A. J. M.; Weerakoon, W. M. W.; Herath, H. M. L. K.; Amaratunga, K. S. P.; Abeywardena, R. M. I. Physiology of Yield Determination of Rice Under Elevated Carbon Dioxide at High Temperatures in Subhumid Tropical Climate. *Field Crops Res.* **2006**, *96*, 336–347.

Erda, L.; Wei, X.; Hui, J.; Yinlong, X.; Yue, L.; Liping, B.; Liyong, X. Climate Change Impacts on Yield Formation of CO_2 Enriched Inter-subspecific Hybrid Rice Cultivar Liangyoupeijiu Under Fully Open-air Field Condition in a Warm Sub-tropical Climate Crop Yield and Quality with CO_2 Fertilization in China. *Philos. Trans. Royal Soc.* B **2005**, *360*, 2149–2154.

Farrell, T. C.; Fox, K. M.; Williams, R. L.; Fukai, S. Genotypic Variation for Cold Tolerance During Reproductive Development in Rice: Screening with Cold Air and Cold Water. *Field Crops Res.* **2006**, *98*, 178–194.

Fischer, G.; Shah, M.; van Velthuizen, H. *Climate Change and Agricultural Vulnerability. International Institute for Applied Systems Analysis,* Report Prepared Under UN Institutional Contract Agreement 1113 for World Summit on Sustainable Development: Laxenburg, Austria, 2002.

Gielen, B.; Scarascia-Mugnozza, G.; Ceulemans, R. Stem Respiration of Populus Species in the Third Year of Free-air CO_2 Enrichment. *Physiol. Plant* **2003**, *117*, 500–507.

Griffin, K. L.; Ball, J. T.; Strain, B. D. Direct and Indirect Effects of Elevated CO_2 on Whole-shoot Respiration in Ponderosa Pine Seedlings. *Tree Physiol.* **1996**, *16*, 33–41.

IISD. *Livelihoods and Climate Change: Combining Disaster Risk Reduction, Natural Resource Management and Climate Change.* Adaptation in a New Approach to the

Reduction of Vulnerability and Poverty; A Conceptual Framework Paper Prepared by the Task Force on Climate Change, Vulnerable Communities and Adaptation, 2003.

IPCC. *Climate Change 2007: The Physical Science Basis*; Solomon, S., Qin, D., Manning, M., Chen, Z., Marquis, M., Averyt, K.B., Tignor, M., Miller, H. L. Eds.; Contribution of Working Group I to the Fourth Annual Assessment Report of the Intergovernmental Panel on Climate Change, Cambridge University Press: Cambridge, UK, 2007; pp 996.

IRRI. *Rice Almanac: Source Book for the Most Important Economic Activity on Earth*, 3rd ed.; CABI: Wallingford, England, 2002.

Jablonski, L. M.; Wang, X. Z.; Curtis, P. S. Plant Reproduction Under Elevated CO_2 Conditions: A Meta-analysis of Reports on 79 Crop and Wild Species. *New Phytol.* **2002,** *156*, 926.

Kanemoto, K.; Yumiko, Y.; Tomoko, O.; Naomi, I.; Nguyen, T.; Nguyen, R. S.; Mohapatra, P. K.; Syunsuke, K.; Reda, E. M.; Junki, I.; Hany, E. S.; Kounosuke, F. Photosynthetic Acclimation to Elevated CO_2 is Dependent on N Partitioning and Transpiration in Soybean. *Plant Sci.* **2009,** *177*, 398–403.

Kasperson, J. X.; Kasperson, R. E.; Turner, B. L.; Hseih, W.; Schiller. Vulnerability to Global Environmental Change. In *The Human Dimensions of Global Environmental Change;* Diekman, A., Dietz, T., Jaeger, C. C., Rosa, E. A., Eds.; MIT Press: Cambridge, MA, 2003.

Kim, H. Y.; Horie, T.; Nakagawa, H.; Wada, K. Effects of Elevated CO_2 Concentration and High Temperature on Growth and Yield of Rice. *Jpn. J. Crop Sci.* **1996,** *65*, 644–651.

Kim, H. Y.; Lieffering, M.; Miura, S.; Kobayashi, K.; Okada, M. Growth and Nitrogen Uptake of CO_2-enriched Rice Under Field Conditions. *New Phytol.* **2001,** *150*, 223–229.

Kurukulasuriya, P.; Rosenthal, S. *Climate Change and Agriculture: A Review of Impacts and Adaptations*. Climate Change Series Paper 91, World Bank, Washington, District of Columbia, 2003, pp 106.

Leakey, A. D. B.; Elizabeth, A. A.; Bernacchi, C. J.; Rogers, A.; P. Lon, S.; Ort, D. R. Elevated CO_2 Effects on Plant Carbon, Nitrogen, and Water Relations: Six Important Lessons from FACE. *J. Exp. Bot.* **2009,** *60*, 2859–2876.

Long, S. P.; Ainsworth, E. A.; Rogers, A.; Ort, D. R. Rising Atmospheric Carbondioxide: Plants FACE the Future. *Ann. Rev. Plant Biol.* **2004,** *55*, 591–628.

Maddison, D.; Manley, M.; Kurukulasuriya, P. *The Impact of Climate Change on African Agriculture: A Ricardian Approach*. World Bank Policy Research Working Paper 4306, 2007.

Mathauda, S. S.; Mavi, H. S.; Bhangoo, B. S.; Dhaliwal, B. K. Impact of Projected Climate Change on Rice Production in Punjab (India). *Trop. Ecol.* **2000,** *41* (1), 95–98.

Mohammed, A. R.; Tarpley, L. High Nighttime Temperatures Affect Rice Productivity Through Altered Pollen Germination and Spikelet Fertility. *Agric. For. Meteorol.* **2009,** *149*, 999–1008.

Monteith, J. L. Climatic Variation and the Growth of Crops. *Quart. J. Royal Meteorol. Soc.* **1981,** *107*, 749–774.

Morita, S.; Yonemaru, J. I.; Takanashi, J. I. Grain Growth and Endosperm Cell Size Under High Night Temperature in Rice. *Ann. Bot.* **2005,** *95*, 695–701.

Nguyen, V. N. Productive and Environmentally Friendly Rice Integrated Crop Management Systems. *IRC Newsl.* **2002,** *51*, 25–32.

Peng, S.; Huang, J.; Sheehy, J. E.; Laza, R. C.; Visperas, R. M.; Zhong, X.; Centeno, G. S.; Khush, G. S.; Cassman, K. G. Rice Yields Decline with Higher Night Temperature from Global Warming. *PNAS* **2004,** *101,* 9971–9975.

Pospisilova, J.; Catsky, J. Development of Water Stress Under Increased Atmospheric CO_2 Concentration. *Biol. Plant* **1999,** *42,* 1–24.

Prasad, P. V.; Boote, K. J.; Allen, L. H.; Sheehy, J. E.; Thomas, J. M. G. Species, Ecotype and Cultivar Differences in Spikelet Fertility and Harvest Index of Rice in Response to High Temperature Stress. *Field Crops Res.* **2006,** *95,* 398–411.

Pritchard, S. G.; Amthor, J. S. *Crops and Environmental Change;* Food Products Press: New York, NY, 2005.

Sakai, H.; Yagi, K.; Kobayashi, K.; Kawashima, S. Rice Carbon Balance Under Elevated CO_2. *New Phytol.* **2001,** *150,* 211–249.

Satake, T.; Yoshida, S. High Temperature-induced Sterilityin Indica Rice at Fowering. *Jpn. J. Crop Sci.* **1978,** *47,* 6–17.

Sheehy, J. E.; Mitchell, P. L.; Ferrer, A. B. Decline in Rice Grain Yields with Temperature: Models and Correlations Can Give Different Estimates. *Field Crops Res.* **2006,** *98,* 151–156.

Shimono, H.; Okada, M.; Inoue, M.; Nakamura, H.; Kobayashi, K.; Hasegawa, T. Diurnal and Seasonal Variations in Stomatal Conductance of Rice at Elevated Atmospheric CO_2 Under Fully Open-air Conditions. *Plant Cell Environ.* **2010,** *33,* 322–331.

Soora, N. K.; Aggarwal, P. K.; Saxena, R.; Rani, S.; Jain, S.; Chauhan, N. An Assessment of Regional Vulnerability of Rice to Climate Change in India. *Clim. Change* **2013,** *118* (3–4), 683–699.

Sreenivasulu, N.; Butardo Jr, V. M.; Misra, G.; Cuevas, R. P.; Anacleto, R.; Kavi Kishor, P. B. Designing Climate-resilient Rice with Ideal Grain Quality Suited for High-temperature Stress. *J. Exp. Bot.* **2015,** *66,* 1737–1748.

Swaminathan, M. S. Rice. *Sci. Am.* **1984,** *250,* 81–93.

Thorlakson, T. *Reducing Subsistence Farmers' Vulnerability to Climate Change: The Potential Contributions of Agroforestry in Western Kenya.* World Agroforestry Centre Occasional Paper 16. World Agroforestry Centre: Nairobi, Kenya, 2011.

Uprety, D. C.; Dwivedi, N.; Jain, V.; Mohan, R.; Saxena, D. C.; Jolly, M.; Paswan, G. Responses of Rice Cultivars to the Elevated CO_2. *Biol. Plant* **2003,** *46,* 35–39.

Verma, D. K.; Srivastav, P. P. Proximate Composition, Mineral Content and Fatty Acids Analyses of Aromatic and Non-aromatic Indian Rice. *Rice Sci.* **2017,** *24* (1), 21–31.

Verma, D. K.; Mohan, M.; Yadav, V. K.; Asthir, B.; Soni, S. K. Inquisition of Some Physico-chemical Characteristics of Newly Evolved Basmati Rice. *Enviorn. Ecol.* **2012,** *30* (1), 114–117.

Verma, D. K.; Mohan, M.; Asthir, B. Physicochemical and Cooking Characteristics of Some Promising Basmati Genotypes. *Asian J. Food Ag.-Ind.* **2013,** *6* (2), 94–99.

Verma, D. K.; Mohan, M.; Prabhakar, P. K.; Srivastav, P. P. Physico-chemical and Cooking Characteristics of Azad Basmati. *Int. Food Res. J.* **2015,** *22* (4), 1380–1389.

Varma, P. *Rice Productivity and Food Security in India.* Centre for Management in Agriculture (CMA), Indian Institute of Management Ahmedabad (IIMA), 2017. DOI 10.1007/978-981-10-3692-7_2.

Wassmann, R.; Dobermann, A. Climate Change Adaptation Through Rice Production in Regions with High Poverty Levels. *SAT J.* **2007,** *4* (1). ejournal.icrisat.org (accessed Oct 23, 2017).

Watanabe, T.; Kume, T. A General Adaptation Strategy for Climate Change Impacts on Paddy Cultivation: Special Reference to the Japanese Context. *Paddy Water Environ.* **2009,** *7,* 313–320.

Xu, D. Q.; Gifford, R. M.; Chow, W. S. Photosynthetic Acclimation in Pea and Soybean to High Atmospheric CO_2 Partial Pressure. *Plant Physiol.* **1994,** *106,* 661–671.

Xu, Z.; Zheng, X.; Wang, Y.; Wang, Y.; Huang, Y.; Zhu, J. Effect of Free-air Atmospheric CO_2 Enrichment on Dark Respiration of Rice Plants (*Oryza sativa* L.) *Agric. Ecosyst. Environ.* **2006,** *115,* 105–112.

Yoshida, S.; Parao, F. T. Climate Influence on Yield and Yield Components of Lowland Rice in the Tropics. *Climate and Rice;* IRRI Publication: Los Banos, Philippines, 1976; pp 471–494.

Yoshimoto, M.; Oue, H.; Kobayashi, K. Energy Balance and Water Use Efficiency of Rice Canopies Under Free-air CO_2 Enrichment. *Agric. For. Meteorol.* **2005,** *133,* 226–246.

Zhong, Li.; Yagi, K.; Sakai, H.; Kobayashi, K. Influence of Elevated CO_2 and Nitrogen Nutrition on Rice Plant Growth, Soil Microbial Biomass, Dissolved Organic Carbon and Dissolved CH4. *Plant Soil* **2004,** *258,* 81–90.

CHAPTER 2

DIRECT-SEEDED VERSUS TRANSPLANTED RICE IN ASIA: CLIMATIC EFFECTS, RELATIVE PERFORMANCE, AND CONSTRAINTS—A COMPARATIVE OUTLOOK

UMAIR ASHRAF[1,2*†], ZHAOWEN MO[1,2], and XIANGRU TANG[1,2*†]

[1]Department of Crop Science and Technology, College of Agriculture, South China Agricultural University, Guangzhou 510642, PR China

[2]Scientific Observing and Experimental Station of Crop Cultivation in South China, Ministry of Agriculture PR China, Guangzhou 510642, PR China

*Corresponding authors. E-mail: umairashraf2056@gmail.com; tangxr@scau.edu.cn

†Authors Umair Ashraf and Zhaowen Mo have equal contributions.

ABSTRACT

Rice—the staple food of the world is largely consumed in Asia where it is the source of calories and livelihood for the rural people. Traditional method of rice production offers higher yield and easy weed control which is being achieved on the cost of intense water and labor usage but this technique is susceptible to the reducing water accessibility. Moreover, the climate fluctuation of this region in addition to the increased temperature and decreased water availability may influence the rice production posing a threat to the food security. Producing rice in water-scare areas is also a great challenge for scientists. Under these conditions, the direct seeding of rice is an alternative technology where less water and labor requirements are needed providing the low production costs. This technique is the hope of future to increase the rice production under extreme conditions, and several Asian countries such as India, Thailand, Philippines, Malaysia, and Pakistan have started to adopt this in dry season, thus offering economic benefits with less labor usage.

This chapter basically compares the transplanted and direct-seeded rice on the basis of climatic conditions, socioeconomic benefits, and limitations.

2.1   INTRODUCTION

Roughly, 60% of the world's population resides in Asia whose ever increasing food demands are threatened by limited water resources. Global production of rice is about 530 million tones whose major part (90–92%) is being consumed only in Asia (IRRI, 1997). It is not only the main source of calories but also a single leading source of income and employment for rural community (FAO, 2004). About 75% of the world's rice production was obtained from 79 million ha of irrigated lowlands (Maclean et al., 2002). Generally, the rice is grown in a traditional way in most of the countries for better crop establishment and easy weed control. Water is getting more and more importance day by day and becoming a precious commodity, as it is used in every scale of life from household to industrial and agricultural purposes. Now it is becoming a challenging task for scientists to develop an efficient rice production system that can cope with water-limiting conditions. It is projected that more than 17 m ha area of transplanted rice (TR) may become vulnerable to "physical water scarcity" and about 22 m ha area may become subject to "economic water scarcity" in Asia by first quarter of this century (Tuong and Bouman, 2002). Hence, it is becoming no longer economically practicable to grow flooded rice for better crop growth and to use it as a better technique for weed control (Johnson and Mortimer, 2005) because sustainability of traditional system of rice cultivation is extremely vulnerable to declining water accessibility (Anwar et al., 2010). Furthermore, nutrient loss in puddled rice production system may also result in ground water contamination as well as puts an additional fertilizer cost for the growers. Therefore, improved water management techniques must be developed to enhance water use efficiency (WUE) to maintain sustainability in rice production system (Bouman et al., 2007).

IPCC (2014) declared the Southeast Asia as the most vulnerable region to climate change that might be a threat to the agricultural productivity of this region. It further projected that climate change impact will fluctuate from region to region which may affect overall agricultural productivity and may become a threat to food security of Asian regions. Increased temperature and decreased water availability may reduce the efficacy of the rice production systems in Asia (ADB, 2009). There exist a correlation between rice yield with temperature and it is projected that 10% yield reduction may occur for every 1°C increase in temperature and case may be more severe especially in rice grown during dry seasons in Asia (Peng et al., 2004). Such adversities in climate may impair world rice production severely (Li and Wassmann, 2011).

Research efforts have so far been intended from last few decades on water saving techniques to maintain rice productivity without any severe loss. To ensure food security, it is necessary to grow more rice with less water (Guerra et al., 1998). In rice production systems, sustainability of water resources has been of major concern (Juraimi et al., 2010). Direct-seeded rice (DSR) has been developed as the most suitable water saving technique of rice cultivation in which crop is grown via direct sowing of rice seeds in unsaturated and nonpuddled conditions (Mahajan et al., 2009; Anwar et al., 2010). According to Wang et al. (2002), approximately 50–60% of irrigation water can be saved and about 200% of water efficiency can be enhanced by adopting DSR system of rice cultivation than lowland flooded rice system. Direct seeding of rice cultivation not only reduces the total water require-ments of the crop but also requires less labor and production cost (Singh et al., 2008). In DSR production systems, normally care should be taken regarding cultivar selection and those cultivars having higher yield potential and well adapted to aerobic soils are used to get better yields; however, in Asia, it is usually grown as a subsistence crop in severe upland situation (Lafitte et al., 2002). In order to save the excess water loss and to promote the water saving technologies in rice cultivation in Asia, the Water-Saving Work Group of the Irrigated Rice Research Consortium is fully dedicated to proliferate the latest promising technologies around the globe regarding rice production (Wang et al., 1998).

Recently, a shift from TR to DSR culture has been observed in South-east Asia which may be due to expensive labor for rice transplantation at peak seasons. In general, lower wage rates and high water availability suits transplanting methods while higher wages and less water availability favors DSR culture shown in Figure 2.1 (Pandey and Velasco, 2002). Pres-ently, proportion of DSR among rice cultivation methods is about 23% over the globe (Rao et al., 2007). In Southeast Asia, direct seeding is mostly preferred in dry season instead of wet season possibly due to better weed control; however, dry-season rice is just a one-quarter of rice production in this region of Asia (Karim et al., 2004). So, DSR culture might be a good alternative to TR (where 1-month-old seedlings transplanted into puddled field with fully anaerobic conditions) in most of the Asian regions which has been successfully adopted in uplands, rainfed lowlands, flood-prone, and

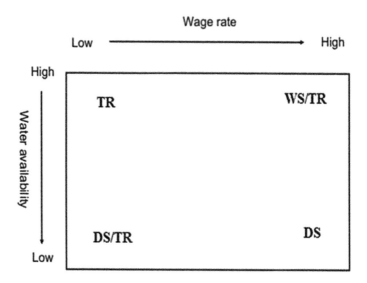

FIGURE 2.1   Factors affecting the selection of rice cultivation methods. TR, transplanted rice; WS, wet seeding; DS, dry seeding.

Source: Adapted from Pandey and Velasco, 2002.

irrigated areas (Azmi et al., 2005; Luat, 2000). Regarding yield, both DSR and TR had similar results (Kukal and Aggarwal, 2002). Even though DSR provided same yields as TR and better economic benefits with less labor usage, it is still not adopted at a large scale. In most of the Asian regions, TR is still preferred over DSR with the constraints what it has. In this review, we compared the relative performance of both DSR and TR regarding climate change effects, comparative economic benefits, and constrains related to their production.

2.2   CLIMATE CHANGE EFFECTS

Increasing world population is pressurizing the agricultural productivity to fulfill the present food demands and future food security. In the upcoming decades, world food security would likely to be facing some greater challenges to fill the demand and supply gap. Future predictions provided by FAO (2009) revealed that world population might go beyond 9 million till 2050 for which 70% more food have to be added in the existing food chain. But, unluckily, agricultural productivity is going unparalleled to food demands.

For instance, about 2–3% reduction has been observed in average annual growth of cereals during 1970–1980 and 1–2% in last decade (World Bank, 2007); however, the situation is going to be even more adverse especially for developing countries which may experience a decline of about 9–21% in overall crop productivity due to climate change (FAO, 2009). Regional and global weather events may become more variable, frequent, and severe than at present (IPCC, 2012) which may result in crop yield fluctuations, variability in food supplies, and may enhance food insecurity risks (FAO, 2009), more specifically in underdeveloped countries (Wang et al., 2013). Asia has a central role in rice production over the globe and contributes more than 90% of world's rice production and its importance lies in the fact that it contributes almost 60% of the three major cereals (rice, wheat, and maize) (FAO, 2005). The overall rice yield variability in Asia over the last 50 years is represented in Table 2.1. From the last 50 years, the rice yield remained more variable in western Asia (>50%) followed by eastern and southern Asia.

Changing climate (especially raising temperatures) significantly affected the rice yield. Present scenario indicates that rice is cultivated in hot regions where temperatures are already to their peaks and above the crop growth requirements, that is, 22–28°C (mean minimum and maximum) during day and night time, respectively (Krishnan et al., 2011). Previously, Baker et al. (1992) reported that rice yields may reduce up to 7–8% with an increase of 1°C within 21–28°C to 34–27°C range in daytime maximum and night minimum temperatures. Along with temperature, effects of elevated levels of CO_2 has also been studied in rice cultivars mostly growing in east and south Asia (Krishnan et al., 2011) while central Asia is little bit ignored. A decrease of about 10% in rice yields with an increase in 1°C under CO_2-enriched conditions has been noticed by Peng et al. (2004). Sheehy et al. (2006)

TABLE 2.1 Rice Yield Variability over the Last 50 Years (from 1963 to 2013).[a]

Region	Rice yield (t ha^{-1})			
	1963	1983	2013	Yield variability (%)
Central Asia	–	–	3.46	–
Eastern Asia	3.18	5.58	7.36	43.21
Southern Asia	1.77	2.45	4.19	42.25
Western Asia	3.60	3.95	7.00	51.40

[a]FAO STAT. http://faostat3.fao.org/home/index.html.

noted 2–6% decrease in rice yield by just an increase in per degree Celsius between 21°C and 28°C mean daily temperatures. However, about 7% yield loss may occur with every unit increase in temperature at present levels of CO_2 while enrichment of CO_2 from 380 ppm to about 700 ppm may have a positive effect (about 31% increase) on rice yield (Krishnan et al., 2007). Moreover, Baker (2004) worked out on US rice cultivars and reported a further increase (40–70%) at 28°C with high CO_2 concentrations. A country-specific variation in rice production due to changing climate has also been reported in China (Bachelet et al., 1995), India (Mall and Aggarwal, 2002), Japan (Horie et al., 2000), and Bangladesh (Karim et al., 1994). Predictions regarding southeast region of Asia showed that rice yield may be lowered up to 3.8% in near future due to climate change with regards to increased temperature and water scarcity (Murdiyarso, 2000). Temperature fluctuations especially in Asian regions largely affect germination, growth behavior, and rice development. Rice seed germination and leaf elongation stages have almost similar critical temperatures. Overall, temperature extremities below 9°C and higher than 45°C might be damaging for rice growth. The critical temperatures for rice development at different growth stages are represented in Table 2.2.

In short, increased temperature and reduced water availability may negatively affect rice productivity. High temperature-induced panicle sterility and disruption in pollen development had severe effects on rice yields (Krishnan et al., 2011). Sowing time is also very crucial in this regard,

TABLE 2.2   Critical Temperature at Different Growth Stages of Rice Development.[a]

Growth stages	Critical temperature (°C)		
	High	Optimum	Low
Germination	≥45	18–40	16–19
Seedling emergence	≥35	25–30	12
Rooting	≥35	25–28	16
Leaf elongation	≥45	31	7–12
Tillering	≥33	25–31	9–16
Panicle primordial initiation	–	–	15
Panicle differentiation	≥30	–	15–20
Anthesis	≥35	30–33	22
Ripening	≥30	20–29	12–18

[a]FAO, 2005. Modified from Redfern et al., 2012.

especially in central Asian region where early seeding may overlap with heat stress at flowering stage while delay in seeding may be exposed to the chilling stress during grain filling stage (Devkota, 2011). So, keeping in view the relationships of seeding time and crop development, rice sowing time must be optimized, especially for DSR cultivation. Simulation and crop models may provide help in order to determine the optimum seeding time and rice cultivars for a specific region under changing climatic conditions (Krishnan et al., 2011). A proper feedback mechanism and technological development should be taken in to account while estimating the rice yields under changing climatic scenario. Moreover, some scientists declared the negative impacts of climate change by using the general circulation models (GCMs) (Matthews et al., 1997). Many rice researchers reported/projected significant changes in rice yield due to climate change in different countries of Asia (Table 2.3).

2.3 RELATIVE PERFORMANCE

2.3.1 GROWTH AND YIELD DYNAMICS

Increasing water deficit in Asia has forced the researchers to develop new techniques to grow rice such as direct seeding as an alternative to TR but due to severe infestation of weeds the final crop yield is reduced in DSR (Akbar et al., 2011). Nonetheless, by growing dry seeded rice rather than flooded, water and labor demands may be reduced (Yadav et al., 2007). In DSR system of rice cultivation, fields stayed unsaturated round the season. In uplands, rice has been grown aerobically in nonflooded soils for centuries, but its average yield was not satisfactory due to severe environmental and agronomic factors such as high weed pressure, low precipitation and poor soil conditions, minimum availability of external inputs, and low yielding rice cultivars. However, development of early maturing and nutrient responsive rice cultivars has encouraged the growers for DSR culture (Lafitte et al., 2002). In DSR, short-duration rice cultivars are preferred over long-duration cultivars. For DSR and TR rice, growth period varied up to reproductive stage, while ripening to harvest period in both types of rice cultivars was almost similar (up to 30 days; Fig. 2.2).

TABLE 2.3 Yield Fluctuations of Rice Under Changing Climate.

Authors	Crop model(s) used	Experimental specifications/details	Remarks	Country
Erda et al. (2005)	Regional climate change model (PRECIS)	Regional crop models were PRECIS driven by (without CO_2 fertilization) in a joint project of DEFRA and MOST to simulate the changes in yields of major Chinese food crops rice, maize, and wheat	Results suggested that by the end of 21st century, the temperature may increase by 3–4°C. Further, modeling predicted that the yields of rice, maize, and wheat could be reduced by 18–37% due to climate change (without CO_2 fertilization) within next 20–80 years	China
Mohandrass et al. (1995)	ORYZA1 rice model	On multilocation trial-based changes in rice yield (%) predicted by the ORYZA1 rice model under three GCMs scenarios, that is, GFDL, GISS, and UKMO in the main and second planting seasons under different scenarios of changing climate (change in CO_2 from 1.5 to 2 times higher) and temperature (1°C, 2°C, and 4°C) higher than prevailed climatic conditions at nine different stations. Normally, in second season, crop is not sown so, results are quoted only for main season (June–November)	The results predicted an increase in rice yield (8.8–40.2%) for each research station under all three scenarios; however, across site analyses predicted for GFDL, GISS, and UKMO with an increase of 25.5%, 28.2%, and 28.4% in rice yields, respectively	India
Shakoor et al. (2015)	VAR model	Data regarding rice production were obtained from Agriculture Statistics of Pakistan and Economic Surveys of Pakistan while annual climatic data from 1980–2013 were collected from Pakistan Meteorological Department	An increase in mean maximum temperature would have positive effects on rice yields while increase in mean minimum temperature may lead to rice yield penalty; however, simulated scenarios for 2030 showed that much increase in temperature and precipitation would affect rice yields negatively. This model further suggested that an increase in temperature for 2–4°C and from 4°C to 5°C would lead to 8% and 11.5% rice yield reduction	Pakistan

TABLE 2.3 (Continued)

Authors	Crop model(s) used	Experimental specifications/details	Remarks	Country
Basak et al. (2009)	DSSAT modeling system	Changes in the yields of rice were predicted for 2030, 2050, and 2070 at 12 different locations of Bangladesh. The weather data to estimate the future yields were generated by regional climate model PRECIS	The model projected a substantial reduction in rice yield about 10%, 20%, and 50% for the years 2030, 2050, and 2070, respectively, due to climate change (increment of temperature). Further, an increase in CO_2 concentrations and incoming solar radiations would lead to increase in rice yields; but these positive effects were not much apparent	Bangladesh
Lizumi et al. (2007)	SIMRIW	The study was conducted to project the influence of global warming on rice production in Japan	The results suggested that global warming mitigated the damages to rice production while enhanced the occurrence of heat stress in hot summers	Japan
Kim et al. (2013)	CERES-Rice 4.0	Data were obtained from experiments conducted using a TGFC with a CO_2 enrichment system during 2009–2010. Based on the empirical calibration and validation, the model was applied to simulate paddy rice production for the region projecting for the years 2050 and 2100	Elevated CO_2 levels would lead to an increase of 12.6% and 22.0% in paddy yield while an increase in temperatures simulated a decrease of 22.1% and 35.0% in rice yield; however, the variations in rice yields depend on cultivars grown	Korea
Peng et al. (2004)	CLICOM-computer system	Climatic data (1979–2003) and relationships among temperature dynamics and rice yields (1992–2003) were drawn at IRRI	An annual increase of 0.35°C and 1.13°C were recorded in mean minimum and mean maximum temperatures. The paddy yield was declined up to 10% with every degree increase in mean minimum temperature	Philippines

TABLE 2.3 (Continued)

Authors	Crop model(s) used	Experimental specifications/details	Remarks	Country
Yuliawana and Handoko (2016)	Shierary Rice Model with GIS	Rice yield decrease was estimated in different regions of Indonesia based on increasing temperature scenarios under climate change	Rise in temperatures has considerable effects on rice yield (both irrigated and rainfed), nonetheless the negative effects were more in rainfed conditions than irrigated. A decrease in rice yield of 11.1% and 14.4% per every degree increase in temperature were observed	Indonesia
Pumijumnong and Arunrat (2013)	i-EPIC model version 0509	Changes in rice yields were estimated from 2007 to 2017 by using the IPCC SRES A2 and the B2	Results revealed that that rice yields were higher under A2 scenario than B2 in both irrigated and rainfed areas	Thailand

CO_2, carbon dioxide; DEFRA, Department for Environment, Food and Rural Affairs, UK; GCMs, general circulation models; GFDL, General Fluid Dynamics Laboratory; GISS, Goddard Institute of Space Studies; IRRI, International Rice Research Institute; MOST, Ministry of Science and Technology, China; SRES, special report on emission scenario; SIMRIW, simulation model for rice and weather relationship; TGFC, temperature gradient field chamber; VAR, vector autoregression; UKMO, United Kingdom Meteorological Office.

Moreover, by comparing the growth and yield efficiency of both DSR and TR, TR is favored pertaining to weed management and crop productivity as Baloch et al. (2006) reported that paddy yield and yield components in TR were substantially higher than that in DSR, whereas weed biomass and its density were higher in rice grown by direct seeding. Normally, yield and yield components of TR are higher than DSR-cultured rice; however, some reports from previous literature claim similar or even higher yields in case of DSR rather than TR. For example, Dingkuhn et al. (1991) obtained higher (3 t ha^{-1}) paddy yield in DSR than TR (2 t ha^{-1}) ascribed to increased panicle numbers, grain size, and less sterility percentage. Further, direct seeding in moistened soil resulted in taller plants, more panicles, and dry biomass, whereas lower chlorophyll and specific leaf weights were observed in TR. Yadav et al. (2007) also reported the rice yield dynamics

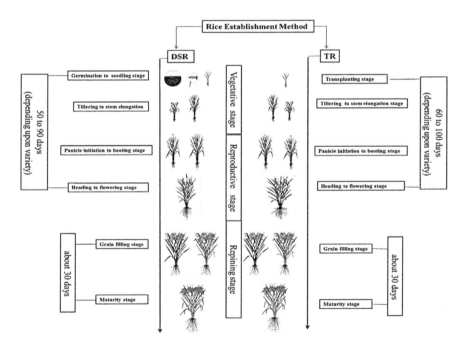

FIGURE 2.2 Growth duration of direct-seeded rice (DSR, left side) and transplanted rice (TR, right side). The growth period of rice can be divided in to three stages, that is, vegetative, reproductive, and maturity. TR often took more days to reach reproductive stage than DSR; however, the growth duration depends on external environmental conditions and cultivar used.

while studying the effect of different sowing techniques, soil types, rice genotypes, and seeding density and reported higher grain yield (30 and 44%) in case of direct drilled puddled and compacted plots compared with unpuddled/uncompacted plots, respectively. Higher paddy yields in trans-planted puddled rice are related to the yield components, that is, effec-tive tillers, filled grains percentage, and grain weight (Ashraf et al., 2014). Comparable rice yields in DSR and TR systems were reported in different on-farm trials in India when weeds were managed properly. However, Johnson et al. (2003) obtained 20% less yield from DSR fields than TR (results from 36 farm trials). Likewise, average yields from both DSR and TR were similar in Philippines when were managed properly (Tabbal et al., 2002). In case of TR, flooded soils prevent direct exposure of light to weeds by making a layer of standing water. It not only suppressed the weed growth effectively but also provided favorable growing conditions for TR (Farooq et al., 2011). Furthermore, transplanting of rice in lines gave better yield than conventional transplantation and direct seeding (Awan et al., 2008). Ramzan (2003) reported 48% and 53% yield reduc-tion in TR and direct seeded, respectively, under flooded conditions and 74% yield reduction in direct seeded in dry soils. Some previous studies arrogated the similar or even higher yields in DSR culture as compared to TR with proper management (Table 2.4). Hence, rice production through direct seeding might be beneficial, as it is quick, less labor intensive with reduced water requirements, and have high benefit to cost ratio in regions where ample water supply is available (Balasubramanian and Hill, 2002), whereas in other aspects, reduced level of irrigation in DSR may also alter physiological features of crop and intensify weed proliferation (Bouman and Thoung, 2001; Warner et al., 2006). Hence, both methods have their own advantages and disadvantages, so more emphasize should be given to resource efficient method for future food security.

2.3.2   *WATER DYNAMICS AND COMPARATIVE WUE*

Area subjected to direct seeding was about the one-fifth of the total area of Asia as estimated in late 1990s (Pandey and Velasco, 2002) as it saves about 73% of irrigation water for land preparation and 56% during growing season. Transplantation under flooded conditions (field is completely satu-rated with water about 5–10 cm during growing period) is the principle method for rice cultivation in Asia. This practice encompasses growing of

TABLE 2.4 Comparative Grain Yield of Direct-Seeded and Transplanted Rice in Different Asian Regions.

Direct-seeded rice (t ha⁻¹)	Transplanted rice (t ha⁻¹)	[a]Difference (%)	Experimental site	References
5.38	5.32	1.13	Southeastern Korea	Ko and Kang (2000)
4.14	4.79	13.57	Muda area (Malaysia)	Cabangon et al. (2002)
3.70	3.69	0.27	Mahanadi delta (India)	Sarkar and Sanjukta (2003)
3.78	4.81	21.41	Rasht (Iran)	
2.71	3.51	22.79	Faisalabad (Pakistan)	Farooq et al. (2006, 2009)
4.60	4.14	11.11	IRRI (Philippines)	Ali et al. (2006)
≤4.00	≥7.00	~42.86	IRRI (Philippines)	Peng et al. (2006)
5.31	5.28	0.57	IRRI (Philippines)	Ali et al. (2007)
3.18	2.31	37.66	Ubon Ratchathani (Thailand)	Hayashi et al. (2007)
7.20	6.60	9.09	Indo-Gangetic Plains	Bhushan et al. (2007)
7.81	8.10	3.58	Uttar Pradesh (India)	Gathala et al. (2011)
4.96	4.61	7.59	Uttar Pradesh (India)	Naresh et al. (2013)
3.58	4.43	19.19	Faisalabad (Pakistan)	Maqsood et al. (2013)
7.69	8.58	10.37	Zhejiang (China)	Chen et al. (2014)
7.91	8.32	4.93	Jiangsu (China)	Hang et al. (2014)
4.79	6.58	27.20	Haryana (India)	Chhokar et al. (2014)
2.68	4.18	35.89	Bihar (India)	Jat et al. (2014)
6.5	9.8	33.67	IRRI (Philippines)	Horgan et al. (2014)
3.84	4.34	11.52	IRRI (Philippines)	Sudhir-Yadav et al. (2014)
4.86	4.25	14.35	Jiangsu (China)	Liu et al. (2014b)
4.57	4.81	4.99	Hubei (China)	Liu et al. (2014)
2.35	3.44	31.69	Faisalabad (Pakistan)	Jabran et al. (2015)
6.94	7.05	1.56	Fukuoka (Japan)	Li et al. (2016b)
8.30	8.68	4.57	Uttar Pradesh (India)	Singh et al. (2016)

[a]Difference (%) = (higher value − lower value/lower value) × 100.

rice seedlings in a nursery bed and then transplanting them to the main field with puddled conditions. Primarily, soaking, plowing, and then puddling is done until a soft muddy, 10–15-cm layer is created in water ponded conditions to prepare the main field where seedlings are going to be transplanted. The main sources of water outflow from the rice field are evapotranspiration (ET), percolation, seepage, and bund overflow, whereas transpiration is the only fruitful way of water usage because it directly links with total dry matter accumulation. Evaporation, "the direct loss of water from the soil surface," and percolation, "the water loss vertically from the water layer beyond the root zone," add nothing to the rice plants (Woperies et al., 1994). In DSR, seepage, percolation, and evaporative losses are less compared to TR. Moreover, DSR system also helps to enhance the water productivity with an effective use of rainfall and by discouraging the attendant loss of soil sediments, silt, fertility, and productivity. Further, DSR culture saved almost 45% of irrigation water as compared to transplanted flooded rice (Lampayan and Bouman 2005). It is well estimated that total water required to prepare the field for transplanting is theoretically about 150–200 mm, but it may go up to 650–900 mm under high temperature conditions (De Datta, 1981; Bhuiyan et al., 1995). Moreover, during crop growth period, field water input may change from 500 to 800 mm (De Detta 1973). Research trials in Indo-Gangetic Plain (IGP) explored the potential water savings up to 60% in DSR on beds compared with TR under puddled conditions (Gupta et al., 2003). Total water required for puddled rice largely depends on the extent of the water outflow, field preparation, and crop growth duration. Currently, most of the short-duration rice cultivars, having 100-days crop growth period, total ET losses are about 400–500 mm and 600–700 mm in the wet and dry seasons, respectively, are being grown in most of the Asian regions. Water requirement may be reduced up to 100–150 mm for land preparation with a short turnover time between land soaking and transplanting (De Datta, 1981). Seepage and percolation losses are significantly higher in TR than DSR at preestablishment phase and whole crop growth season while lower at crop growth phase (Cabangon et al., 2002). Total water saving in DSR mainly results from several factors such as no loss of water during land preparation, low percolation and seepage losses from the field, and less evaporative losses (Bouman et al., 2005) which reduced 51% of water usage with 32–88% higher as compared to WUE in rice (articulated as gram of grain/kg of water). Direct seeding of rice with microirrigation practices might be helpful for efficient use of water with minimum losses (Parthasarathi et al., 2012). Still, in some cases, percolation and seepage losses were found higher in DSR as compared to rice cultivated by puddling, but such cases are very few (Singh et al., 2001). Overall, TR showed higher seepage and percolation losses than wet seeded and dry seeded rice and the trend of water loss was linear with crop growth period depicted in Figure 2.3 (Cabangon et al., 2002).

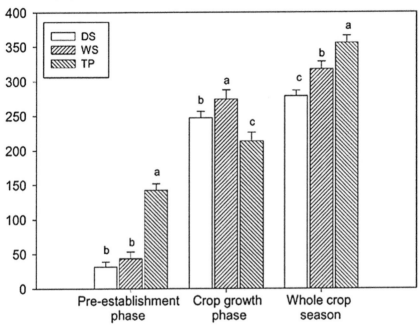

FIGURE 2.3   Comparative seepage and percolation losses in irrigations provided to various rice establishment methods. Water losses are more from pre-establishment phase to whole crop season. Columns with different letters are significantly different from each other ($p \leq$ 0.05). DS, dry seeded; WT, wet seeded; TP, transplanted rice.

Source: Reprinted from Cabangon, R. J.; Tuong, T. P.; Abdullah, N. B. Comparing Water Input and Water Productivity of Transplanted and Direct-seeded Rice Production Systems. *Agric. Water Manag.* **2002,** *57*, 11–31. © 2002, with permission from Elsevier.

In a field trial, Sarkar and Singh (2007) sown rice seeds in moist soil by maintaining a water level of 20 ± 5 cm after 5 days of sowing and compared with 30 days old seedlings which have been transplanted in traditional way and observed taller plants, more dry matter accumulation, numbers of panicles, and less sterility percentage in TR than DSR. Moreover, Bouman (2001) revealed that rainfall is inadequate to maintain productivity of lowland TR (which necessitate about 1200–1500 mm) but enough for DSR (approximately 800 mm). In many Asian regions, farmers often postpone plowing and puddling while waiting for the seedlings being raised in seedbeds, which may lead to extended duration and enhanced water needs during land preparation (IRRI, 1978). Wickham and Sen (1978) also stated that percolation may account up to 40% of total water applied to grow TR during land preparation. Hence, by lowering duration of land preparation, the water input of rice cultivation can potentially be decreased (Tuong, 1999). Moreover, direct seeding either wet seeding (sowing of pregerminated seeds directly into the puddled field) or dry seeding (broadcasting of dry seeds onto dry or moist soil) could be an opportunity to curtail the land preparation period. Bhuiyan et al. (1995) concluded that direct drilled rice abridged water requirement compared with TR which might be due to less seepage, percolation, and evaporative losses. Table 2.5 indicated that values for irrigation service units (ISU) are higher in case of TR than wet-direct-seeded (WDS) and dry-direct-seeded (DDS) rice with maximum yield in TR.

Hence, to save water losses with higher use efficiencies in sustainable rice production systems, water management practices should be integrated with germplasm/cultivar selection and other crop/resource management practices at micro- and macrolevel. System-level management (water management at farm level) might also be a good option for consumptive use of water in paddy fields; however, consequences of reduced water inputs on nutrient acquisition, weed growth, rice sustainability, and crop productivity in rice production systems are another set of barriers that needs to be broken/crossed over in future.

TABLE 2.5   Paddy Yield (t ha^{-1}) and Water Productivity (kg rice m^{-3} Water) with Respect to Different Irrigation Service Units (ISU).[a]

Parameters	Dry-direct-seeded (ISU)	Wet-direct-seeded (ISU)	Transplanted (ISU)
Yield	4.14 ± 0.17 b	4.50 ± 0.23 ab	4.79 ± 0.23 a
WP$_I$	1.48 ± 0.26 a	0.62 ± 0.30 b	1.00 ± 0.30 b
WP$_{I+R}$	0.28 ± 0.02 a	0.26 ± 0.02 a	0.26 ± 0.02 a
WP$_{ET+E}$	0.40 ± 0.02 a	0.44 ± 0.02 a	0.42 ± 0.02 a
WP$_{ET}$	0.48 ± 0.03 b	0.54 ± 0.04 b	0.62 ± 0.04 a

Row values (means $\pm$ SE) followed by the same letter do not differ significantly at $p \leq 0.05$ by LSD. WP$_I$, water productivity under irrigation; WPI+R, total water input; WPET+E, evapotranspiration from rice area + evaporation from nonrice area; WPET, evapotranspiration from rice area.

[a]Reprinted from Cabangon, R. J.; Tuong, T. P.; Abdullah, N. B. Comparing Water Input and Water Productivity of Transplanted and Direct-seeded Rice Production Systems. *Agric. Water Manag.* **2002**, *57*, 11–31. © 2002, with permission from Elsevier.

2.3.3   RELATIVE RESOURCE UTILIZATION

DSR culture utilizes fewer resources such as water, nutrition, labor with high efficiency thus incurs less production cost than TR (Bhuiyan et al., 1995). Substantial reduction in labor requirement was observed when TR culture was transferred to DSR (Pandey and Velasco 1999). DSR also requires less water to mature than TR as it utilizes less water with more efficiency (Dawe, 2005). For instance, up to 57% water savings were reported in experiments conducted at Northwest India when rice was sown directly in nonpuddled soils (at field capacity) (Singh et al., 2002; Sharma et al., 2002), while up to 20% reduction in water requirements was recorded in short-term field trials (Gupta et al., 2003). Further, compared to TR, a reduction in water use up to 60% and 10% increase in yield in DSR culture on raised beds was found. Gill et al. (2006) recorded higher water productivity (0.35 and 0.76) in DSR than TR (0.31 and 0.51) in a 2 years field experiment during 2002 and 2003, signifying high WUE in case of DSR. When compared with nonpuddled and zero tilled rice cultivation, DSR used 40% less water under no-till conditions than TR, whereas 30% water savings were recorded with DSR on raised beds coupled with 25% yield loss (Gupta et al., 2006). High economic prof-itability associated with no-tilled and DSR cultivation on raised beds may replace TR under puddled conditions with DSR culture (Bhushan et al., 2007).

Further, nutritional requirements of DSR are totally different than TR coupled with land preparation and water management strategies. In DSR, micronutrient deficiencies are of major concern such as Fe, S, Zn, Mn that might be due to improper N fertilization (Gao et al., 2006). High soil pH and carbonates and low redox potential might be the main reasons of Zn deficiency in paddy fields (Mandal et al., 2000). Fe oxidation through root released oxygen in aerobic soils lowers soil pH, thus restricts Zn release from insoluble fractions and leads to Zn-starved rice plants in DSR culti-vation (Kirk and Bajita, 1995); however, high soil pH enhances P and Zn solubility in rhizosphere (Saleque and Kirk, 1995). So, minimum resource utilization with maximum efficiency must be kept in mind to harvest more with less expense.

2.3.4   LABOR AND NET RETURN

No doubt, direct seeding of rice with chemical weed control saves labor consumption but is it beneficial for Asia, especially in South Asian region, where many people rely on the daily wages? Actually, it depends on the economic value of labor of a region. Under unemployment conditions, labor is not too much costly or may be lower than its market value but on the other hand it may be more costly at peak seasons and may be unavailable for most of the time. This uncertainty of the labor availability lead to adoption of farm mechanization as a technology-oriented approach for precision farming but its adoption largely depends upon the economic conditions of the farmers. In general, DSR requires less labor, less cost of production as compared to TR which needs more labor puddling, transplanting, and irrigation activities (Wang et al., 2002). Lack of trained labor is also a major constraint experienced by the rice growers for nursery transplanting and weeds management (WWF-ICRISAT Project, 2010). In total, labor scarcity necessitates the adoption of direct seeding of rice in Asia as it reduces labor requirement and may be adopted as a mechanized and precise operation for rice cultivation.

Direct seeding is a cost-effective method of rice cultivation and provides higher returns as compared to all other transplanting techniques, if weeds are controlled timely through herbicides, which require higher initial costs (Pathak et al., 2011). Although yield may be same or higher (in some cases) in TR but saving of irrigation water, reduced labor, easy and quick sowing, and reduced use of farm machinery give a leading edge to DSR cultivation and helps to raise the benefit:cost ratio (Ehsanullah et al., 2007). Naresh et al. (2013) supported DSR as a resource conservative approach than TR and further stated that although TR has even more yield but reduced production cost (about 20–30%) in DSR compensated this gap. Ali et al. (2013) compared different methods of rice cultivation and found the rice yield and cost:benefit ratio in following order: pregerminated seed broadcast < direct seeding < conventional rice plantation < line transplantation. Likewise, DSR produced same as TR under both dry and wet seasons, while dry seeding had an edge over TR regarding benefit:cost ratio (Ali et al., 2006).

2.3.5   COMPARATIVE ECONOMIC ASSESSMENT

Both direct seeded and TR have their own economic benefits and constraints; however, direct seeded have some additional benefits than TR in most of the Asian countries. Recently, a shift to DSR from TR in many Asian countries has been brought due to high water requirements, labor, and high input costs, as well as labor shortage at transplanting time for TR, which not only cause delay in rice sowing but also reduce profit margins (Pandey and Velasco 1999; Pandey and Velasco 2002). Higher labor demands for paddy field preparation, nursery uprooting, and transplantation pushed farmers to adopt DSR culture. Data/information presented in Table 2.6 showed that DSR provided almost equal yields to TR with lower production costs and higher net benefits. In this regard, Bhushan et al. (2007) also argued that higher net returns with increased economic profitability in DSR culture can easily replace TR. Moreover, early harvest in case of DSR may earn fair prices in the market than late mature TR (Shrestha, 2004; Dhakal et al., 2015). Even though DSR required less operational costs, farmers of most of the Asian countries still adopt TR rice culture method as it provides higher paddy yield with good quality which have higher market demands.

2.4 CONSTRAINTS

2.4.1 MAJOR CONSTRAINTS WITH DSR CULTURE

2.4.1.1 WEED INFESTATION

The major hurdle in successful production of DSR is the presence of intensive weed flora compared to TR (Rao et al., 2007; Ehsanullah et al., 2014). In conventional irrigated lowland rice systems, rice has a 15–20 days head start over weeds, which encourages rice to compete efficiently against unemerged weeds at transplanting time. The water layer after transplanting efficiently provokes the emergence and growth of various types of weeds including upland and semiaquatic weeds. Upland and aerobic rice face severe weed pressure and competition than flooded and lowland rice. Mahajan et al. (2010) recorded almost double weed infestation and biomass in DSR field than TR. In TR, weeds are discouraged by transplanting rice seedlings in stagnant water conditions, which have a head start over emerging weeds (Ashraf et al., 2014). Conversely, dry-tillage and alternate wetting and drying in DSR culture are favorable for luxurious weed growth causing yield loss up to 50–90%. Furthermore, weed-related losses in rice grain yield are about 10% and they may be more severe with higher weed density (Rao et al., 2007). Weed reduced grain yield up to 53% in wet-seeded and 74% in dry-seeded rice

TABLE 2.6 Comparative Economic Assessment of Direct-Seeded and Transplanted Rice Production Systems in Asian Countries.

Rice growing countries	Treatment description/research specifications/general information	Findings/remarks/consequences	Reference
Central Asia			
Afghanistan	A report was published regarding production and economical aspects of rice production in different regions of Afghanistan	Both direct-seeded and transplanted rice are widely practiced in Afghanistan, whereas Afghan-98, Basmati 385, Baghlan-98, Swat 2, and JP 5 are the most popular rice growing cultivars. The rice planted areas have shrinked due to lack of labor and due to a state of armed conflict. Furthermore, the production cost of rice in eastern regions was about 490 USD ha^{-1} with an average yield of 3.5 t ha^{-1} and 150 USD t^{-1} of paddy	GRiSP (2013)
Turkmenistan	Overall, rice is a minor crop of Turkmenistan, whereas cotton and wheat are the major ones. Rice is planted in April–May and harvested in August–September	The annual consumption rate of 13 kg milled rice per person was recorded whilst the average paddy yield was 2.43 t ha^{-1}	GRiSP (2013)
Uzbekistan	IWD method of rice establishment was compared with CT-I and CT-II systems in a field trial during 2008–2009	The maximum yield with higher net returns was obtained with the following trend: CT-I > CT-II > IWD. About 67–73% water was saved in case of IWD compared with CT-I, whereas CT-II showed the similar water productivity as IWD	Devkota et al. (2010)
Kazakhstan	The area under rice cultivation is 94 m ha out of the 23,400 m ha of arable land with an arid to semiarid type of climate	Rice is not a major item to consume, nevertheless the per head usage of milled rice was recorded up to 8.6 kg. The average paddy yield per annum was 3.97 t ha^{-1}	GRiSP (2013)
East Asia			
China	Classical transplanted rice cultivation system was compared with DSR. Indica rice cultivars, that is, "Lvhan1, Huanghuazhan, and Yangliangyou 6" were used as experimental material in a 2-year field experiment	The grain yield of DSR was identical as harvested from TR. Further, DSR used 15.3% less water with higher nitrogen use efficiency (~11–20%) than TR. Higher resource use efficiency of DSR proved better alternative to TR with higher net return and economic margins	Liu et al. (2015)

TABLE 2.6 (Continued)

Rice growing countries	Treatment description/research specifications/general information	Findings/remarks/consequences	Reference
Japan	Rice is extensively grown in the ranged from latitudes, including temperate, subtropical, and subfrigid area all around the country. It is widely cultivated in the plains of the major river basins, terraces, and valleys. Almost 85% of the farmer community is associated with rice cultivation	Japan has some serious constraints regarding rice cultivation, nevertheless by reducing production cost, applying advance technologies to enhance resource use efficiency and net benefits, use of paddy fields for multipurpose, as well as crop rotations and diverse cropping systems are important for sustaining rice yields. The average annual rice productions were recorded up to 6.51 t ha^{-1}	GRiPS (2013)
South Asia			
India	A 7-year field trial on rice–wheat cropping system with various establishment methods of rice and wheat, that is, CTR followed by CTW, CTR followed by ZTW; PBDSR followed by sowing PBW; ZTDSR followed by CTW; ZTDSR followed by ZTW (with and without residues) and unpuddled TR followed by ZTW was performed to evaluate the economic feasibility of rice–wheat rotation in eastern Gangetic plains of South Asia	The net returns were remained higher in CA-based systems including DSR (≥1175 USD) than CT (puddled transplanted rice) (≥1044 USD); however, CA-based economic benefits were realized after 2–3 years	Jat et al. (2014)
Bangladesh	Economic productivity of direct-seeded and transplanted "*Aman*" and "*Boro*" rice were evaluated in a field experiment. Direct seeding was performed by using drum seeder on wet soil while seedlings were manually transplanted in wet puddled field in transplanted rice production system	The production cost of DSR were remained up to 417–446 USD with net returns of 1042–1126 USD in comparison to TR where cost of production was 403–505 USD with net returns of 993–1125. DSR gave 4–6% higher gross margins, nonetheless input cost remained 12–20% higher than TR	Rashid et al. (2009)

TABLE 2.6 (Continued)

Rice growing countries	Treatment description/research specifications/general information	Findings/remarks/consequences	Reference
Pakistan	Economic feasibility, water productivity, and yield dynamics in transplanted rice (with continuous and alternate drying and flooding conditions) and DSR were assessed in a 2-year field trial (2008–2009) with following local rice cultivars: "Super Basmati," "Shaheen Basmati," and "Basmati 2000"	Both transplanted rice with alternate wetting and drying and direct seeding proved effective in water saving with higher water productivity than transplanted rice with continuous flooding. Direct-seeded "Shaheen Basmati" and "Basmati 2000" in transplanted rice with continuous flooding provided the highest economic returns during 2008 and 2009, respectively. Moreover, in 2008, Shaheen Basmati and Super Basmati grown as direct seeding had the highest benefit–cost ratio (2.76 and 2.69, respectively), while Basmati 2000 grown under continuous flooding had the lowest (2.03). In 2009, "Basmati 2000" established under continuous flooding had the highest benefit–cost ratio (1.83)	Jabran et al. (2015)
Sri Lanka	Farmers' perceptions about rice planting methods, cultural practices adopted, and yield gaps in DSR culture compared with TR was surveyed by using a structured questionnaire, published/unpublished data/reports and field surveys in different regions	About 99% of the farmers switched to direct seeding from transplanted rice because of (1) high cultivation cost, (2) low return to invest, and (3) less time for crop establishment, more than 92% farmers complained for reduced labor requirements in DSR and labor shortage for transplanting for TR. Furthermore, reasons of 76% farmers for DSR adoption was the same returns as in TR while 45% farmers claimed less water availability was the cause to shift to DSR culture	Weerakoon et al. (2011)
Nepal	A group of 60 farmers (selected from the master list of CSISA) were chosen randomly in 2010 from different areas of Nepal, who shifted to DSR cultivation from the last 1 year (2009) to TR cultivation system, were interviewed to know about the general perception and economic assessment of both rice cultivation systems	DSR culture required less cost of cultivation than TR, whereas net returns per hectare also remained lower in DSR than TR. Similarly, benefit–cost ratio was also remained higher (2.00) in DSR compared with TR (1.63). Less labor, reduced irrigation water, higher paddy price at early harvest were the principle components of higher profitability in DSR	Dhakal et al. (2015)

TABLE 2.6 (Continued)

Rice growing countries	Treatment description/research specifications/general information	Findings/remarks/consequences	Reference
Bhutan	Direct seeding and nursery transplanting are the most common rice establishment methods in Bhutan; however, latter is the most common among Bhutanese farming community. Normally, transplanting is practiced in all zones while direct seeding is confined to mid and low zones only	Poor stand establishment and weeds are the most common problems in DSR while more labor is required for uprooting and transplanting of the rice seedlings in TR. Overall, less labor, early harvest, and similar yields in DSR provided better economic returns than TR. The cost of rice production is higher in high altitudes than mid and low altitudes while farmers who adopted the improved farm practices with latest varieties earned 110% more cash income than nonadopters	Shrestha (2004)
Southeast Asia			
Vietnam	To compare the labor usage in direct-seeded and transplanted rice, farm survey data were collected from Long An, Vietnam	Rice cultivation through conventional tillage (flooded transplantation) required 68 persons-days ha^{-1} while wet/dry DSR only required 38 persons-days ha^{-1}. Hence, DSR saved up to 40 labor and reduced overall production cost	Pandey et al., 2002
Myanmar	Two types of rice cultivation methods are widely adopted in Myanmar, that is, dry upland (includes direct/dry seeding) and wet cultivation in monsoon (nursery transplantation in partially submerged and fully submerged conditions)	Direct seeding was reported to require 1.5-fold higher cost ha^{-1} as compared to wet cultivated rice in monsoon due to higher labor, fertilizer, diesel, and water application costs in case of DSR; however, dry seeding needs extra irrigations to get optimum yields	Young et al. (1998)
Philippines	A partial budget analysis, labor cost at the expense of crop-establishment method was estimated in a study titled "Benchmarking the Philippine rice economy relative to major rice-producing countries in Asia"	Direct seeding was found economically effective with a labor cost of 3.32 P/kg of rice with 2 man-days h^{-1} as compared to TR where labor cost was 34.75% higher (4.42 P/kg of rice) with 25 man-days h^{-1}. Overall, total expenses, labor requirements, and time consumption was significantly higher in TR compared with DSR	Bordey et al. (2015)

TABLE 2.6 (Continued)

Rice growing countries	Treatment description/research specifications/general information	Findings/remarks/consequences	Reference
Cambodia	A joint project "Cambodia–IRRI-Australia" was initiated to compare the efficiency of different rice production systems, that is, traditional transplanted and improved direct seeded, economic models, and constraints in rice production systems in Cambodia	In traditional TR production system the net revenue was remained lower (131.00 USD) with a return person^{-1} day of 1.191 USD compared with DSR where 180.70 and 1.505 USD were recorded as net revenue and return person^{-1} day, respectively	Rickman et al. (2001)
Thailand	Cooperative performance of DSR (broadcasted) and TR was evaluated in various regions of Thailand	In many areas of Thailand, especially northeastern part, the conventional transplanting of rice has been majorly replaced by direct seeding due to higher production costs, low net returns and more labor requirements	Naklang et al. (1997)
Indonesia	Rice is cultivated in lowlands (mostly transplanted flooded rice) and uplands (direct seeded + wet seeded). Major portion of rice is produced from two islands "Java" and "Sumatra," while about 60% total production only from Java Island. Lowland rice is artificially irrigated while upland is totally dependent on rainfall	Lowland rice received less irrigations, low fertilizer rates, and labor requirements which keep its production costs low, whereas upland rice mainly depends on rainfall	GRiSP (2013)
Korea	Cost and labor usage in mechanized transplanted rice was compared with wet DSR	The cost of production and labor usage was remained higher in TR than DSR where only 27 person-days ha^{-1} was used compared with 42 person-days ha^{-1} in TR	Lee et al. (2002)
Malaysia	Relative cost of production and rice productivity of both direct and transplanted (mechanized/manual) rice cultivation systems has been assessed	The total cost of direct seeding ha^{-1} was lower (329.09 USD) than mechanized and manual transplanting, that is, 402.42 and 504.54 USD, respectively, nevertheless, the profit percentage were remained maximum in mechanized transplanting (164%) followed by and direct seeding (68%) and manual (64%) transplantation of rice	Wah (1998)

TABLE 2.6 (Continued)

Rice growing countries	Treatment description/research specifications/general information	Findings/remarks/consequences	Reference
Southwest Asia			
Iran	A field experiment with split-plot design was conducted where cultivars were randomized in main-plots while rice planting methods in subplots to assess the comparative performance of direct-seeded and transplanted rice	The both rice cultivation methods are significantly different regarding rice yields where TR gave higher yields than DSR, nevertheless DSR proved cost-effective with 20–30% less production cost. The reduced production cost was attributed to less farm operations, less labor usage, and reduced water requirements in DSR	Akhgari and Kaviani (2011)

CA, conservation agriculture; CT, conventional tillage; CT-I, conventional transplanted rice with flooded irrigation systems; CT-II, conventional transplanted rice with intermittent irrigation systems; CTR, conventional transplanted rice; CTW, conventional tilled wheat; DSR, direct-seeded rice; IWD, Intermittent wet and dry; PBDSR, permanent bed direct-seeded rice; PBW, permanent bed wheat; ZTDSR, zero-tilled direct-seeded rice; ZTW, zero-tilled wheat.

(Ramzan, 2003) and up to 68–100% in DSR (Mamun, 1990). DSR fields were found to accomplish by diverse weed species than TR (Tomita et al., 2003).

Canopy of TR captures more space and have leverage over flourishing weeds by establishing their roots earlier than germinating weeds. Moreover, instant flooding after transplanting restricts the establishment of various weeds (Johnson, 2002). In the same way, by using the high seeding rates, weed density in DSR can be reduced to a large extent (Hayat et al., 2003). Further, Mahajan et al. (2009) reported 28.3% more weed pressure in DSR as compared to TR. Thus, it seems that weeds are the major restraint to DSR production and hence, accomplishment of this technology generally depends on successful weed management. Major rice-infested weeds, especially in Asia, have been represented in Table 2.7.

TABLE 2.7   Most Common Rice Weeds in Asia.[a]

S. N.	Botanical names	Category	Family
1	*Paspalum distichum*	Grass	Poaceae
2	*Echinochloa colona*	Grass	Poaceae
3	*Leptochloa chinensis*	Grass	Poaceae
4	*Echinochloa crus-galli*	Grass	Poaceae
5	*Oryza sativa* (weedy rice)	Grass	Poaceae
6	*Digitaria setigera*	Grass	Poaceae
7	*Digitaria ciliaris*	Grass	Poaceae
8	*Eleusine indica*	Grass	Poaceae
9	*Ischaemum rugosum*	Grass	Poaceae
10	*Digitaria ciliaris*	Grass	Poaceae
11	*Cyperus rotundus*	Sedge	Cyperaceae
12	*Fimbristylis miliacea*	Sedge	Cyperaceae
13	*Cyperus difformis*	Sedge	Cyperaceae
14	*Cyperus iria*	Sedge	Cyperaceae
15	*Commelina benghalensis*	Broadleaved	Commelinaceae
16	*Monochoria vaginalis*	Broadleaved	Pontederiaceae
17	*Eclipta prostrata*	Broadleaved	Asteraceae
18	*Ipomoea aquatica*	Broadleaved	Convolvulaceae
19	*Ludwigia octovalvis*	Broadleaved	Onagraceae
20	*Sphenoclea zeylanica*	Broadleaved	Sphenocleaceae
21	*Ludwigia adscendens*	Broadleaved	Onagraceae

[a]IRRI, 2003. Modified from Juraimi et al., 2013.

2.4.1.2   NEMATODE INVASION

In partially and fully aerobic conditions (in case of DSR), some soilborne pests and diseases were reported to invade the yield, particularly root knot nematodes (RKN). About 12–80% higher rice yield was reported when actions were taken against RKN in upland rice or rice grown in temporarily submerged conditions (Prot and Matias, 1995). Flooded soils, alternatively, may lower the chances of root damage caused by RKN because they cannot attack under water-saturated conditions (Bridge et al., 2005). Kreye et al. (2009) conducted an experiment to ascertain an interactive effect of water and nitrogen on phenology of the rice under tropical environmental conditions and concluded invasion of RKN is also a significant factor of rice yield reduction. Furthermore, they found the similar results in another experimental modeling approach which includes both biotic (RKN) and abiotic (water and nutrients) stress factors.

2.4.1.3   NUTRIENT ACCESSIBILITY

Transferring from TR to DSR, nutrient acquisition is the most important factor. Paddy fields in saturated and unsaturated conditions differ in their physicochemical properties and nutritional status. Belder et al. (2005) reported comparatively low nitrogen uptake (about 22%) and nitrogen recovery under aerobic soil in comparison with puddled conditions where 49% N-uptake was recorded. Moreover, about 47% nitrogen has been volatilized under aerobic field conditions due to rapid nitrification–denitrification processes. Increased nitrification–denitrification processes may be explicated by differences in redox potential in aerobic and flooded soils. Application of N-fertilizer as basal dose (just before transplanting) showed the lowest recovery of nitrogen, however, when applied at the rate of 150 kg N ha^{-1}, only 31% of applied concentration was retained in the soil and roots after harvest. Hence, high N recoveries (0.6–0.7 kg kg^{-1}) in rice give an idea that increased N recoveries in DSR might be achievable by adjusting N doses with its application time and crop requirement. Furthermore, nutrient management such as deep placement and the use of slow release fertilizers might be used in aerobic soils (Zeigler and Puckridge, 1995). However, under flooded conditions, N should be converted and assimilated in ammonia (NH$_4^+$) form for better nitrogen use efficiency (NUE) which can be increased up to two-fold with improved management practices (Cassman et al., 1998). Application of nitrification inhibitors, that is, dicyandiamide can also be applied with fertilizer in paddy fields. This chemical is commercially marketed in Japan. Improving NUE with reduced volatilization losses, N-recovery may be enhanced under aerobic soil conditions. Rice yield associated with NUE can be enhanced finding best nitrogen–water combinations (Bouman et al., 2006).

2.4.1.4 PANICLE STERILITY AND STAGNANT YIELD

Both direct seeded and TR are sensitive to drought; however, effects are more obvious at flowering (the reproductive stage) (Liu and Bennett, 2011). Thus, any short incidence of drought period may prove devastating especially in DSR than TR. Reduced water potential lessens time to anthesis, and thus lead to panicle sterility and ultimately yield reduction (Farooq et al., 2006). Drought conditions also diminish pollen viability by reducing starch content thus resulted in hampered yield by reducing panicle fertility (Lalonde et al., 1997). Further, Farooq et al. (2009) observed more sterility percent as well as opaque, chalky, and futile kernels in DSR than TR.

Furthermore, reduced yield in DSR (Kreye et al., 2009; Farooq et al., 2007) might have a correlation or interaction with changing macro or microclimate around the rhizosphere, for example, soil degradation and structural imbalance, nutrient diminution, soil sickness due to soilborne pathogens, and allopathic interventions (Ventura and Watanabe, 1978). A substantial reduction in the subsequent rice crop was reported in Philippines due to presence of allelochemicals in the soil from the preceding rice stubbles (Olofsdotter, 2001). In Taiwan, phytotoxins (released from previous rice residues) caused 25% yield reduction to the succeeding rice crop (Chou, 1980).

2.4.1.5 CULTIVAR DEVELOPMENT AND LODGING ISSUES

Cultivar development for DSR was not emphasized properly that is a core reason behind its low popularity among farming community. Development of lodging and blast resistant, deep rooted, early vigor, short duration, and short statured varieties is still a question to be answered (Ikeda et al., 2008). Varieties having improved resistance against blast along with resistance against adverse soil conditions such as mineral deficiency and mineral toxicity and better herbicides tolerance with short mesocotyls are need of time. Early vegetative vigor accompanied with early maturity is very important to lessen weeds competition and for getting higher yields in double cropping systems (Coffman and Nanda, 1982). Short duration and drought-tolerant rice varieties are suitable for DSR, such as IR36 with 105-day duration and good drought tolerance (Gines et al., 1978). Farooq et al. (2011) observed that lodging, the permanent vertical displacement of stem of crop plant (Berry et al., 2004), was more common in DSR fields as compared to transplanted rice (TPR) fields. Yun et al. (1997) compared growth and yield of cultivars of japonica–indica crosses under DSR upland fields and found that indica cultivars performed better in terms of yield than japonica cultivars and must be included in breeding programs for DSR. Rice cultivars having thick band of sclerenchyma are less prone to lodging (Ramaiah and Mudaliar, 1934). Moreover, lodging-resistant rice cultivars have more vascular bundles, both peripheral and in the inner section of the outer layers (Chaturvedi et al., 1995). During ripening phase, DSR is more prone to lodging as compared to TPR (Setter et al., 1997). Lodging reduces yield as photosynthesis is decreased as a result of self-shading, and grain quality is hampered due to increased coloring and decreased taste (Matsue et al., 1991; Setter et al., 1997). Additionally, a lodged crop is extremely burdensome to be harvested mechanically. In this context, thick stem with broad stem walls, intermediary plant heights, and high lignin contents are the characteristics for lodging-resistant cultivars (Mackill et al., 1996).

2.4.1.6   DISEASE RISK

Rice, in both conditions; transplanted or direct seeded is prone to several diseases; among which, rice blast is most devastating one (Bonman and Leung, 2004; Farooq et al., 2011). However, an aerobic condition intensifies the severity of rice blast (Bonman 1992; Mackill and Bonman, 1992). Blast spread is favored by deficiency of water and move to direct seeding culture from transplanting. Different processes such as germination, spore formation and release, and rice blast epidemics are affected by level of water supply as revealed by some studies (Kim, 1987). Pathogen's life cycle is influenced by crop microclimate, especially dew, which increases the host susceptibility which is related to water management practices (Sah and Bonman, 2008). In DSR, unlike anaerobic culture, soil is found in moist or dry condition because of poor management of irrigation practices which favors dew depositions and makes the conditions favorable for blast development (Savary et al., 2005). High incidence of diseases such as sheath blight, dirty panicle, ragged stunt virus, and yellow orange leaf virus has been observed in DSR (Pongprasert, 1995). But, at times, more attack of arthropods insects was noticed in TR as compared with DSR (Oyediran and

Heinrichs, 2001). Shifting toward dry seeded culture is also influenced by RKN (Prot et al., 1994).

2.4.2 MAJOR CONSTRAINTS WITH TR

2.4.2.1 WATER AND SOIL RESOURCE-BASED DEGRADATION

Conventionally, 3000–5000 L of water is used to produce 1 kg of rice. Rice crop demands two times more water in comparison with maize or wheat, so the productivity of water in rice is much less. Globally, shares of rice in fresh water consumption are about 85%. Rice crop consumes five to six times more water as compared to wheat in some regions of IGP. In many parts of South Asia, the diminishing resources of water have augmented competition between industrial and domestic sector, and increasing costs are already affecting the sustainability of flooded rice culture. For example, in the upper transect of the IGP, rice cultivation resulted in a decline in water tables and water quality. Furthermore, in India, during last two decades, water table declined in the range of 3–10 m in several districts in the rice–wheat growing regions of Haryana. In central Punjab only, the groundwater table has fallen at about 23 cm y^{-1}. Excessive pumping depletes ground water and causes pollution such as arsenic contamination as has been observed in many parts of West Bengal (Pathak et al., 2011). Water application in rice production systems needs to be reduced by improving WUE through techniques such as laser land leveling, crack plowing to reduce bypass flow and bund maintenance to reduce losses caused by seepage, percolation, and evaporation. In general, DSR has potential to improve the WUE. However, extensive research is needed on this aspect.

Concerns about sustainability are arising throughout tropical rice ecosystems because of decreasing soil fertility as most countries move into the postgreen revolution era. Recent trends of yield decline/stagnation observed in long-term experiments in South Asia were mostly due to soil-related causes such as the decline in soil carbon (C) and macro- and micronutrients in rice–rice and rice–wheat cropping systems; accumulation of phenolic compounds, Fe^{2+}, and sulfides in the rice–rice system; and the increase in soil salinity (Neue et al., 1998; Erenstein, 2009; Pathak et al., 2011; Jat et al., 2014, Ahmed et al., 2015; Singh et al., 2016). In short term, salinity buildup leads to the reduced yields, whereas in the long term, it can lead to abandoning of crop lands. Farmers are also using poor-quality water for irrigation in several areas of the IGP for rice and run the risk of further aggravating soil degradation. The soil quality of rice systems therefore needs to be continuously monitored and efforts should be made to improve the soil health.

2.4.2.2 RICE RESIDUE MANAGEMENT

Rice residues are often not managed or disposed off in a proper way. For example, in the states of Haryana and Punjab (India), about 60–80% of rice straws are burnt in the field (Pathak et al., 2006). In Punjab (India) only, about 12 million t of rice residues are burnt annually. In order to prepare lands for the succeeding crop, rice residue has to be burned, removed, or incorporated into the soil. Residue burning and removal from the rice fields are the two mostly adopted practices in Punjab (Pakistan) (Ahmad et al., 2015). This practice of burning rice residues is not acceptable environmentally as it not only releases smoke and soot particles which will lead to human respiratory disorders but also results in emission of greenhouse gases (GHG) such as nitrous oxide (N_2O), methane (CH_4), and carbon dioxide (CO_2) leading to global warming and nutrient loss (Le Mer and Roger, 2001; Zhang et al., 2011). Furthermore, it results in loss of plant nutrients such as N, P, K, and S. Almost the entire amounts of C and N, 25% of P, 50% of S, and 20% of K present in straw are lost due to burning. According to Timsina and Connor (2001), about 85% of India's cereal production comes from Rice–wheat cropping system that has been spread over an area of about 10 m ha in IGP of India (Kumar et al., 1998). Residues burning and removal practices adopted during the whole crop season accelerated problems such as environmental pollution, soil erosion, and soil degradation (Montgomery, 2007) and disturbing ecosystem stability (Srinivasan et al., 2012). In Pakistan, farmers use rice straw predominantly to feed animals, when it is not burnt. Each practice has different cost of implications. Complete removal of residues on average costs 4586 PKR (~55 USD) per acre, whereas on average basis, complete removal of residue is 34% more costly to farmers as compared to their full burning. Thus, farmers would need to be subsidized to avoid residue burning practices. Farmers' residue management decisions are influenced by a number of socioeconomic factors (Ahmad et al., 2015). In two different rice cropping systems in China, residues retention increased stock of soil organic carbon and also increased yields in comparison with residue removal from the paddy fields (Chen et al., 2015). Contrarily, in DSR, by using "Happy Seeder" machine, seeds can be sown directly on residue retained fields which may discourage the practice of burning of rice residues.

2.4.2.3   GREENHOUSE GAS EMISSION

Methane (CH_4) and nitrous oxide (N_2O) emissions from rice production systems contribute to global climate change. A small but highly specific bacteria group called methanogens produce methane through the metabolic activities. Submerged, anaerobic condition developed in wetland rice fields increases their activity which limits the transport of oxygen into the soil, and activities of microbes make the water-saturated soil practically free of oxygen. Methane is not produced in upland, dry soil. Therefore, water management plays a vital role in methane emission. In anaerobic rice, methane emission can be controlled by alternate wetting and drying, especially aeration during mid-season. During off-season, incorporation of organic matter into soil or composting by boosting the microbial activities can be an effective tool. Pathak et al. (2010) reported that 3.37 Mt of CH_4 was emitted from Indian rice fields spreading on area of 43.86 m ha in 2007. Different rice cultures, that is, irrigated multiple aeration rice, deep water, single aeration, rainfed drought prone, rainfed flood prone rice, and irrigated continuously flooded rice contributed toward methane emission 4%, 8%, 16%, 17%, 21%, and 34% of CH_4, respectively. For example, only rice production systems (DSR and TR) accounts up to 8.7–28% of the CH_4 emitted by anthropogenic activities (Mosier et al., 1998); however, rice might be the first most target of GHS emitted from the paddy fields (Wassmann et al., 2004). The maximum emission was from TR culture (34%) followed by rainfed flood prone rice fields (21%). Nitrous oxide, 298 times more effective than CO_2, is also potentially released from paddy fields. In comparison to DSR, TR emits higher GHGs depicted in Figure 2.4 (Pathak et al., 2012). The major sources are burning of fossil fuels and organic material, fertilizer and manure application, and soil cultivation. Higher rates of GHGs emission were also recorded in TR compared with DSR (Corton et al., 2000). Appropriate crop-management practices with improved NUE hold the key to reduce nitrous oxide emission. GHGs emissions can be reduced by site-specific nutrient management, biochar application, fertilizer placement, and nutrient management in a better accordance with plant demands (Schlesinger, 1999; Ahmad et al., 2009; Li et al., 2013; Nayak et al., 2013). Alternatively, DSR culture limits the emission of GHGs from paddy fields.

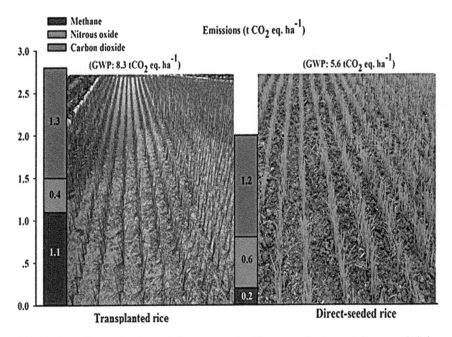

FIGURE 2.4 **(See color insert.)** GHGs emission from transplanted and direct-seeded rice production systems.

Source: Reprinted from Hussain, S.; Peng, S.; Fahad, S.; Khaliq, A.; Huang, J.; Cui, K.; Nie, L. Rice Management Interventions to Mitigate Greenhouse Gas Emissions: A Review. *Environ. Sci. Pollut. Res.* **2015**, *22*, 3342–3360. Springer-Verlag: Berlin, Heidelberg, 2014. With permission from Springer.

2.4.2.4 PESTICIDE- AND FERTILIZER-RELATED POLLUTION

The pesticides usage has become an integral part of puddled rice cultivation as it has enhanced productivity of agricultural production systems. Intensive use of these pesticides is posing serious threats to natural resources by causing water, air, and soil pollution in rice production systems. Whatever the method, formulation, or applicator used, some part of these pesticides escapes into environment and enters into food chain causing health problems such as decreased fertility, high cholesterol level, high infant mortality, carcinogenicity, and various genetic and metabolic disorders in human beings. The use of pesticides is also disturbing natural balances of ecosystem as it kills the natural beneficial insects of rice crops which feed upon harmful insects. Integrated pest management should be adopted to lessen pesticides use and to reduce pesticide-related pollution.

Fertilizer-related pollution is caused mainly due to excessive application of nitrogenous fertilizer and at inappropriate times in many intensively flooded rice systems, which results in poor recovery of fertilizer recovery by the rice crop. Only one-third of applied nitrogen (N) is taken up by crop and the remaining two-third is lost from soil–plant systems into the environment as a result of volatilization, denitrification, leaching, and runoff causing fertilizer pollution (Ladha et al., 2005). The major pathways resulting in losses are leaching; mainly nitrate (NO^{3-}) and to some extent NH_4+, and denitrification of soluble organic N resulted in emissions of dinitrogen (N_2) gases, nitrous oxide (N_2O), nitric oxide (NO), and ammonia (NH_3) volatilization (Xing et al., 2009; Qin et al., 2010). NUE can be improved significantly by adopting proper nutrient management practice, that is, in time and balanced use of fertilizer keeping in view the demand and supply factor (Abid et al., 2015). It will maximize the uptake of nutrients by crops and will reduce the losses through volatilization, denitrification, etc.

2.5   CONCLUSIONS AND FUTURE DIRECTIONS

World's changing climatic scenario and future projections of climate change indicate that agricultural productivity may be affected in both ways (positively and negatively). However, negative effects may be more severe than positive. Increasing temperature may cause drought, water scarcity, heat stress, can change rainfall pattern, seasonal variations and overlaps, crop sowing and harvesting periods to a large extent. Specifically in rice, chances of yield reduction are more than yield increase but options are always there to combat the challenges. Regarding rice cultivation method, rice grown by transplanting of nursery gives better yield and grows well under standing water but world's fresh water resources are not sufficient to cultivate rice in this traditional method, predominating water crises warrant the sustainability of TR and demanded the implementation of other water saving technologies. Direct seeding of rice may be proved an efficient technology regarding better production with minimal water losses, reduced GHGs emission, higher water and NUEs, reduced labor requirements, and cost effective. In order to get promising yield, suitable cultural practices, weed-free environment, and proper crop care on regular basis should be carried out with integrated weed management programs.

On the other hand, unscientific water management and imbalanced fertilizer application put a terrific pressure on the rice growers to make rice farming more feasible and sustainable regarding economically and ecologically. Hence, rice yield penalty and yield sustainability have to keep in mind for adoption in the farmer community. From the literature that we have reviewed, it is unequivocal that direct seeding of rice has been recognized as an impending new technology of rice cultivation having low water requirements and economically attractive, nevertheless, constraints with weed management and development of prominent rice genotypes suitable to grow under aerobic conditions have to be aloof to make it more proficient and to get maximum returns with limited water resources.

In future, further improvements in crop modeling (techniques, tools, and methodologies) and development of simulation crop models may help to assess the future climate scenarios with respect to crop improvement. Introduction of new and cost-effective rice growing techniques, planting methods, and farm machinery which may reduce the water losses effectively are the needs of the day. Developing new ideotypes for rice by analyzing the signaling mechanisms under aerobic environment and by knowing the activation mechanism of aquaporins in the root system of aerobic rice may be quite helpful for suitable rice selection. Assessment of physiological diversity in plants by quantitative trait loci may quite be helpful to understand the plants' adaptive response to a specific environment and development of rice genotypes suitable for direct-seeding environments. There is also a need to ensue stay green mutant plants, climate smart rice, and microirrigation systems for future prospective in rice cultivation. Moreover, drought-tolerant lines having high yield potential should be selected for aerobic cultivation.

2.6 SUMMARY

Rice is one of the most important cereals which serves as staple food for more than half of the world. Undoubtedly, the conventional method (via transplantation of rice seedlings in puddled fields) to grow rice is generally ascribed with higher yield that could be attributed to better weed control. Nonetheless, this yield increment is on the expense of employing intensive labor and water usage for its cultivation which leads researchers to think about improved rice cultivation techniques under current scenario of fresh water scarcity. DSR has been considered as an alternative method of rice cultivation that encompasses low input demands and is being practiced successfully in developed countries and being adopted in many Asian countries such as Pakistan, Thailand, Philippines, Malaysia, and India. However, shift from transplantation to direct seeding may substantially lessen water requirements, nutrient demands, emissions of GHGs from paddy fields, and economic costs, but weed infestation, nutrient accessibility, nematode invasion, panicle sterility and stagnant yield, cultivar development and lodging, and disease risk are still the major challenges in this regard. In this review, we discussed the climatic effects on rice, yield and water dynamics and comparative WUE, resource utilization, labor requirements and net returns of both DSR and TR, and some common problems related to their cultivation. In this perspective, potential approaches, techniques, and practices associated with rice cultivation were also suggested as future directions in the end.

ACKNOWLEDGMENTS

The funding was provided by National Natural Science Foundation of China (31271646), Natural Science Foundation of Guangdong Province (8151064201000017), Agricultural Research Projects of Guangdong Province (2011AO20202001), and Agricultural Standardization Project of Guangdong Province (4100 F10003) in completing this review is highly acknowledged. Authors are also thankful to Dr. Shakeel Ahmad Anjum and Dr. Imran Khan from Department of Agronomy, University of Agriculture, Pakistan for their critical comments and useful suggestions during preparation of this chapter.

KEYWORDS

- **direct-seeded rice**
- **transplanted rice**
- **chilling stress**
- **greenhouse gases**
- **rice cropping systems**
- **rice yield variability**
- **yield dynamics**

REFERENCES

Abid, M.; Khan, I.; Mahmood, F.; Ashraf, U.; Imran, M.; Anjum, S. A. Response of Hybrid Rice to Various Transplanting Dates and Nitrogen Application Rates. *Philippine Agr. Sci.* **2015,** *98* (1), 98–104.

ADB (Asian Development Bank). *The Economics of Climate Change in Southeast Asia: A Regional Review*; Manila, Philippine, 2009.

Ahmad, S.; Li, C. F.; Dai, G. Z.; Zhan, M.; Wang, J. P.; Pan, S. G.; Cao, C. G. Greenhouse Gas Emission from Direct Seeding Paddy Field Under Different Rice Tillage Systems in Central China. *Soil Tillage Res.* **2009**, *106*, 54–61. DOI:10.1016/j.still.2009.09.005

Ahmad, P.; Hashem, A.; Abd_Allah, E. F.; Alqarawi, A. A.; John, R.; Egamberdieva, D.; Gucel, S. Role of *Trichoderma harzianum* in Mitigating NaCl Stress in Indian Mustard (*Brassica juncea* L) Through Antioxidative Defense System *Front. Plant Sci.* **2015**, *6*, 868, DOI: 10.3389/fpls.2015.00868.

Ahmed, T.; Ahmed, B.; Ahmed, W. Why do Farmers Burn Rice Residue? Examining Farmers' Choices in Punjab, Pakistan. *Land Use Policy* **2015**, *47*, 448–458.

Akbar, N.; Ehsanullah, Jabran, K.; Ali, M. A. Weed Management Improves Yield and Quality of Direct Seeded Rice. *Aust. J. Crop Sci.* **2011**, *5*, 688–694.

Akhgari, H.; Kaviani, B. Assessment of Direct Seeded and Transplanting Methods of Rice Cultivars in the Northern Part of Iran. *Afr. J. Agric. Res.* **2011**, *6* (31), 6492–6498

Ali, M. A.; Ladha, J. K.; Rickman, J.; Lales, J. S. Comparison of Different Methods of Rice Establishment and Nitrogen Management Strategies for Lowland Rice. *J. Crop Improv.* **2006**, *16*, 173–189.

Ali, M. A.; Ladha, J. K.; Rickman, J.; Lales, J. S. Nitrogen Dynamics in Lowland Rice as Affected by Crop Establishment and Nitrogen Management. *J. Crop Prod.* **2007**, *20*, 89–105.

Ali, Q. M.; Ahmad, A.; Ahmed, M.; Arain, M. A.; Abbas, M. Evaluation of Planting Methods for Growth and Yield of Paddy (*Oryza sativa* L.) Under Agro-Ecological Conditions of District Shikarpur. *Am.-Eurasian J. Agric. Environ. Sci.* **2013**, *13*, 1503–1508.

Anwar, M. P.; Juraimi, A. S.; Man, A.; Puteh, A.; Selamat, A.; Begum, M. Weed Suppressive Ability of Rice (*Oryza sativa* L.) Germplasm Under Aerobic Soil Conditions. *Aust. J. Crop Sci.* **2010**, *4*, 706–717.

Ashraf, U.; Anjum, S. A.; Ehsanullah, Khan, I.; Tanveer, M. Planting Geometry-induced Alteration in Weed Infestation, Growth and Yield of Puddled Rice. *Pak. J. Weed Sci. Res.* **2014**, *20*, 77–89.

Awan, T. H.; Ali, I.; Safdar, M. E.; Ahmad, M.; Akhtar, M. S. Comparison of Parachute, Line and Traditional Rice Transplanting Methods at Farmer's Field in Rice Growing Area. *Pak. J. Agric. Sci.* **2008**, *45*, 432–438.

Azmi, M.; Chin, D. V.; Vongsaroj, P.; Johnson, D. E. In Emerging Issues in Weed Management of Direct-seeded Rice in Malaysia, Vietnam, and Thailand. *Rice is Life: Scientific Perspectives for the 21st Century*, Proceedings of the World Rice Research Conference, Tsukuba, Japan, Nov 4–7, 2004; 2005; International Rice Research Institute, Manila, pp 196–198.

Bachelet, D.; Kern, J.; Tolg, M. Balancing the Rice Carbon Budget in China Using Spatially-distributed Data. *Ecol. Model.* **1995**, *79*, 167–177.

Baker, J. T. Yield Responses of Southern US Rice Cultivars to CO_2 and Temperature. *Agric. For. Meteorol.* **2004**, *122*, 129–137.

Baker, J. T.; Allen Jr, L. H.; Boote, K. J. Response of Rice to Carbon Dioxide and Temperature. *Agric. For. Meteorol.* **1992**, *60*, 153–166.

Balasubramanian, V.; Hill, J. E. In Direct Seeding of Rice in Asia: Emerging Issues and Strategic Research Needs for 21st Century. *Direct Seeding Research Strategies and Opportunities*; International Rice Research Institute: Manila, Philippines, 2002; pp 15–39

Baloch, M. S; Awan, I. U.; Hussan, G.; Khakwani, A. A. Effect of Establishment Methods and Weed Management Practices on some Growth Attributes of Rice. *Rice Sci.* **2006,** *13,* 131–140

Basak, J. K.; Ali, M. A.; Islam, M. N.; Alam, M. J. B. Assessment of the Effect of Climate Change on Boro Rice Production in Bangladesh Using CERES-Rice in a Temperate Climate. *Glob. Change Biol.* **2009,** *19,* 548–562.

Belder, P.; Bouman, B. A. M.; Spiertz, J. H. J.; Peng, S.; Castaneda, A. R.; Visperas, R. M. Crop Performance, Nitrogen and Water Use in Flooded and Aerobic Rice. *Plant Soil* **2005,** *273,* 167–182.

Berry, P. M.; Sterling, M.; Spink, J. H.; Baker, C. J.; Sylvester-Bradley, R.; Mooney, S. J.; Tams, A. R.; Ennos, A. R. Understanding and Reducing Lodging in Cereals. *Adv. Agron.* **2004,** *84,* 217–271.

Bhuiyan, S. I.; Sattar, M. A.; Khan, M. A. K. Improving Water Use Efficiency in Rice Irrigation Through Wet Seeding. *Irrigation Sci.* **1995,** *16,* 1–8.

Bhushan, L.; Ladha, J. K.; Gupta, R. K.; Singh, S.; Tirol-Padre, A.; Saharawat, Y. S.; Gathala, M.; Pathak, H. Saving of Water and Labor in a Rice–Wheat System with No-tillage and Direct Seeding Technologies. *Agron. J.* **2007,** *99,* 1288–1296.

Bonman, J. M. Durable Resistance to Rice Blast Disease-environmental Influences. *Euphytica* **1992,** *63,* 115–123.

Bonman, J. M.; Leung, H. Breeding for Durable Resistance to Rice Blast Disease Dream or Reality? *Phytopathology* **2004,** *93* (Suppl 113). Publication No. P-2003-0110-SSA. American Phytopathological Society Annual Meeting.

Bouman, B. A. M. Water-efficient Management Strategies in Rice Production. *Int. Rice Res. Notes,* **2001,** *16,* 17–22.

Bouman, B. A. M.; Toung T. P. Field Water Management to Save Water and Increase Productivity in Irrigated Low Land Rice. *Agric. Water Manag.* **2001,** *49,* 11–30.

Bouman, B. A. M.; Peng, S.; Castaneda, A.; Visperas, R. M. Yield and Water Use of Irrigated Tropical Aerobic Rice Systems. *Agric. Water Manag.* **2005,** *74,* 87–105.

Bouman, B. A. M.; Yang, X.; Wang, H.; Wang, Z.; Zhao, J.; Chen, B. Performance of Aerobic Rice Varieties Under Irrigated Conditions in North China. *Field Crops Res.* **2006,** *97,* 53–65.

Bouman, B. A. M.; Humphreys, E.; Tuong, T. P.; Barker, R. Rice and Water. *Adv. Agron.* **2007,** *92,*187–237.

Bridge, J.; Luc, M.; Plowright, R. A. In Nematode Parasites of Rice. *Plant Parasitic Nematodes in Subtropical and Tropical Agriculture;* Luc, M, Sikoraand, R. A., Bridge, J. Eds.; CABI Publishing: Cambridge, 2005; pp 87–130

Bordey, F. H.; Launio, C. C.; Beltran, J. C.; Litonjua, A. C.; Manalili, R. G.; Mataia, A. B.; Moya, P. F. In Game Changer: Is PH Rice Ready to Compete at Least Regionally? *Rice Science for Decision Makers*; Philippine Rice Research Institute Maligaya, Science City of Muñoz: Nueva Ecija, 2015.

Cabangon, R. J.; Tuong, T. P.; Abdullah, N. B. Comparing Water Input and Water Productivity of Transplanted and Direct-seeded Rice Production Systems. *Agric. Water Manag.* **2002,** *57*, 11–31.

Cassman, K. G.; Peng, S.; Olk, D. C.; Ladha, J. K.; Reichardt, W.; Dobermann, A.; Singh, U. Opportunities for Increased Nitrogen-use Efficiency from Improved Resource Management in Irrigated Rice Systems. *Field Crops Res.* **1998,** *56*, 7–39.

Chaturvedi, G. S.; Misra, C. H.; Singh, C. N.; Pandey, C. B.; Yadav, V. P.; Singh, A. K.; Divivedi, J. L.; Singh, B. B.; Singh, R. K. Physiological Flash Flooding. International Rice Research Institute: Los Ban~os, Philippines, 1995; pp 79–96.

Chen, S.; Wang, D.; Xu, C.; Ji, C.; Zhang, X.; Zhao, X.; Zhang,. Chauhan, B. S. Responses of Super Rice (*Oryza sativa* L.) to Different Planting Methods for Grain Yield and Nitrogen Use Efficiency in the Single Cropping Season. *PLoS One* **2014,** *9* (8), e104950. DOI: 10.1371/journal.pone.0104950.

Chen, Z. D.; Zhang, H. L.; Dikgwatlhe, S. B.; Xue, J. F.; Qiu, K. C.; Tang, H. M. Soil Carbon Storage and Stratification Under Different Tillage/Residue-management Practices in Double Rice Cropping System. *J. Integr. Agric.* **2015,** *14* (8), 1551–1560.

Chhokar, R. S.; Sharma, R. K.; Gathala, M. K.; Pundir, A. K. Effects of Crop Establishment Techniques on Weeds and Rice Yield. *Crop Prot.* **2014,** *64*, 7–12.

Chou, C. H. Allelopathic Researches in the Subtropical Vegetation in Taiwan. *J. Comp. Physiol.* **1980,** *5*, 222–234.

Coffman, W. R.; Nanda, J. S. In Cultivar Development for Dry Seeded Rice. *Report of a Workshop on Cropping Systems Research in Asia.* International Rice Research Institute: Los Ban~os, Laguna, Philippines, 1982; pp 149–156

Corton, T. M.; Bajita, J. B.; Grospe, F. S.; Pamplona, R. R.; Assis, C. A.; Wassmann, R.; Lantin, R. S.; Buendia, L. V. Methane Emission from Irrigated and Intensively Managed Rice Fields in Central Luzon (Philippines). *Nutr. Cycl. Agroecosys.* **2000,** *58*, 37–53.

Dawe, D. Increasing Water Productivity in Rice-based Systems in Asia: Past Trends, Current Problems, and Future Prospects. *Plant Prod. Sci.* **2005,** *8*, 221–230.

De Datta, S. K. Principles and Practices of Rice Production. International Rice Research Institute: LosBan~os, Philippines, 1981; pp 618.

Devkota, K. P. Resource Utilization and Sustainability of Conservation Based Rice–Wheat Cropping Systems in Central Asia. Ph.D. Dissertation, ZEF, Rheinische Friedrich-Wilhelms University of Bonn, 2011; p 181. http://hss.ulb.uni-bonn.de/2011/2594/2594.pdf (accessed Dec 6, 2017).

Devkota, K. P.; Manschadia, A. M.; Egamberdiev, O.; Guptac, R. K.; Devkota, M. K.; Lamersb, J. P. A. In *Effect of Water-saving Irrigation, Tillage and Residue Management on Yield and Water Productivity of Rice in Khorezm, Uzbekistan,* Conference on International Research on Food Security, Natural Resource Management and Rural Development, ETH Zurich, Tropentag, Germany, Sept 14–16, 2010.

Dhakal, M.; Sah, S. K.; McDonald, A.; Regmi, A. P. Perception and Economics of Dry Direct Seeded Rice in Terai of NEPAL. *J. Food Agric. Environ.* **2015,** 16, 103–111.

Dingkuhn, M.; Penning de Vries, F. W. T.; De Datta, S. K.; van Laar, H. H. In Concepts for a New Plant Type for Direct Seeded Flooded Tropical Rice. *Direct-Seeded Flooded Rice in the Tropics.* International Rice Research Institute: Manilla, 1991; pp 17–38.

Ehsanullah, Akbar, N.; Jabran,. Habib, T. Comparison of Different Planting Methods for Optimization of Plant Population of Fine Rice (*Oryza sativa*) in Punjab (Pakistan). *Pak. J. Agric. Sci.* **2007**, *44*, 597–599.

Ehsanullah, Anjum, S. A.; Ashraf, U.; Tanveer, M.; Khan, I. Effect of Sowing Dates and Weed Control Methods on Weed Infestation, Growth and Yield of Direct-seeded Rice. *Philippine Agric. Sci.* **2014**, *97* (3), 307–312.

Erda, L.; Xiong, W.; Ju, H.; Xu, Y.; Li, Y.; Bai, L.; Xie, L. Climate Change Impacts on Crop Yield and Quality with CO_2 Fertilization in CHINA. *Philos. Trans. Royal Soc.* **2005**, *360*, 2149–2154.

Erenstein, O. Comparing Water Management in Rice–Wheat Production Systems in Haryana, India and Punjab, Pakistan. *Agric. Water Manag.* **2009**, *96* (12), 1799–1806.

FAO (Food and Agriculture Organization). Rice and Us. (Food and Agriculture Organization of the United Nations), 2004. http://www ao.org/rice2004/en/aboutrice.htm (accessed Nov 23, 2017).

FAO (Food and Agriculture Organization). Global Climate Changes and Rice Food Security. FAO: Rome, 2005. http://www.fao.org/climatechange/15526-03ecb62366f779d-1ed45287e698a44d2e.pdf (accessed Nov 23, 2017).

FAO (Food and Agriculture Organization). *Increasing Crop Production Sustainably, the Perspective of Biological Processes.* FAO: Rome, 2009.

Farooq, M.; Basra, S. M. A.; Wahid, A. Priming of Field-sown Rice Seed Enhances Germination, Seedling Establishment, Allometry and Yield. *Plant Growth Regul.* **2006**, *49*, 285–294.

Farooq, M.; Basra, S. M. A.; Ahmad, N. Improving the Performance of Transplanted Rice by Seed Priming. *Plant Growth Regul.* **2007**, *51*, 129–137.

Farooq, M.; Basra, S. M. A.; Ahmad, N.; Murtaza, G. Enhancing the Performance of Transplanted Coarse Rice by Seed Priming. *Paddy Water Environ.* **2009**, 7, 55–63.

Farooq, M.; Siddique, K. H. M.; Rehman, H.; Aziz, T.; Lee, D. J.; Wahid, A. Rice Direct Seeding: Experiences, Challenges and Opportunities. *Soil Tillage Res.* **2011**, *111*, 87–98.

Gao, X. P.; Zou, C. Q.; Fan, X. Y.; Zhang, F. S.; Hoffland, E. From Flooded to Aerobic Conditions in Rice Cultivation: Consequences for Zinc Uptake. *Plant Soil* **2006**, *280*, 41–47.

Gathala, M. K.; Ladha, J. K.; Kumar, V.; Saharawat, Y. S.; Kumar, V.; Sharma, P. K.; Sharma,. Pathak, H. Tillage and Crop Establishment Affects Sustainability of South Asian Rice–Wheat System. *Agron. J.* **2011**, *103* (4), 961–971.

Gill, M. S.; Kumar, P.; Kumar, A. Growth and Yield of Direct-seeded Rice (*Oryza sativa*) as Influenced by Seeding Technique and Seed Rate Under Irrigated Conditions. *Indian J. Agron.* **2006**, *51*, 283–287.

Gines, H.; Lavapiez, L.; Nicolas, J.; Torralba, R.; Morris, R. A. In *Dry-seeded Rice: Agronomic Experiences in a Rainfed and Partially Irrigated Area.* Paper Presented at the Ninth Annual Meeting Annual Scientific Meeting of Crop Science Society of Philippines, Illio City, Philippines, May 11–13, 1978.

GRiSP (Global Rice Science Partnership). *Rice Almanac*, 4th ed.; International Rice Research Institute: Los Baños, Philippines, 2013; p 283.

Guerra, L. C.; Bhuiyan, S. I.; Tuong, T. P.; Barker, R. *Producing more Rice with less Water from Irrigated Systems.* SWIM Paper 5, IWMI/IRRI: Colombo, Sri Lanka, 1998; pp 24.

Gupta, R. K.; Naresh, R. K.; Hobbs, P. R.; Jiaguo; Ladha, J. K. In Sustainability of Post-green Revolution Agriculture. The Rice–Wheat Cropping Systems of the Indo-Gangetic Plains and China. *Improving the Productivity and Sustainability of Rice–Wheat Systems: Issues and Impacts,* ASA Special Publication 65: Wisconsin, USA, 2003.

Gupta, R. K.; Ladha, J. K.; Singh, S.; Singh, R. J.; Jat, M. L.; Saharawat, Y.; Singh, V. P.; Singh, S. S.; Sah, G.; Gill, M. S.; Alam, M.; Mujeeb, H.; Singh, U. P.; Mann, R.; Pathak, H.; Singh, B. S.; Bhattacharya,. Malik, R. K. In *Production Technology for Direct Seeded Rice.* Rice Wheat Consortium Technical Bulletin 8. New Delhi, India, 2006.

Hang, X.; Zhang, X.; Song, C.; Jiang, Y.; Deng, A.; He, R.; Lu,. Zhang, W. Differences in rice yield and CH_4 and N_2O Emissions Among Mechanical Planting Methods with Straw Incorporation in Jianghuai Area, China. *Soil Tillage Res.* **2014,** *144,* 205–210.

Hayashi, S.; Kamoshita, A.; Yamagishi, J.; Kotchasatit, A.; Jongdee, B. Genotypic Differences in Grain Yield of Transplanted and Direct-seeded Rainfed Lowland Rice (*Oryza sativa* L.) in Northeastern Thailand. *Field Crops Res.* **2007,** 102, 9–21.

Hayat, K.; Awan, I. U.; Hassan, G. Impact of Seeding Dates and Varieties on Weed Infestation, Yield and Yield Components of Rice (*Oryza sativa* L.) Under Direct Wet Seeded Culture. *Pak. J. Weed Sci. Res.* **2003,** *9,* 59–65.

Horgan, F. G.; Figueroa, J. Y.; Liberty, M.; Almazan, P. Seedling Broadcasting as a Potential Method to Reduce Apple Snail Damage to Rice. *Crop Prot.* **2014,** *64,* 168–176.

Horie, T.; Baker, J. T.; Nakagawa; Matsui, T. In Crop Ecosystem Responses to Climatic Change: Rice. *Climate Change, Plant Productivity and Global Implications;* Reddy, K. R., Hodges, H. F., Ed.; CABI Publishing: New York, USA, 2000; pp 81–106.

Hussain, S.; Peng, S.; Fahad, S.; Khaliq, A.; Huang, J.; Cui, K.; Nie, L.; Rice Management Interventions to Mitigate Greenhouse Gas Emissions: A Review. *Environ. Sci. Pollut. Res.* **2015,** *22,* 3342–3360. DOI: 10.1007/s11356-014-3760-4.

Ikeda, H.; Kamoshita, A.; Yamagishi, J.; Ouk, M.; Lor, B. Assessment of Management of Direct Seeded Rice Production Under Different Water Conditions in Cambodia. *Paddy Water Environ.* **2008,** *6,* 91–103.

IPCC (Intergovernmental Panel on Climate Change). *Managing the Risks of Extreme Events and Disasters to Advance Climate Change Adaptation;* Special Report. Cambridge University Press: Cambridge, UK, 2012.

IPCC, 2014. Summary for Policymakers. In: *Climate Change 2014: Impacts,Adaptation, and Vulnerability. Part A: Global and Sectoral Aspects.* Contribution of Working Group II to the Fifth Assessment Report of the Intergovernmental Panel on Climate Change, Cambridge University Press, Cambridge, 2014.

IRRI (International Rice Research Institute). *Annual Report for 1977;* IRRI: Los Ban ˜os, Philippines, 1978; p 548.

IRRI (International Rice Research Institute). *Rice Almanac,* 2nd ed.; IRRI: Los Banos, Philippines, 1997; p 181.

Jabran, K.; Ehsanullah, Hussain, M.; Farooq, M.; Haider, N.; Chauhan, B. S. Water Saving, Water Productivity and Yield Outputs of Fine-grain Rice Cultivars Under Conventional and Water-saving Rice Production Systems. *Exp. Agric.* **2015,** *51* (4), 567–581.

Jat, R. K.; Sapkota, T. B.; Singh, R. G.; Jat, M. L.; Kumar, M.; Gupta, R. K. Seven Years of Conservation Agriculture in a Rice–Wheat Rotation of Eastern Gangetic Plains of South Asia: Yield Trends and Economic Profitability. *Field Crops Res.* **2014**, *164*, 199–210.

Johnson, D. E. Weed Management in Small Holder Rice Production in the Tropics. Natural Resources Institute, University of Greenwich, Chatham, Kent, UK; 2002.

Johnson, D. E.; Mortimer, A. M. In *Issues for Weed Management in Direct Seeded Rice and the Development of Decision-support Frameworks.* Workshop on Direct Seeded Rice in the Rice–Wheat System of the Indo-Gangetic Plains; University of Agriculture and Technology, Pantanagar, Uttaranchal, India, Feb 1–2, 2005; pp 8.

Johnson, D. E.; Mortimer, A. M.; Orr, A.; Riches, C. *Weeds, Rice and Poor People in South Asia*; Natural Resources Institute: Chatham, UK, 2003. http://www.nri.org/work/weed-sricepoor.pdf (accessed Dec 5, 2017).

Juraimi, A. S.; Begum, M.; Yusuf, M. N. M.; Man, A. Efficacy of Herbicides on the Control Weeds and Productivity of Direct Seed Rice Under Minimal Water Conditions. *Plant Prot. Q.* **2010**, *25*, 19–25.

Juraimi, A. S.; Uddin, K.; Anwar, P.; Muda, M.; Mahmud, T.; Ismail, M. R.; Man, A. Sustainable Weed Management in Direct Seeded Rice Culture: A Review. *Aust. J. Crop Sci.* **2013**, *7* (7), 989–1002.

Karim, Z.; Ahmed, M.; Hussain, S. G.; Rashid, K. B. Impact of Climate Change on the Production of Modern Rice in Bangladesh. In *Implications of Climate Change for International Agriculture: Crop Modeling Study*; US Environmental Protection Agency: Washington D.C, 1994.

Karim, R. S. M.; Man, A. B.; Sahid, I. B. Weed Problems and Their Management in Rice Fields of Malaysia: An Overview. *Weed Biol. Manag.* **2004**, *4*,177–186.

Kim, C. K. Disease Dispersal Gradients of Rice Blast from Point Source. *Korean J. Plant Prot.* **1987**, *3*, 131–136.

Kim, H. Y.; Ko, J.; Kang, S.; Tenhunen, J. Impacts of Climate Change on Paddy Rice Yield in a Temperate Climate. *Glob. Change Biol.* **2013**, *19*, 548–562.

Kirk, G. J.; Bajita, J. B. Root Induced Iron Oxidation, pH Changes and Zinc Solubilization in the Rhizosphere of Lowland Rice. *New Phytol.* **1995**, *131*, 129–137.

Ko, J. Y.; Kang, H. W. The Effects of Cultural Practices on Methane Emission from Rice Fields. *Nutr. Cycl. Agroecosyst.* **2000**, *58*, 311–314.

Kreye, C.; Bouman, B. A. M.; Faronilo, J. E.; Llorca, L. Causes for Soil Sickness Affecting Early Plant Growth in Aerobic Rice. *Field Crops Res.* **2009**, *114*, 182–187.

Krishnan, P.; Swain, D. K.; Bhaskar, B. C.; Nayak, S. K.; Dash, R. N. Impact of Elevated CO_2 and Temperature on Rice Yield and Methods of Adaptation as Evaluated by Crop Simulation Studies. *Agric. Ecosyst. Environ.* **2007**, *122*, 233–242.

Krishnan, B.; Ramakrishnan, K.; Reddy, R.; Reddy, V. R. High-temperature Effects on Rice Growth, Yield, and Grain Quality. *Adv. Agron.* **2011**, *111*, 87–206.

Kukal, S. S.; Aggarwal, G. C. Percolation Losses of Water in Relation to Puddling Intensity and Depth in a Sandy Loam Rice (*Oryza sativa*) Field. *Agric. Water Manag.* **2002**, *57*, 49–59.

Kumar, P.; Joshi, P. K.; Johansen, C.; Asokan, M. Sustainability of Rice–Wheat Based Cropping Systems in India: Socio Economic and Policy Issues. *Econ. Political Wkly.* **1998**, *33*, 152–158.

Ladha, J. K.; Pathak, H.; Krupnik, T. J.; Six, J.; Van Kessel, C. Efficiency of Fertilizer Nitrogen in Cereal Production: Retrospects and Prospects. *Adv. Agron.* **2005,** *87,* 85–156.

Lafitte, R. H.; Courtois, B.; Arraudeau, M. Genetic Improvement of Rice in Aerobic Systems: Progress from Yield to Genes. *Field Crops Res.* **2002,** 75, 171–190.

Lalonde, S.; Beebe, D.; Saini, H. S. Early Signs of Disruption of Wheat Anther Development Associated with the Induction of Male Sterility by Meiotic-stage Water Deficit. *Sex. Plant Reprod.* **1997,** *10,* 40–48.

Lampayan, R. M.; Bouman, B. A. M. In *Management Strategies for Saving Water and Increase Its Productivity in Lowland Rice-based Ecosystems,* Proceedings of the First Asia-Europe Workshop on Sustainable Resource Management and Policy Options for Rice Ecosystems (SUMAPOL), Hangzhou, Zhejiang Province, PR China, May 11–14, 2005. On CDROM: Altera, Wageningen, Netherlands, 2005.

Le Mer, J.; Roger, P. Production, Oxidation, Emission and Consumption of Methane by Soils: A Review. *Eur. J. Soil Biol.* **2001,** *37,* 25–50. DOI: 10.1016/S1164-5563 (01)01067-6.

Lee, M. H.; Kim, J. K.; Kim, S. S.; Park, S. T. In Status of Dry-seeding Technologies for Rice in Korea. *Direct Seeding: Research Strategies and Opportunities*; Pandey, S., Mortimer, M., Wade, L., Tuong, T. P., Lopez, K., Hardy, B., Eds.; International Rice Research Institute: Los Ban~os, Philippines, 2002; pp 3–14.

Li, T.; Wassmann, R. Modeling Approaches for Assessing Adaptation Strategies in Rice Germplasm Development to Cope with Climate Change, 2011. http://www.fao.org/fileadmin/templates/agphome/documents/IRRI_website/Irri_workshop/LP_16.pdf (accessed Nov 18, 2017).

Li, C. F.; Zhang, Z. S.; Guo, L. J.; Cai, M. L.; Cao, C. G. Emissions of CH_4 and CO_2 from Double Rice Cropping Systems Under Varying Tillage and Seeding Methods. *Atmos. Environ.* **2013,** *80,* 438–444.

Li, M.; Ashraf, U.; Tian, H.; Mo, Z.; Pan, S.; Anjum, S. A.; Duan, M.; Tang, X. Manganese-induced Regulations in Growth, Yield Formation, Quality Characters, Rice Aroma and Enzyme Involved in 2-Acetyl-1-Pyrroline Biosynthesis in Fragrant Rice. *Plant Physiol. Biochem.* **2016a,** *103,* 167–175.

Li, D.; Nansek, T.; Matsue, Y.; Chome, Y.; Yokota, S. Variation and Determinants of Rice Yields Among Individual Paddy Fields: Case Study of a Large-scale Farm in the Kanto Region of Japan. *J. Fac. Agric. Kyushu U.* **2016b,** *61* (1), 205–214.

Liu, J. X.; Bennett, J. Reversible and Irreversible Drought-induced Changes in the Anther Proteome of Rice (*Oryza sativa* L.) Genotypes IR64 and Moroberekan. *Mol. Plant* **2011,** *4* (1), 59–69. DOI: 10.1093/mp/ssq039.

Liu, H.; Hussain, S.; Peng, S.; Huang, J.; Cui, K.; Nie, L. Potentially Toxic Elements Concentration in Milled Rice Differ Among Various Planting Patterns. *Field Crops Res.* **2014a,** *168,* 19–26.

Liu, S.; Zhang, Y.; Lin, F.; Zhang, L.; Zou, J. Methane and Nitrous Oxide Emissions from Direct-seeded and Seedling-transplanted Rice Paddies in Southeast China. *Plant Soil* **2014b,** *374,* 285–297.

Liu, H.; Hussain, S.; Zheng, M.; Peng, S.; Huang, J.; Cui, K.; Nie, L. Dry direct-Seeded Rice as an Alternative to Transplanted-flooded Rice in Central China. *Agron. Sustain. Dev.* **2015,** *35* (1), 285–294.

Lizumi, T.; Hayashi, Y.; Kimura, F. Influence on Rice Production in Japan from Cool and Hot Summers After Global Warming. *J. Agric. Meteorol.* **2007**, *63* (1), 11–23.

Luat, N. V. In Integrated Weed Management and Control of Weeds and Weedy Rice in Vietnam. *Wild and Weedy Rice in Rice Ecosystems in Asia: A Review, Limited Proceedings Number 2;* Baki, B. B., Chin, D. V., Mortimer, M., Eds.;. International Rice Research Institute: Los Bano ˜s, Philippines, 2000; pp 1–3.

Mackill, D. J.; Bonman, J. M. Inheritance of Blast Resistance in Near-isogenic Lines of Rice. *Phytopathology* **1992**, *82*, 746–749.

Mackill, D. J.; Coffman, W. R.; Garrity, D. P. *Rainfed Lowland Rice Improvement*; International Rice Research Institute: Los Bano˜s, Philippines, 1996.

Maclean, J. L.; Dawe, D.; Hardy, B.; Hettel, G. P. (Eds.). *Rice Almanac*. International Rice Research Institute: Los Banos, Philippines, 2002; p 253.

Mahajan, G.; Chauhan, B. S.; Johnson, D. E. Weed Management in Aerobic Rice in Northwestern Indo-Gangetic Plains. *J. Crop Improv.* **2009**, *23*, 366–382.

Mahajan, G.; Gill, M. S.; Singh, K. Optimizing Seed Rate to Suppress Weeds and to Increase Yield in Aerobic Direct-seeded Rice in Northwestern Indo-Gangetic Plains. *J. New Seeds* **2010**, *11*, 225–238.

Mall, R. K.; Aggarwal, P. K. Climate Change and Rice Yields in Diverse Agro-environments of India. I. Evaluation of Impact Assessment Models. *Clim. Change* **2002**, *52*, 315–330.

Mamun, A. A. Weeds and Their Control: A Review of Weed Research in Bangladesh. *Agricultural and Rural Development in Bangladesh*; Japan International Co-operation Agency: Dhaka, Bangladesh, 1990; Vol. 19, pp 45–72.

Mandal, B.; Hazra, G. C.; Mandal, L. N. Soil Management Influences on Zinc Desorption for Rice and Maize Nutrition. *Soil Sci. Soc. Am. J.* **2000**, *64*, 1699–1705.

Maqsood, M.; Shehzad, M. A.; Ali, S. N. A.; Iqbal, M. Rice Cultures and Nitrogen Rate Effects on Yield and Quality of Rice (*Oryza sativa* L.). *Turk. J. Agric. Forstr.* **2013**, *37* (6), 665–673.

Matthews, M.; Kropff, J.; Horie, T.; Bacheletd, D. Simulating the Impact of Climate Change on Rice Production in Asia and Evaluating Options for Adaptation. *Agric. Syst.* **1997**, *54*, 399–425.

Matsue, Y.; Mizuta, K.; Furuno, K.; Yoshida, T. Studies on Palatability of Rice Grown in Northern Kyushu. *Jpn. J. Crop Sci.* **1991**, *60*, 490–496.

Mohandrass, S.; Kareem, A. A.; Ranganathan, T. B.; Jeyaraman, S. In Rice Production in India Under the Current and Future Climate. *Modeling the Impact of Climate Change on Rice Production in Asia*; Mathews, R. B., Kroff, M. J., Bachelet, D., van Laar, H., Eds.; CAB International: United Kingdom, 1995; pp 165–181.

Mosier, A.; Kroeze, C.; Nevison, C.; Oenema, O.; Seitzinger, S.; Cleemput, O. Closing the Global N_2O Budget: Nitrous Oxide Emissions Through the Agricultural Nitrogen Cycle. *Nutr. Cycl. Agroecosyst.* **1998**, *52*, 225–248.

Montgomery, D. R. Soil Erosion and Agricultural Sustainability. Proc. *Natl. Acad. Sci.* **2007**, *104*, 13268–13272.

Murdiyarso, D. Adaptation to Climatic Variability and Change: Asian Perspectives on Agriculture and Food Security. *Environ. Monit. Assess.* **2000**, *61*, 123–131.

Naklang, K. In *Direct Seeding for Rainfed Lowland Rice in Thailand*. Breeding Strategies for Rainfed Lowland Rice in Drought-prone Environments; Proceedings of an International

Workshop, Ubon Ratchathani, Thailand, Nov 5–8, 1996; Australian Center for International Agricultural Research: Canberra, 1997; pp 126–136.

Naresh, R. K.; Misra, A. K.; Singh, S. P. Assessment of Direct Seeded and Transplanting Methods of Rice Cultivars in the Western Part of Uttar Pradesh. *Int. J. Pharm. Sci. Bus. Manag.* **2013,** *1*, 1–8.

Nayak, D.; Cheng, K.; Wang, W.; Koslowski, F.; Yan, X.; Guo, M.; Newbold, J.; Moran, D.; Cardenas, L.; Pan, G.; Smith, P. *Technical Options to Reduce Greenhouse Gas Emissions from Croplands and Grasslands in China.* UK-China Sustainable Agriculture Innovation Network-SAIN. Policy Brief No. 9, Oct 2013.

Neue, H. U.; Quijano, C.; Senadhira, D.; Setter, T. Strategies for Dealing with Micronutrient Disorders and Salinity in Lowland Rice Systems. *Field Crops Res.* **1998,** *56* (1–2), 139–155.

Olofsdotter, M. Rice—A Step Toward Use of Allelopathy. *Agron. J.* **2001,** *93*, 3–8.

Oyediran, I. O.; Heinrichs, E. A. Arthropod Populations and Rice Yields in Directseeded and Transplanted Lowland Rice in West Africa. *Int. J. Pest Manag.* **2001,** *47*, 195–200.

Pandey, S.; Velasco, L. E. In Economics of Alternative Rice Establishment Methods in Asia: A Strategic Analysis. *Social Sciences Division Discussion Paper;* International Rice Research Institute: Los Bano˜s, Philippines, 1999.

Pandey, S.; Mortimer, M.; Wade, L.; Thong, T. P.; Lopez, K.; Hardy, H. (Eds.). In *Direct Seeding: Research Issues and Opportunities.* Proceedings of the International Workshop on Direct Seeding in Asian Rice Systems: Strategic Research Issues and Opportunities, Jan 25–28, 2000. Bangkok, Thailand. Los Banos (Philippines) (*Intl. Rice Res. Inst.*) 2002, p 383.

Pandey, S.; Velasco, L. In Economics of Direct Seeding in Asia: Patterns of Adoption and Research Priorities. *Direct Seeding: Research Strategies and Opportunities;* Pandey, S., Mortimer, M., Wade, L., Tuong, T. P., Lopes, K., Hardy, B., Eds.; International Rice Research Institute: Los Ban˜os, Philippines, 2002.

Parthasarathi, T.; Vanitha, K.; Lakshamanakuma, P.; Kalaiyarasi, D. Aerobic Rice-mitigating Water Stress for the Future Climate Change. *Int. J. Agron. Plant Prod.* **2012,** *3*, 241–254.

Pathak, H.; Singh, R.; Bhatia, A.; Jain, N. Recycling of Rice Straw to Improve Crop Yield and Soil Fertility and Reduce Atmospheric Pollution. *Paddy Water Environ.* **2006,** *4*, 111–117.

Pathak, H.; Bhatia, A.; Jain, N.; Aggarwal, P. K. In *Greenhouse Gas Emission and Mitigation in Indian Agriculture: A Review.* Ing Bulletins on Regional Assessment of Reactive Nitrogen, Bulletin No. 19, (Ed. Bijay Singh), SCON-ING, New Delhi, 2010; pp 1–34.

Pathak, H.; Tewari, A. N.; Sankhyan, S.; Dubey, D. S.; Mina, U.; Singh, V. K.; Jain, N.; Bhatia, A. Direct-seeded Rice: Potential, Performance and Problems: A Review. *Curr. Adv. Agric. Sci.* **2011,** *3* (2), 77–88.

Pathak, H.; Aggarwal, P. K.; Singh, S. D. (Eds.). *Climate Change Impact, Adaptation and Mitigation in Agriculture: Methodology for Assessment and Applications.* Indian Agricultural Research Institute, New Delhi, 2012; pp xix + 302.

Peng, S.; Huang, J.; Sheehy, J. E.; Laza, R. C.; Visperas, R. M.; Zhong, X.; Centeno, G. S.; Khush, G. S.; Cassman, K. G. Rice Yields Decline with Higher Night Temperature from Global Warming. *Proc. Natl. Acad. Sci.* **2004,** *101*, 9971–9975.

Peng, S.; Bouman, B. A. M.; Visperas, R. M.; Castan˜eda, A.; Nie, L.; Park, H. K. Comparison Between Aerobic and Flooded Rice: Agronomic Performance in a Long-term (8-season) Experiment. *Field Crops Res.* **2006**, *96*, 252–259.

Pongprasert, S. In Insect and Disease Control in Wet-seeded Rice in Thailand. *Constraints, Opportunities, and Innovations for Wet-seeded Rice*, Discussion Paper Series No. 10; Moody, K., Ed.; International Rice Research Institute: Los Bano˜s, Philippines, 1995; pp 118–132.

Prot, J. C.; Matias, D. M. Effects of Water Regime on the Distribution of *Meloidogyne graminicola* and Other Root Parasitic Nematodes in a Rice Field Topo Sequence and Pathogenicity of *M. graminicola* on Rice Cultivar UPLR15. *Nematology* **1995**, *41*, 219–228.

Prot, J. C.; Soriano, I. R. S.; Matias, D. M. Major Root-parasitic Nematodes Associated with Irrigated Rice in the Philippines. *Fund. Appl. Nematol.* **1994**, *17*, 75–78.

Pumijumnong, N.; Arunrat, N. Simulating the Rice Yield Change in Thailand Under SRES A2 and B2 Scenarios with the EPIC Model. *J. Agri-Food Appl. Sci.* **2013**, *1* (4), 119–125.

Qin, Y.; Liu, S.; Guo, Y.; Liu, Q.; Zou, J. Methane and Nitrous Oxide Emissions from Organic and Conventional Rice Cropping Systems in Southeast China. *Biol. Fertil. Soils* **2010**, *46*, 825–834. DOI:10.1007/s00374-010-0493-5.

Ramaiah, K.; Mudaliar, S. D. Lodging of Straw and Its Inheritance in Rice (*O. sativa*). *Indian J. Agric. Sci.* **1934**, *4*, 880–894.

Ramzan, M. *Evaluation of Various Planting Methods in Rice–Wheat Cropping System, Punjab, Pakistan*; Rice Crop Report, 2003–2004; 2003; pp 4–5.

Rao, A. N.; Jhonson, D. E.; Sivaprasad, V.; Ladha, J. K.; Mortimer, A. M. Weed Management in Direct Seeded Rice. *Adv. Agron.* **2007**, *93*, 153–255.

Rashid, M. H.; Alam, M. M.; Khan, M. A.; Ladha, J. K. Productivity and Resource Use of Direct-(Drum)-seeded and Transplanted Rice in Puddled Soils in Rice–Rice and Rice–Wheat Ecosystems. *Field Crops Res.* **2009**, *113* (3), 274–281.

Redfern, S. K.; Nadine, A.; Jesie, S. B. Rice in Southeast Asia: Facing Risks and Vulnerabilities to Respond to Climate Change. In *Building Resilience for Adaptation to Climate Change in the Agriculture Sector*; Food and Agriculture Organization of the United Nations: Rome, 2012; p 295.

Rickman, J. F.; Meas, P.; Som, B.; Poa, S. In *Direct Seeding of Rice in Cambodia*. Increased Lowland Rice Production in the Mekong Region; Proceedings of an International Workshop, Vientiane, Laos, Oct 30–Nov 2, 2000. Australian Centre for International Agricultural Research: Canberra, 2001; pp 60–65.

Sah, D. N.; Bonman, J. M. Effects of Seedbed Management on Blast Development in Susceptible and Partially Resistant Rice Cultivars. *J. Phytopathol.* **2008**, *136*, 73–81.

Saleque, M. A.; Kirk, G. J. D. Root Induced Solubilization of Phosphate in the Rhizosphere of Lowland Rice. *New Phytol.* **1995**, *129*, 325–336.

Sarkar, R. K.; Sanjukta, D. Yield of Rainfed Lowland Rice with Medium Water Depth Under Anaerobic Direct Seeding and Transplanting. *Trop. Sci.* **2003**, *43*, 192–198.

Sarkar, S. C.; Singh, V. K. In *Response of Kharif Paddy, Wheat and Mustard to Soil Application of ZnSO₄ and Borax in Farmer Field of Project Areas in West Bengal*; Proceedings of Workshop on Micronutrients, Bhubaneswar, India, Jan 22–23, 1992. 2007.

Schlesinger, W. H. Carbon Sequestration in Soils. *Science* **1999**, *284*, 2095. DOI:10.1126/science.284.5423.2095.

Shakoor, U.; Saboor, A.; Baig, I.; Afza, A.; Rahman, A. Climate Variability Impacts on Rice Crop Production in Pakistan. *Pak. J. Agric. Res.* **2015**, *28* (1), 19–27.

Sharma, P. K.; Bhushan, L.; Ladha, J. K.; Naresh, R. K.; Gupta, R. K.; Balasubramanian, B. V.; Bouman, B. A. M. In *Crop–Water Relations in Rice–Wheat Cropping Under Different Tillage Systems and Water Management Practices in a Marginally Sodic Medium Textured Soil*; Proceedings of the International Workshop on Water-wise Rice Production, Los Bano˜s, Philippines; Bouman, B. A. M., Hengsdijk, H., Hardy, B., Bihdraban, B., Toung, T. P., Ladha, J. K. Eds.; International Rice Research Institute: Los Bano˜s, Philippines, 2002; pp 223–235.

Shrestha, S. *An Economic Impact Assessment of the Rice Research Program in Bhutan*; International Rice Research Institute: Los Banos, Philippines, 2004.

Savary, S.; Castilla, N. P.; Elazegui, F. A.; Teng, P. S. Multiple Effects of Two Drivers of Agricultural Change, Labour Shortage and Water Scarcity, on Rice Pest Profiles in Tropical Asia. *Field Crops Res.* **2005**, *91*, 263–271.

Setter, T. I.; Laureles, E. V.; Mazaredo, A. M. Lodging Reduces Yield of Rice by Self Shading and Reduction of Photosynthesis. *Field Crops Res.* **1997**, *49*, 95–106.

Sheehy, J. E.; Mitchell, P. L.; Ferrer, A. B. Decline in Rice Grain Yields with Temperature: Models and Correlations can Give Different Estimates. *Field Crops Res.* **2006**, *98*, 151–156.

Singh, V. P.; Singh, G.; Singh, R. K. Integrated Weed Management in Direct Seeded Spring Sown Rice Under Rainfed Low Valley Situation of Uttaranchal. *Indian J. Weed Sci.* **2001**, *33*, 63–66.

Singh, A. K.; Choudhury, B. U.; Bouman, B. A. M. In Effects of Rice Establishment Methods on Crop Performance, Water Use and Mineral Nitrogen; Proceedings of the International Workshop on Water-wise Rice Production, Los Bano˜s, Philippines; Bouman, B. A. M., Hengsdijk, H., Hardy, B., Bihdraban, B., Toung, T. P., Ladha, J. K., Eds.; International Rice Research Institute: Los Bano˜s, Philippines, 2002; pp 237–246.

Singh, S.; Ladha, J. K.; Gupta, R. K.; Bhushan, L.; Rao, A. N. Weed Management in Aerobic Rice Systems Under Varying Establishment Methods. *Crop Prot.* **2008**, *27*, 660–671.

Singh, V. K.; Dwivedi, B. S.; Singh, S. K.; Majumdar, K.; Jat, M. L.; Mishra, R. P.; Rani, M. Soil Physical Properties, Yield Trends and Economics After Five Years of Conservation Agriculture Based Rice–Maize System in North-Western India. *Soil Tillage Res.* **2016**, *155*, 133–148.

Srinivasan, V.; Maheswarappa, H. P.; Lal, R. Long Term Effects of Topsoil Depth and Amendments on Particulate and Non Particulate Carbon Fractions in a Miamian Soil of Central Ohio. *Soil Tillage Res.* **2012**, *121*, 10–17.

Sudhir-Yadav; Evangelista G.; Faronilo, J.; Humphreys, E.; Henry, A.; Fernandez, L. Establishment Method Effects on Crop Performance and Water Productivity of Irrigated Rice in the Tropics. *Field Crops Res.* **2014**, *166*, 112–127.

Tabbal, D. F.; Bouman, B. A. M.; Bhuiyan, S. I.; Sibayan, E. B.; Sattar, M. A. On-farm Strategies for Reducing Water Input in Irrigated Rice; Case Studies in the Philippines. *Agric. Water Manag.* **2002**, *56*, 93–112.

Timsina, J.; Connor, D. J. The Productivity and Management of Rice–Wheat Cropping Systems: Issues and Challenges. *Field Crops Res.* **2001,** *69,* 93–132.

Tomita, S.; Miyagawa, S.; Kono, Y.; Noichana, C.; Inamura, T.; Nagata, Y.; Sributta, A.; Nawata, E. Rice Yield Losses by Competition with Weeds in Rainfed Paddy Fields in North–East Thailand. *Weed Biol. Manag.* **2003,** *3,* 162–171.

Tuong, T. P. Productive Water Use in Rice Production: Opportunities and Limitations. *J. Crop Prod.* **1999,** *2,* 241–264.

Tuong, T. P.; Bouman, B. A. M. In *Rice Production in Water–Scarce Environments*; Proceedings of the Water Productivity Workshop, Colombo, Sri Lanka, Nov 12–14, 2001; International Water Management Institute: Colombo, Sri Lanka, 2002.

Yuliawana, T.; Handoko, I. In *The Effect of Temperature Rise to Rice Crop Yield in Indonesia Uses Shieary Rice Model with Geographical Information System (GIS) Feature.* The 2nd International Symposium on LAPAN-IPB Satellite for Food Security and Environmental Monitoring 2015, LISAT-FSEM 2015; Procedia Environmental Sciences, 2016, Vol. 33, 214–220..

Ventura, W.; Watanabe, I. Growth Inhibition due to Continuous Cropping of Dryland Rice and Other Crops. *Soil Sci. Plant Nutr.* **1978,** *24,* 375–389.

Wah, C. A. *Direct Seeded Rice in Malaysia-A Success Story.* Asia-Pacific Association of Agricultural Research Institutions, FAO Regional Office for Asia and Pacific: Bangkok, 1998; pp 15–33.

Wang, C.; Prinn, R. G.; Sokolov, A. A Global Interactive Chemistry and Climate Model: Formulation and Testing. *J. Geophys. Res.* **1998,** *103,* 3399–3418.

Wang, H. Q.; Bouman, B. A. M.; Zhao, D. L.; Wang, C. G.; Moya, P. F. In Aerobic Rice in Northern China: Opportunities and Challenges. *Water-wise-rice Production*; Proceedings of the International Workshop on Water-wise Rice Production, Los Banos, Philippines, April 8–11, 2002; Bouman, B. M. A., et al., Eds.; International Rice Research Institute: Los Banos, Philippines, 2002; pp 143–154

Wang, J. X.; Huang, J. K.; Yan, T. T. Impacts of Climate Change on Water and Agricultural Production in Ten Large River Basins in China. *J. Integr. Agric.* **2013,** *12,* 101–108.

Warner, J. F.; Bindraban, P. S.; Keulen, H. V. Introduction: Water for Food and Ecosystems: How to Cut Which Pie? *Int. J. Water Resour. Dev.* **2006,** *22,* 3–13.

Wassmann, R.; Neue, H. U.; Ladha, J. K.; Aulakh, M. S. Mitigating Greenhouse Gas Emissions from Rice–Wheat Cropping Systems in Asia. *Environ. Dev. Sust.* **2004,** *6,* 65–90.

Weerakoon, W. M.; Mutunayake, M. M.; Bandara, C.; Rao, A. N.; Bhandari, D. C.; Ladha, J. K. Direct-seeded Rice Culture in Sri Lanka: Lessons from Farmers. *Field Crops Res.* **2011,** *121* (1), 53–63.

Wickham, T. H.; Sen, L. N. In Water Management for Lowland Rice: Water Requirements and Yield Response. *Soils and Rice*; International Rice Research Institute: Manila, Philippines, 1978; pp 649–669.

Wopereis, M. C. S.; Bouma, J.; Kropff, M. J.; Sanidad, W. Reducing Bypass Flow Through a Dry and Cracked Previously Puddled Rice Soil. *Soil Tillage Res.* **1994,** *29,* 1–11.

World Bank. *World Development Report 2008: Agriculture for Development*; The World Bank: Washington, DC, USA, 2007.

Xing, G. X.; Zhao, X.; Xiong, Z. Q.; Yan, X. Y.; Xua, H.; Xie, Y. X.; Shi, S. L. Nitrous Oxide Emission from Paddy Fields in China. *Acta. Ecol. Sin.* **2009,** *29,* 45–50. DOI: 10.1016/j. chnaes.2009.04.006.

Yadav, D. B.; Yadav, A.; Malik, R. K.; Gurjeet, G. In *Efficacy of PIH 2023, Penoxsulam and Azimsulfuron for Post-emergence Weed Control in Wet Direct Seeded Rice*; Biennial Conference, Indian Society of Weed Science, Nov 2–3, 2007. Department of Agronomy, CCS Haryana Agricultural University, Hisar, India, 2007.

Young, K. B.; Wailes, E. J.; Cramer, G. L. Economic Assessment of the Myanmar Rice Sector: Current Developments and Prospects. Arkansas Agricultural Experiment Station, Division of Agriculture University of Arkansas, Research Bulletin 958, Fayetteville, AR 72701, USA, 1998.

Yun, S. I.; Wada, Y.; Maeda, T.; Miura, K.; Watanabe, K. Growth and Yield of Japonica × Indica Hybrid Cultivars Under Direct Seeding and Upland Cultivation Condition. *Jpn. J. Crop Sci.* **1997,** *66,* 386–393.

Zhang, W.; Yu, Y.; Huang, Y.; Li, T.; Wang, P. Modeling Methane Emissions from Irrigated Rice Cultivation in China from 1960 to 2050. *Glob. Change Biol.* **2011,** *17,* 3511–3523. DOI:10.1111/j.1365-2486. 2011.02495.x.

Zeigler, R. S.; Puckridge, D. W. Improving Sustainable Productivity in Ricebased Rainfed Lowland Systems of South and Southeast Asia. *J. Geol.* **1995,** *35,* 307–324.

CHAPTER 3

TEMPERATE RICE CULTURE: CONSTRAINTS AND INTERVENTIONS THEREOF TO MITIGATE THE CHALLENGES

ASIF B. SHIKARI[1*], S. NAJEEB[2], GAZALA H. KHAN[1], SHABIR H. WANI[3], ZAHOOR A. BHAT[3], G. A. PARRAY[2], and SHAFIQ A. WANI[1]

[1]Centre for Plant Biotechnology, Sher-e-Kashmir University of Agricultural Sciences and Technology, Shalimar, Srinagar 191121, Jammu and Kashmir, India

[2]Mountain Research Centre for Field Crops, Sher-e-Kashmir University of Agricultural Sciences and Technology, Shalimar, Srinagar 191121, Jammu and Kashmir, India

[3]Division of Genetics and Plant Breeding, Sher-e-Kashmir University of Agricultural Sciences and Technology, Shalimar, Srinagar 191121, Jammu and Kashmir, India

*Corresponding author. E-mail: asifshikari@gmail.com

ABSTRACT

Rice (*Oryza sativa* L.), an important source of dietary calories in diet, can be grown from purely tropical through subtropical to temperate cold region. In order to meet the fast-increasing population demands, the rice production must be enhanced especially in the temperate rice-growing areas which are deprived of the green revolution. Several issues related to temperate ecologies such as cold stress, leaf and panicle blast, sheath blight, etc. lead to the low rice production which must be solved by developing the suitable cold-tolerant varieties besides being photosynthetically efficient. The breeding is another option to improve the rice quality traits that would make rice cultivation more economic and competitive in nature. Therefore, the main challenge of short growing season and peculiarities of weather encountered in the hilly areas must be solved by breeders and agronomists with the help of genetic engineering and omics-based techniques.

3.1   INTRODUCTION

Rice (*Oryza sativa* L.) is grown worldwide and provides 20% of dietary calories to human population. Rice ecology ranges from purely tropical (below mean sea level), through subtropical to temperate cold region, where it is grown at altitudes ranging up to 2300 m amsl. Globally, rice is cultivated on an area of 158 million ha with a total production of 700 million t of paddy or 470 million t of milled rice annually (FAO, 2014). Presently, Asia produces the food for 27 persons per hectare against an estimated demand of 43 people per hectare of land by the year 2050 (Skamnioti and Gurr, 2009). Temperate rice occupies 20% of the area under rice cultivation worldwide and is grown in countries such as Bhutan, Australia, Central Asia, China, Japan, Chile, Korea, Nepal, Philippines, Russia, Turkey, the United States, Uruguay, and few other places. The fact that world needs to produce additional 8–10 million t of rice after each passing year to meet out the increasing population demand, to safeguard the productivity in temperate rice-growing areas is of equal importance. Major production constraints, however, daunt the rice yields in such ecologies which are some ways similar and somehow different than what is experienced in subtropical or tropical regions. Temperate ecologies generally have internal niche specific problems which can range from genetic to technological in nature. One of the major production constraints in temperate areas happens to be cold stress. Also, low temperature limits rice production in subtropical areas and also in high-altitude areas of tropics. Therefore, development of suitable cold-tolerant varieties with acceptable grain quality and early maturity would help to make rice cultivation more economic and competitive in nature.

3.2   MAJOR CONSTRAINTS AND CHALLENGES UNDER TEMPERATE CONDITIONS

There are various constraints and challenges in rice cultivation under the temperate conditions which limit production and productivity. Some of them are described below.

3.2.1   COLD STRESS TOLERANCE

A multitude of stresses influence the potential yield in temperate areas and higher elevations of subtropical regions most important of which happens to be stress due to cold/freezing temperatures. Rice is a crop of tropics and inherently lacks tolerance against cold stress. In parts of South and Southeast Asia, an estimated 7 million ha cannot be planted with modern varieties because of low temperature stress (Sthapit and Witcombe, 1998). Cold stress is encountered at various crop growth stages and jeopardizes the yield through different genetic and physiological mechanisms. Depending upon the crop growth stage which is exposed to cold, the different genotypes/varieties are expected to show variable responses to cold stress. This is because there are different genomic regions/quantitative trait loci (QTL) governing cold tolerance at specific growth stages of a plant. This makes cold tolerance on the whole a very complex character which cannot be defined out of context without the reference to the particular developmental stage of crop and obviously the environment. Rice plants experience injury at seedling stage if grown in early spring in temperate regions. This results in seedling mortality and weak seedling which may easily be predisposed to fungal infections thereby making the situation worse. Cold stress results in low initial germination particularly in case of direct sowing which leads to reduced plant stand in upland conditions. Seedling damaged with cold bites take longer time to get established which causes poor tillering and loss in yield.

Cold-tolerant genotypes possess high photosynthetic efficiency than cold-susceptible genotypes under stress environments. Cold stress is not only met at seed/seedling stage but also can be more challenging at flowering. Cold snaps at booting and anthesis cause shriveled anthers and nonviable pollen grains. This has been related to inefficient transfer of sugars to pollen in cold susceptible genotypes. Drop in pollen viability has got drastic effect on final yield. Apart from direct cold injury at seedling and flowering, prevalence of cold irrigation water and air tends to prolong vegetative growth in cold susceptible genotypes, which is the cause for delayed maturity. Cold snaps at terminal stages of crop-growth result in shriveled grains and deterioration of quality. The highest altitude where rice is grown is about 2800 m in Bumthang, Bhutan. During the rice-cropping season, the high-altitude environment has a low-high-low temperature pattern where low temperature is a problem in the early growth stage and also in the reproductive and ripening stages. The minimum temperatures below 15°C combined with low water temperature at seedling and tillering stage cause cold damage. In Korea, rice growth periods span 6 months from mid-April to mid-October, with lowest temperature around 13°C in April and October to the warmest of 25°C in August. Thus, rice cultivation can easily face cold injury in spring and autumn. Due to low degree of tolerance to cold, cultivation of the high-yielding "Tongil-type" rice cultivars declined rapidly and mostly high-yielding japonicas have been grown in farmers' fields since last two and a half decades. 93% of rice area (891,493 ha) is grown under 20 medium to long duration japonica varieties. In Nepal, the hills and high hills of temperate regions cover 28.8% of the total rice area but they have lower yield than tropical and subtropical areas (*Terai*) of the country. Indica rice is mostly grown in these regions with lower productivity than japonica rice. Many of the landraces having ample degree of cold tolerance include *Chhomrong, Jumli Marshi*, and *Jurneli*, etc. In India, temperate rice is grown in northwestern Himalayan states of Himachal Pradesh, Jammu and Kashmir, and Uttarakhand and northeastern hill region. About 1 million ha, which accounts for 2.3% of total area under rice in these regions are grown under extreme cold temperatures. Grain yield averages around 1.1 t per hectare as against the mean yield of 1.9 t per hectare at national level. In Kashmir, the northern most State of India, the rice grows up to 2500 m amsl under assured irrigation. The source of irrigation happens to be melting glaciers, which when reaches to fields is still at 15–18°C. At the same time crop is surrounded with low atmospheric temperatures (7–12°C) during seedling growth. The temperature fluctuations at nursery stage result in delayed seedling emergence from water surface, slow root growth, etiolation, and seedling mortality. In Kashmir, temperatures at transplanting range between 14.7°C and 28.7°C. The optimum temperature for rooting of transplanted seedlings ranges between 28°C and 31°C with little or no increase of new roots at 15°C. Major cause of nonadaptability of most of the exotic materials happen to be stunting of growth due to low night temperatures which at the mid-tillering stage may touch to 12°C in times of brief rainy week. Temperature during the tillering phase in Kashmir (18–30°C) is suitable because the tillering rate rises at 18–28°C (Hamdani et al., 1979).

Any unprecedented dip in temperatures may cause severe decline in yield as the cultivation window is narrow and the crop cannot be delayed due to terminal onset of winter immediately after optimum harvest time. Cold naps at flowering drastically cause white tips at higher altitudes and also result in high degree of sterility and eventually low yields. Considerable progress has been achieved in breeding of varieties such as *K-332, K-84, Barkat* (for very high altitudes ranging from 1800 to 2200 m amsl), *K-39-96-3-1-1-1-2* (for mid-altitudes), *Jehlum, Shalimar Rice-1*, and *Shalimar Rice-2* (for altitudes ranging from 1500 to 1650 m amsl) (Parray and Shikari, 2008). Indica/japonica program was initiated to address the issues pertaining two target ecologies in Philippines that included, irrigated lowland areas, the major rice-producing region where the aim was to achieve excellent grain quality and resistance to insect pests and diseases, lodging resistance, shattering tolerance, and no on-the-panicle sprouting. The other ecology covered cool elevated areas, where low temperature causes crucial injury to rice plants and stunted growth or sterility, leading to unstable rice production. This type of rice is targeted for Cordillera Administrative Region, where two types of rice are required: (1) short-duration (6 months) improved *bulu* with tolerance of low temperature at the seedling stage during the dry season and (2) short-duration (5 months or less) modern varieties with tolerance of low temperature at the reproductive stage in the wet season.

Cold tolerance itself is a complex trait which demands the improvement in component attributes such as seedling vigor, germinability, seedling growth, degree of yellowing, root growth, and photosynthetic ability. Cold tolerance at reproductive stage is controlled altogether by different set of genes and much of this relies on pollen viability under cold air and water temperatures. The genes *Cts1* and *Cts2* have been reported to be responsible for conferring cold tolerance at vegetative stage and were found to influence the degree of leaf yellowing (Kwak et al., 1984) and withering (Nagamine, 1991). While these authors suggest the major effect of such genes, cold tolerance is believed to be a complex trait involving multiple genes (Andaya and Tai, 2006). Mapping of genes/QTLs should form an important strategy as has been practiced by various workers (Qian et al., 2000; Andaya and Mackill, 2003; Jena et al., 2010; Fujino and Sekiguchi, 2011). Mostly, the screening procedures for cold tolerance are based on the stringency of cold temperatures and duration of exposure but usually they lead to a destructive process. Seedling survival percentages have been suggested to be reliable as survival per se is related to normal metabolic rates under cold temperatures. (Morsy et al., 2007). Though numerous genotyping platforms are available for delineating QTLs for cold tolerance, phenotyping protocols need to be properly standardized vis-a-vis the growth stages. This will help in mining of more effective alleles in natural populations. Reverse genetics approaches such as targeting induced local lesions in genomes and insertional mutagenesis would help to characterize the major genes for tolerance for their functional domains. Proteomics and other metabolomics approaches are capable of bringing forth molecular mechanism and cross-talk between cold and related stresses. Xu et al. (2012) reported the resequencing of 50 cultivated and wild rice accessions. Such megascale genomic platforms have helped recently to sequence thousands of germplasm in a very short time period and at a very low cost. The information generated may be of immense value in understanding genomic regions associated with cold stress tolerance in rice. Various genes, namely, miR-171, miR-444a, OsCDPKs, OsDREBs, OsMAPs, and OsNACs and the products of such genes are functioning in cold sensing and transcriptional regulation, and the posttranscriptional processing of the response to cold have been identified (Zhang et al., 2013).

3.2.2   NARROW GENETIC BASE

More than 250,000 rice germplasm accessions, including both cultivated types and their wild relatives, are maintained in crop gene banks worldwide. However, a startling fact is that 95% of these repositories of important genes for array of economically important traits have never been used in any breeding programs. Japonica ecotype in rice is inherently more adaptable to cold temperate climate due to its high photosynthetic efficiency and inbuilt tolerance to cold stress. However, the genetic base is considerably narrow as compared to indica which puts a ceiling on genetic enhancement of yield per unit area. One such example is of India, where it seems that varietal profile in temperate regions is much narrow with the number of varieties released through State and Central agencies limiting to just 1% of total when tallied against subtropical and tropical regions of the country. In Australia, much attention has been now diverted to breed for varieties for cold regions which apart from having cold tolerance must have better water use efficiency or can be said as need for aerobic rice systems to combat moisture deficit conditions. In Bhutan, the varietal expansion is for some other needs ranging from dietary food to rice suitable for alcohol brewing and snacks such as *sip* and *zaw*. By and large the proportion of traditional landraces grown in the country is more than 50% which leads to low productivity. China National Rice Molecular Breeding Network (CNRMBN) focuses on component of japonica rice improvement. The CNRMBN has established two gene pools, an elite gene pool (EGP) and a donor gene pool (DGP). The EGP consists of 30 commercially grown inbred varieties and parents of the best hybrid cultivars that were predominantly commercial varieties in different rice ecosystems of China. The DGP consists of 169 lines that were selected to represent the maximum geographic and genetic diversity within *Oryza sativa*. Based on rice SSR marker analysis, they had been found to contain several dozen landraces that have never been used in any previous breeding programs. Yu et al. (2003) classified the parents as indica (68.2%), japonica (30.3%), and intermediate types derived from indica/japonica. Besides, a deepwater rice,

Jalmagna, from India, which forms a single solitary group. From Nepal, 13 japonica varieties were identified based on isozyme analysis (Joshi and Bimb, 2004) and included *Chainung 242, Chandannath 1, Chandannath 3, Chhormmrong, Khumal 11, Khumal 5, Khumal 6, Khumal 7, Khumal 9, Machhapuchhre 3, Manjushree 2, Palung 2*, and *Taichung 176*. The japonica rice varieties grown in the United States are based on two very small pools of introductions in the Southern United States (tropical japonicas) and California (temperate japonicas). This narrow germplasm base represents a major constraint to the genetic improvement of japonica rice. In India, important earlier varieties released for altitudes greater than 2000 m amsl include *Barkat (K-78), K-332, Kohsar (K-429)* in Kashmir and *Norin 8* and *Norin 18* for mid-hills of Himachal Pradesh.

3.2.3   IMPORTANT RICE DISEASE

3.2.3.1   RICE BLAST

Blast of rice, caused by *Magnaporthe oryzae*, is a devastating disease of rice (Figs. 3.1 and 3.2) and a major production constraint across all the rice growing ecologies. The disease appears on different parts of plant at various growth stages and causes huge losses to both straw and grain (Fig. 3.2). The fungus best perpetuates at maximum temperatures around 25°C with drastic diurnal changes during the growing period. Therefore, temperate hill ecologies are more conducive to establishment and spread of the disease. Yield losses could be much serious in upland conditions leading to total crop failures. Fungicides often used to contain blast, add to the cost of production. The cost incurred by using blast effective fungicides in Japan in the year 2000 was estimated at 160 million euro (Yamaguchi, 2004). In China, during the 10-year period from 1980 to 1990, the crop area treated with chemical fungicides increased from 2.4 to 8.9 million ha (Shen and Lin, 1994). In Bhutan, the 1995 blast epidemic affected the entire temperate rice region and resulted in losses estimated at 1099 t of paddy. More than 100 genes have been reported to confer genetic resistance against disease (Table 3.1). However, important genes which provide meaningful resistance against most virulent isolates are limited. At the same time not all the genes are effective all over mostly because of enormous diversity in nature of pathogen virulence which gives rise to new races and strains across length and breadth of rice production systems. For instance, genes such as *Pib* were found effective against races in Japan. In Southeast Asian countries, *Pi2, Pizt, Pi40*, and *Pi9* connote broad spectrum resistance. *Pita* had been a success in the United States for very long time. In India and Nepal, genes such as *Pi9*, *Pi2*, and *Pi40* have been found effective. In Northwestern Himalayan region

FIGURE 3.1   Infected field with rice blas disease in Jammu and Kashmir, India. (a) Heavey incidence of rice blast at early transplanting stage and (b) field view of rice blast at maturity stage in a crop grown at Tangdhar, Kashmir, Jammu and Kashmir, India.

FIGURE 3.2 Symptoms of rice blast disease caused by *Magnaporthe oryzae* appear at different plant growth stages. (a) Leaf blast, (b) collar blast, (c) nodal blast, and (d) panicle blast.

under the temperate rice producing areas, *Pi54* has shown durable resistance to blast. The important japonica varieties known to possess blast resistance genes *Pi9*, *Pia*, *Pib*, *Pii*, *Pii*, *Pii*, *Pik*, *Pik*, *Piz*, and *Piz* include Unbong, Suweon, Sinunbong, Sangjuchal, Sangju, Junghwa, Joan, Jinbuchal, Haepyeong, and Goun.

Rice blast is a global problem. Korea reported the yield losses of 30–40% in the 1970s (Khush, 1989). About 10–30% annual yield losses due to rice blast have been reported (Talbot, 2003). Different lines of approaches have been researched and suggested by various workers to neutralize devastating effects of blast fungus. Some key methods that can serve the purpose include the development of resistance gene (R-gene) pyramids (Hittalmani et al. 2000); R-gene rotation; cultivar mixture (Zhu et al., 2000); use of multilines (Koizumi, 2001); transgenic crop resistance (Narayanan et al., 2002); biological control agents; silicon application; nitrogen and water management; field sanitation; and fungicides (Mew et al., 2004). The most effective option lies in the exploitation and deployment of genetic resistance to blast. The use of multilines and cultivar mixtures has been advocated; however, these are practicable only under subsistence agriculture. Narayanan et al. (2002) demonstrated the transgenic breeding approach to combine bacterial leaf blight and blast resistance in IR50. However, in the backdrop of regulatory issues surrounding transgenic crops, the conventional or advanced breeding methodologies to incorporate genetic resistance into susceptible backgrounds seem much more plausible. One such methodology entails the transfer of multiple R-genes into the genetic background of popular cultivars. Targeting few of these genes into the adaptable backgrounds could be

TABLE 3.1 Worldwide Study on Genes Employed Different Rice Cultivars to Develop Blast Resistance.

Country	Author	Gene	Ch. No.	Physical position (bp)	Source rice cultivars
China	Ashikawa et al. (2008)	*Pikm*	11	27314916–27532928	Tsuyuake (J)
	Causse et al. (1994)	*Pi11*	8	–	Zhai-Ya-Quing8 (I)
		Pizh		4372113–21012219	
	Chen et al. (2004)	*Pid(t)1*	2	20143072–22595831	Digu (I)
	Chen et al. (2006)	*Pid2*	6	17159337–17163868	
	Deng et al. (2006)	*Pigm(t)*	6	10367751–10421545	Gumei4 (I)
	Hayashi et al. (2006)	*Pikp*	11	27314916–27532928	HR22 (I)
	Huang et al. (2011)	*Pi47*	11	–	Xiangzi 3150 (I)
		Pi48	12		
	Lei et al. (2005)	*Piy1(t)*	2	–	Yanxian No 1 (I)
		Piy2(t)			
	Li et al. (2009)	*Pi37*	1	33110281–33489931	St-No 1 (J)
	Liu et al. (2004)	*PiGD1*	8	–	Sanhuangzhan 2 (I)
		PiGD-2	10		
		PiGD3	12		
	Liu et al. (2005)	*Pi36*	8	2870061–2884353	Q61 (I)
	Liu et al. (2007)	*Pi39(t)*	4, 12	–	Chubu 111 (J) Q15(I)
	Pan et al. (1996)	*Pi15*	9	9641358–9685993	GA25 (J)
	Qu et al. (2006)	*Pi9*	6	10386510–10389466	O. minuta (W)
	Wang et al. (2009)	*Pik*	11	27314916–27532928	Kusabue (I)
	Wu et al. (2005)	*Pi26*	6	8751256–11676579	Gumei 2 (I)
	Yang et al. (2009)	*Pi41*	12	33110281–34005652	93-11 (I)
	Zhou et al. (2004)	*Pig(t)*	2	34346727–35135783	Guangchangzhan (I)
	Zhuang et al. (2001)	*Pi25*	6	18080056–19257588	Gumei 2 (I)
France	Sallaud et al. (2003)	*Pi24(t)*	1	5242654–5556378	Azuenca (J)
		Pi25(t)	2	34360810–37725160	IR64 (I)
		Pi26(t)	5	2069318–2760202	Azucena (J)
		Pi27	1	5556378–744329	Q14 (I)
		Pi27(t)	6	6230045–6976491	IR64 (I)
		Pi28(t)	10	19565132–22667948	
		Pi29(t)	8	9664057–16241105	
		Pi30(t)	11	441392–6578785	
		Pi31(t)	12	7731471–11915469	
		Pi32(t)		13103039–18867450	
		Pi33	8	5915858–6152906	

TABLE 3.1   (Continued)

Country	Author	Gene	Ch. No.	Physical position (bp)	Source rice cultivars
India	Gowda et al. (2006)	*Pi38*	11	19137900–21979485	Tadukan (I)
	Barman and Chattoo (2004)	*Pitp(t)*	1	25135400–28667306	Tetep (I)
	Kumar et al. (2010)	*Pi42(t)*	12	19565132–22667948	DHR9 (I)
	Naqvi et al. (1996)	*Pi10*	5	14521809–18854305	Tongil (I)
		Pi157	12	8826555–18050447	Moroberekan (J)
	Sharma et al. (2005)	*Pikh (Pi54)*	11	24761902–24762922	Tetep (I)
Indonesia	Dwinita et al. (2008)	*Pirf2-1(t)*	2	–	*O. rufipogan* (W)
		Pir2-3(t)			IR64 (I)
Japan	Fjellstrom et al. (2004)	*Piks*	11	27314916–27532928	Shin 2 (J)
	Fujii et al. (1995)	*PBR*	11	–	St-No 1 (J)
	Fujii et al. (1999)	*Pb1*	11	21711437–21361768	Modan (I)
	Fukuoka and Okuno (2001)	*pi21*	4	5242654–5556378	Owarihatamochi (J)
	Goto (1970)	*Piis1*	11	2840211–19029573	Imochi Shiraz (J)
		Piis2	–	–	
		Piis3			
		Pikur1	4	24611955–33558479	Kuroka (J)
		Pise	11	5740642–16730739	Sensho (J)
		Pise2	–	–	
		Pise3			
	Goto (1976)	*Mpiz*	11	4073024–16730739	Zenith (J)
		Piz	6	10155975–10517612	Zenith (J), Tadukan (I), Toride 1 (J), Fukunishiki (J)
	Goto (1981)	*Pia*	11	4073024–8078510	Aichi Asahi (J)
	Goto (1988)	*Pikur2*	11	2840211–18372685	Kuroka (J)
	Hayashi et al. (2006)	*Pi19(t)*	12	8826555–13417087	Aichi Asahi (J)
		Pit	1	2270216–3043185	Tjahaja (I), K59 (I)
	Hayashi et al. (2010b)	*Pb1*	11	21711437–21361768	Modan (I)
	Imbe et al. (1997)	*Pish*	1	33381385–35283446	Shin 2 (J)
			11		Nipponbare (J)

TABLE 3.1   (Continued)

Country	Author	Gene	Ch. No.	Physical position (bp)	Source rice cultivars
	Inukai et al. (1996)	*Pi12*	12	6988220–15120464	K80-R-Hang (J) Moroberekan (J)
	Ise et al. (1991)	*Pii*	9	2291804–28431560	Ishikari Shiroke (J) Fujisaka 5 (J)
	Kinoshita and Kiyosawa (1997)	*Pii2*	9	1022662–7222779	Ishikari Shiroke (J)
	Nakamura et al. (1997)	*Pita2*	12	10078620–13211331	Shimokita (J)
	Nguyen et al. (2006)	*Pi35(t)*	1	–	Hokkai 188 (J)
	Pan et al. (1996)	*Pikg*	11	27314916–27532928	GA20 (J)
		Pii1	6	2291804–28431560	Fujisaka 5 (J)
		Pi8		6230045–8751256	Kasalath (I)
	Pan et al. (1998)	*Pi14(t)*	2	1–6725831	Maowangu
	Pan et al. (1999)	*Pi16(t)*	2	1–6725831	Aus373 (I)
	Sallaud et al. (2003)	*Pi62(t)*	12	2426648–18050026	Yashiro-mochi (J)
	Shinoda et al. (1971)	*Pif*	11	24695583–28462103	Chugoku 31-1 (J)
	Wang et al. (1999)	*Pib*	2	35107768–35112900	Tohoku IL9 (J)
	Zenbayashi et al. (2002)	*Pi34*	11	19423000–19490000	Chubu32 (J)
Korea	Ahn et al. (1996)	*Pi22(t)*	6	4897048–6023472	Suweon365 (J)
		Pi23	5	10755867–19175845	
	Ahn et al. (2000)	*Pi18(t)*	11	26796917–28376959	Suweon365 (J)
Philippines	Amante et al. (1992)	*Pi13(t)*	6	12456009–16303608	O. minuta (W), Kasalath (I), Maowangu
	Hua et al. (2012)	*Pi1*	11	26498854–28374448	LAC23 (J)
	Imbe et al. (1997)	*Pi20*	12	6988220–10603823	IR24 (I)
	Jeon et al. (2003)	*Pi5(t)*	9	–	Moroberekan (J)
	Jeung et al. (2007)	*Pi40(t)*	6	16274830–17531111	O. australiensis (W)
	Mackill and Bonman (1992)	*Pi3(t)*	6	–	Pai-kan-tao (J)
	Pan et al. (1995)	*Pi17*	7	22250443–24995083	DJ123 (I)
	Sallaud et al. (2003)	*Pi67*	–	–	Tsuyuake
	Tabien et al. (1996)	*Pib2*	11	26796917–28376959	Lemont (J)
	Yu et al. (1991)	*Pi1*	11	26498854–28374448	LAC23 (J)

TABLE 3.1 (Continued)

Country	Author	Gene	Ch. No.	Physical position (bp)	Source rice cultivars
The United States	Chen et al. (1999)	*Pi44*	11	20549800–26004823	Moroberekan (J)
	Bryan et al. (2000)	*Pita*	12	10603772–10609330	Tadukan (I)
	Chauhan et al. (2002)	*PiCO39(t)*	11	6304007–6888870	CO39 (I)
	McCouch et al. (1994)	*Pi6(t)*	12	4053339–18867450	Apura (I)
	Tabien et al. (1996)	*Pitq2*	2	–	Teqing (I)
		Pitq3	3		
		Pitq4	4		
	Tabien et al. (2000)	*Pilm2*	11	13635033–28377565	Lemont (J)
		Pitq1	6	28599181–30327854	Tequing (I)
		Pi-tq5	2	34614264–35662091	
		Pitq6	12	5758663–7731471	

Ch. No., chromosome number; I, indica; J, japonica.

[a]Modified from Sharma, T. R.; Rai, A. K.; Gupta, S. K.; Vijayan, J.; Devanna, B. N.; Ray, S. Rice Blast Management Through Host-plant Resistance: Retrospect and Prospects. *Agric. Res.* **2012,** *1* (1), 37–52. © NAAS (National Academy of Agricultural Sciences), 2012. With permission from Springer.

achieved precisely by marker-assisted selection. Fu et al. (2012) incorporated the blast resistance genes *Pi1* and *Pi2* (from donor BL122) into the elite maintainer line Rongfeing. Singh et al. (2012) performed the marker-assisted backcrossing procedure for incorporation of genes such as *Pi54* and *Piz5* in Basmati rice. We for the first time developed a set of near-isogenic lines in the background of Basmati variety (Pusa Basmati 1) (Khanna et al., 2015). The seven different NILs were formed with the help of marker-assisted foreground and background selection and carried the genes *Pi54*, *Pi1*, *Pita*, *Pib*, *Pi2*, *Pi5*, and *Pi9* (Inukai et al., 1994). The two- and three-gene pyramided lines were also developed. Gene pyramiding strategies offer to provide durable resistance in susceptible cultivars. High-throughput genotyping platforms make it easy to map QTLs and genes linked to minor or major gene resistance. Novel alleles of already known genes that perform much more effectively than previously available ones could be discovered and transferred from wild to cultivated background as has been recently suggested for gene *Pi54* (Das et al., 2012). Our group discovered the novel allelic sources of blast resistance which could be used as potential donors in breeding programs (Shikari et al., 2013, 2014). Quantitative resistance has deep rooted governance principle than major gene resistance and is thought to be much more effective in long term. However, little efforts have been done to map and exploit such resistance into practical breeding programs. This can be achieved through integration of approaches such as genomic selection or genotyping by sequencing technology accompanied with phenotyping for minor gene resistance. One of the important and least studied areas in rice blast resistance research is that against panicle blast in rice. Though there are 100 odd genes reported for blast resistance, only Pb1 locus has been identified to control panicle blast (Hayashi et al 2010b). The sources of germplasm are too limited and unavailable which calls to identify novel sources of durable resistance to panicle blast.

3.2.3.2   BROWN SPOT DISEASE

Brown spot is one of the serious diseases caused by *Cochliobolus miyabeanus* (Fig. 3.3), particularly in northwestern Himalayan region. As such, the disease does not cause economic loss when reported in its typical form of scattered dense brownish spots on leaves; however, it reduces the market value of rice when brown spots appear on glumes. Good fertility management can avoid the onset of disease. The disease seems to be more serious under low N conditions, poor management, and appears mostly on late season varieties.

3.2.3.3   BAKANAE DISEASE

Bakanae is also known as "foolish seedling" disease. It is another upcoming disease caused by *Gibberella fujikuroi* that is a serious constraint in temperate areas (Fig. 3.4). We reported the foot rot symptoms at more than 2200 m amsl in temperate varieties such as *K-332, K-429*, etc. Very few resistance donors or QTLs are available so far which can be brought into practical breeding of resistant cultivars. Nevertheless, scope lies in identification of novel donors for bakanae resistance across diverse germplasm. Fiyaz et al. (2014) reported some novel sources of resistance against bakanae isolates and mapped (Fiyaz et al., 2016) QTLs for resistance, namely, qBK1.1, qBK1.2, and qBK1.3 which explained 4.76%, 24.74%, and 6.49% of phenotypic variation, respectively. Our center has identified several promising lines showing broad resistance to *Gibberella fujikuroi* (Lone et al., 2016).

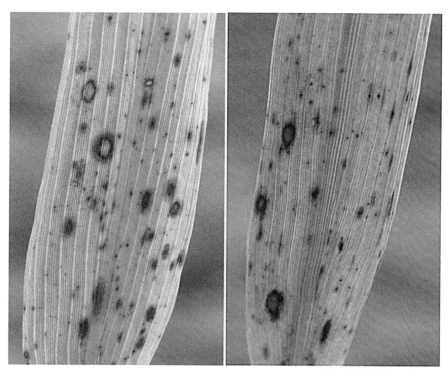

FIGURE 3.3 Symptoms of rice brown spot disease caused by *Cochliobolus miyabeanus.* Spots and lesions on leaves.

FIGURE 3.4 Bakanae or foolish seedling disease of rice caused by *Gibberella fujikuroi.* (a) Symptoms of excessive elongation at tillering stage and (b) foot rot symptoms.

3.2.4   EARLY MATURITY

Moisture stress or less water availability has sometimes been a problem in temperate regions such as in Australia, where the areas prefer to use short duration varieties to avoid dry patches. Short duration varieties at the same time need to have good seedling vigor for rapid attainment of vegetative phase which form an indirect component to early maturity. Short duration varieties fit into multiple cropping patterns and also provide more crop yield per hectare per day.

3.2.5   PHYSIOLOGICAL ISSUES

Straighthead a physiological disorder resulting in malformed and empty florets, which is aggravated under cold underlying waters, has been reported. In Kashmir, the rice variety, *Shalimar Rice-1* has been reported to be doing exceptionally well with plump and filled terminal spikelets even under cold stagnant waters.

3.2.6   WEEDY RICE

Temperate rice fields in Bhutan are highly infested with wild forms of rice, *Oryza sativa* f. *spontanea*. Difficult to identify at early growth stage, weedy rice is highly shattering trait which deteriorates both quality and yield. Molecular markers may help to purify the nucleus seed which after multiplication controlled plots may be dissimilated. Also, volunteer plants in field may be identified through distinguishing morphological feature of a released variety.

3.2.7   WATERLOGGING

In low lying areas where rice field happens to be lower than riverbed, waterlogging is a problem that results in low tillering ability, poor response to fertilizer application, and crop lodging which results in on-field sprouting of matured seeds. Under waterlogged conditions Al, Mn, Zn, and Fe toxicity is common.

3.3 INTERVENTIONS AND TECHNOLOGICAL DEVELOPMENT TO MITIGATE THE CHALLENGES

3.3.1 RICE QUALITY IMPROVEMENT

3.3.1.1 IN BASMATI RICE

Basmati rice native to Himalayan foothills of Indian subcontinent is a cultural heritage of the region, a legacy, and valuable trade commodity. This is a unique varietal group, which has found worldwide acceptance for its exquisite quality traits and harmonious blend of the features comprising extra long, slender superfine grains, length-wise kernel elongation with minimum breadth-wise expansion on cooking, fluffy texture of cooked rice, pleasant aroma, delightful mouth feel, taste, and appearance. Aromatic rices have a special place in world rice markets as they are highly priced. Among the aromatic rices, long-grain basmati types fetch highest premium and India is one of the major exporters of basmati rice in the world. Super-fine long slender grain (>6.61-mm long and with LBR > 4), characteristic pleasant aroma, high elongation ratio on cooking (>1.7 times), and soft texture of cooked rice are the distinctive features of basmati rices. Basmati cultivation is confined to Northwest Indian States—Haryana, Punjab, Utta-ranchal, and Western Uttar Pradesh and to a limited area in Delhi, Himachal Pradesh, Jammu and Kashmir, and Rajasthan. It has touched foreign earn-ings exceeding a whooping figure of INR 30,000 crore for India. Success story of Indian Basmati rice culture is really big and may not be dealt here as most of this contribution comes from subtropical regions near Himalayas which cultivate most recently bred Basmati rices varieties apart from tradi-tional types such as Basmati 370, Type 3, Ranbir Basmati, etc. It is pertinent to mention though the characteristic features of aroma and taste are best expressed under cool moist conditions near foothills. The problem in typi-cally higher elevations (>1500 m amsl) is that full season Basmati varieties do not flower or mature. The grain quality is a function of starch filling and packaging prior to harvest. The erratic temperatures in cold temperate regions may result in chalky grains, loose starch, and bad quality, thereby deteriorating the essence in final produce. Very long-grained types may therefore, not be a choice in regions with limited cropping season which may provide us next option of introducing semifine grained types with aroma. In one such example, we proved that atypical Basmati rice Pusa Sungandh-3 was introduced from Indian Agricultural Research Institute, New Delhi and tried in Kashmir at 1560 m amsl. The level of aroma under temperate conditions increased many folds and better expression of quality was pronounced.

With the depleting natural resource base, deteriorating soil health, declining input use efficiency, and plateauing of yield in irrigated ecology to achieve the targeted production would be an uphill task in the coming decades. Moreover, the rising cost of cultivation, various production constraints, competitive marketing scenario, and low benefit to cost ratio has made rice cultivation less economic. This has caused people to shift toward cultivation of more remunerative horticultural and other crops. If this trend continues, the lack of motivation will result in declining acreage under rice at the cost of alternative crop species. In the backdrop of above said consistent technical and economical bottlenecks, the rice culture in temperate regions needs to be revived and revisited in its policy and scientific intervention, which can be achieved by several means and the most important of those seems to be the introduction and popularization of value-added rice which naturally includes fine-grained scented varieties. This underlines the idea of using second-line Basmati varieties in temperate regions to make rice production more economic and as a remunerative enterprise.

3.3.1.2   IN HYBRID RICE

India is a highest net exporter of rice which caters 30% of net exports with the foreign exchange earnings of $6.4 billion. In order to sustain self-sufficiency and global competitive advantage, India must grow with 2 million t of rice every year. Peoples Republic of China is a leading producer of hybrid rice and countries such as India, Bangladesh, Indonesia, Vietnam, and Philippines have commercialized the hybrid rice technology in past years. By adopting hybrid rice technology, China has witnessed enhanced yield levels which can feed additionally, an estimated 60 million people per year (Li et al. 2010). Given the high labor to land ratio and more than 50% of area under assured irrigation, the hybrid rice technology can be grand successes in pushing further the already achieved yield levels. The temperate areas are more suited to this concept which already show great success in having yield levels in inbreds as high as 7 t/ha. In such conditions, hybrids have capacity to surpass 10 t mark. However, the biggest hurdle in such areas (citing example from high altitudes of Jammu and Kashmir province of India), much more needs to be done as traditional cytoplasmic male sterility (CMS) lines IR58025A, IR68888A, and others do not suit to the climate. It was not possible to achieve the promised yield levels with most of the rice hybrids released for higher hills. The line conversion thus, was started in early 2000's and as of now there is a set of CMS lines and restorers developed to satisfy much needed dimension to rice breeding and its practical utilization in mountainous regions. Initiatives were successfully concluded with development of temperate fertility restorers and sufficient realization of heterosis levels before commercial rice hybrids may be out for field release (Khan et al., 2016).

3.3.2 MOLECULAR BREEDING AS NOVEL APPROACHES

Given the fact that 120,000 rice accessions are maintained in the T.T. Chang Genetic Resources Center of the International Rice Research Institute (IRRI), such a huge diversity would be advantageous once the accessions are characterized and utilized for breeding new varieties. However, only 5% of this diversity has been tapped which calls for sequential phenotype-based classification followed by characterization at molecular level before the important QTLs are finally discovered. Next-generation sequencing platforms have made it possible to resequence the rice germplasm (Subbaiyan et al., 2012) for identification of the allelic variation found in key rice quality genes. This revolutionary approach would ultimately pave a way toward discovery of agronomically useful loci based on expression analysis (eQTLs) or metabolic fingerprinting to capture mQTLs. The genome-wide association studies offer an encouraging tool for fine mapping of the major QTLs underlying complex traits of grain quality from untapped germplasm (Bandillo et al., 2013). Genomic selection on the other hand is a special kind of selection process, wherein genomics-assisted breeding value is fixed for all uncorrelated markers and statistically relate it to target trait, and sum of all the predictions thus, is used to identify small-effect QTLs. The process has been detailed by various authors such as Heffner et al. (2009); Jannink et al. (2010). Carefully selected superior lines based on yield- and quality-related traits or for stress tolerance may be used as parents to generate multiparent advanced generation intercross or nested association mapping populations to break very tight linkages in a short time period. This would help in dissecting complex loci and more so high-resolution mapping of genes/QTLs.

In parallel with advances made in rice genome sequencing and discovery of enormous number of markers, rice molecular breeding has reached a stage where numerous deliverables have come to limelight and also to use in farmer's fields. In many cases, the incorporation of genes has brought into mega varieties, a novel combination of traits which guarantee to uphold rice production system above the vagaries of climate change. The most prominent ones include the rice tolerant to submergence Sub1 (Sarkar et al., 2006; Xu et al., 2000, 2006; Neeraja et al., 2007; Perata and Voesenek, 2007; Septiningsih et al. 2009; Singh et al. 2010), drought, for example, qtl12.1 (Bernier et al. 2009a, 2009b; Swamy et al. 2011), salinity SalTol (Kaushik et al., 2003; Thomson et al. 2010; Pandit et al., 2010), and P-deficiency toler-ance (Kottearachchi and Wijesekara, 2013; Chin et al., 2011).

Before the onset of the year 2025, the global food demand needs to be met by increasing food grain production further by another 50%. (Khush, 2001). Scientists across the world have to be indomitable to achieve this herculean task. The target trait though, is grain yield; however, it is a much complex trait simultaneously under the control of multiple genes/QTLs. To target the improvement of such a trait is possible after resolving the deficien-cies with favorable alleles at all component loci across the genome. Grain yield is a complex agronomic trait, determined by multiplicative action of three component traits, namely, number of effective tillers per plant, number of grains per panicle, and grain weight. Grain number that has been reported to be guided by grain number 1a (Gn1a), a major QTL which was mapped in backcross inbred lines of Koshihikari (lower grain number) and Habataki (high number) (Ashikari et al., 2005) (Table 3.2). Gn1a encodes an enzyme, namely, cytokinin oxidase/dehydrogenase (OsCKX2), which degrades the cytokinin. The reduced expression of Gn1a results in accumulation of cyto-kinin in the primitive inflorescence and results enhanced number of grains per panicle. Ashikari et al. (2005) developed Near-isogenic line (NIL-Gn1) in the genetic background of Koshihikari with 45% increase in grain number per panicle and 35% increase in grain number per plant, as compared to recurrent parent, Koshihikari. Another determinant of grain yield is grain size which along with filling rate influences the final test weight. GS3 was the first major QTL in this category to be isolated and it has been found that it contributes to both grain length and weight (Fan, et al., 2006). GS3 was identified in a cross between long grained, Minghui 63 and small grained, Chuan7. Fan et al. (2006) used BC_3F_2 population derived from Chuan7 as recurrent parent and selected homozygous plants with the grain length of 10.2 mm (contributed by Minghui 63 allele), that were 39% longer than recurrent parent, Chuan7 (7.32 mm). As a consequence, the 1000-grain weight of homozygous plants carrying the Minghui 63 allele increased to 25.6 g, with 46% more weight than homozygous plants with Chuan7 allele (17.5 g). Later Mao et al. (2010) identified two more alleles, in addition to Minghui63 and Chuan7 and classified four types of GS3 alleles. Grain size QTL, GW2 was discovered in a cross between WY3 with large grain and Fengaizhan-1 having small grain. This influences both grain width and weight (Song et al., 2007). GW2 encodes the RING-type protein having E3 ubiquitin ligase activity. The protein is known to cause degradation in ubiquitin-proteasome pathway. GW2 is a loss-of-function allele which results in increased cell number that causes larger and wider spikelet hull and therefore, accelerated grain filling. Song et al. (2007) reported 26.2% increase in grain width and 49.8% increase in 1000-grain weight for NIL (GW2) in the genetic background of Fengaizhan-1. Although the number of grains per panicle was 29.9% lower in NIL (GW2), the total grain yield per plant increased by 19.9% over Fengaizhan-1.

3.4 FUTURE PERSPECTIVES AND RESEARCH OPPORTUNITIES

The fruits of green revolution did not reach the rice-growing regions in stress ecologies. The narrow options are available to genuinely raise the yield ceiling in mountain and hill regions where limited genetic resource is at disposal to fish out important genes related to agronomical traits and for tolerance to stresses. Major stresses which limit our production potential include cold stress, drought, rice blast, and several niche specific problems which need to be addressed differently. Pathogen population of hills have their distinct racial profile for which queue may not be drawn from much widely established race spectrum prevalent in tropics. On the other hand, breeding for rice quality traits is immensely important to provide competitive advantage to farm produce in local and international markets. It has been indicated that lot much needs to be done

TABLE 3.2 Important Grain-related Traits and Genes Which Contribute to Yield per Plant.

Trait	QTL	Chromosome	Intervals	Protein type	References
Grain size	GS3	3	GS09-MRG5881	Transmembrane protein	Fan et al. (2006)
Grain weight	gw3.1	3	JL8-RM3180	RING-type ubiquitin E3 ligase	Li et al. (2004)
Grain weight	gw8.1	8	RM531-RM42	RING-type ubiquitin E3 ligase	Xie et al. (2006)
Grain weight	gw9.1	9	RM5661-RM215	RING-type ubiquitin E3 ligase	Xie et al. (2008)
Grain number	Gn1a	1	3A28-3A20	Cytokinin oxidase	Ashikari et al. (2005)

Gn1a, grain number 1a; gw, grain weight; QTL, quantitative trait locus.

beyond classification based on amylose and gelling properties in rice. This is evident from studies on IRRI rice stocks where more than 10,000 lines have both intermediate GC and AC, yet they differ in organoleptic qualities. This proves that all other background sensory and nonsensory compounds take part in shaping final rice quality and hence guide the consumer acceptability at wide range. Advance in rice molecular genetics and genomics tools have broadened our knowledge from simple understanding of gene structure and its function to metabolic cross talk between various complex pathways which shape most important traits favoring tolerance to stresses in rice.

3.5   SUMMARY AND CONCLUSION

Temperate rice is grown in considerable area across the globe, which includes the countries such as Bhutan, Australia, Central Asia, China, Japan, Chile, Korea, Nepal, Philippines, Russia, Turkey, the United States, and Uruguay. The constraints to rice production besides some specific ones are cold/freezing stress at seedling and reproductive stage, leaf and panicle blast, sheath rot, sheath blight, and few others. Low temperatures in some of these countries result in average yield loss of 1.0–3.9 t/ha. Further, narrow genetic diversity leads to narrow adaptability which is the primary cause of low yield. Hybrid rice has taken up in some the countries such as China and India; however, hybrids suitable for use in cold temperate high altitudes are generally lacking. Upland rices are challenged with both drought and cold and the cultivation is strained at all the growth stages. The expression of quality traits such as aroma happens to be complete under moderate temperatures which open up the scope to introduce consumer preferred quality rice genotypes to provide competitive advantage to rice culture. In hilly areas, the major impediment to production lies with the nonadaptability of varieties over large geographic area. The performance of varieties remains confined to particular ecological niches which calls for breeding of location specific varieties. In the backdrop of unique problems being faced in hill and mountain agriculture, which are characterized by short growing season and vagaries of weather, the real challenge to breeders and agronomists would be to breed for varieties resilient to abiotic and biotic stresses, besides being photosynthetically efficient, in order to realize high yield per unit area per day. These issues could be met out by tools of modern biology and omics technologies.

KEYWORDS

- temperate rice fields
- blast-resistance genes
- marker-assisted selection
- water use efficiency

- physiological disorder
- resistant cultivars
- biological control

REFERENCES

Ahn, S. N.; Kim, Y. K.; Hong, H. C.; Han, S. S.; Kwon, S. J.; Choi, H. C.; Moon, H. P.; McCouch, S. R. Molecular Mapping of a Gene for Resistance to a Korean Isolate of Rice Blast. *Rice Genet. Newslett.* **1996,** *13,* 74.

Ahn, S. N.; Kim, Y. K.; Hong, H. C.; Han, S. S.; Kwon, S. J.; Choi, H. C.; Moon, H. P.; McCouch, S. Molecular Mapping of a New Gene for Resistance to Rice Blast. *Euphytica* **2000,** *116,* 17–22.

Amante-Bordeos, A.; Sitch, L. A.; Nelson, R.; Damacio, R. D.; Oliva, N. P.; Aswidinnoor, H.; Leung, H. Transfer of Bacterial Blight and Blast Resistance from the Tetraploid Wild Rice (*Oryza minuta*) to Cultivated Rice (*Oryza sativa* L.). *Theor. Appl. Genet.* **1992,** *84,* 345–354.

Andaya, V. C.; Mackill, D. J. Mapping of QTLs Associated with Cold Tolerance During the Vegetative Stage in Rice. *J. Exp. Bot.* **2003,** *54,* 2579–2585.

Andaya, V. C.; Tai, T. H. Fine Mapping of the qCTS12 Locus, a Major QTL for Seedling Cold Tolerance in Rice. *Theor. Appl. Genet.* **2006,** *113,* 467–475.

Ashikari, M.; Sakakibara. H.; Lin. S. Y.; Yamamoto. T.; Takashi, T. Cytokinin Oxidase Regulates Rice Grain Production. *Science* **2005,** *309,* 741–45.

Ashikawa, I.; Hayashi, N.; Yamane, H.; Kanamori, H.; Wu, J.; Matsumoto, T.; Ono, K.; Yano, M. Two Adjacent Nucleotide- binding Site-leucine-rich Repeat Class Genes Are Required to Confer Pikm-specific Rice Blast Resistance. *Genetics* **2008,** *180,* 2267–2276.

Bandillo, N.; Raghavan, C.; Muyco, P. A.; Sevilla, M. A.; Lobina, I. T.; Dilla-Ermita, C. J.; Tung, C. W.; McCouch S.; Thomson, M.; Mauleon, R.; Singh, R. K.; Gregorio, G.; Redona, E.; Leung, H. Multi-parent Advanced Generation Inter-cross (MAGIC) Populations in Rice: Progress and Potential for Genetics Research and Breeding. *Rice* **2013,** *6,* 11.

Barman, S. R.; Chattoo, B. B. Identification of a Major Blast Resistance Gene in the Rice Cultivar Tetep. *Plant Breed.* **2004,** *123,* 300–302.

Bernier, J.; Serraj, R.; Kumar, A.; Venuprasad, R.; Impa, S.; Gowda, R. P. V.; Oane, R.; Spaner, D.; Atlin, G. The Large-effect Drought-resistance QTL qtl12.1 Increases Water Uptake in Upland Rice. *Field Crops Res.* **2009a,** *110,* 139–146.

Bernier, J.; Kumar, A.; Venuprasad, R.; Spaner, D.; Verulkar, S.; Mandal, N. P.; Sinha, P. K.; Peeraju, P.; Dongre, P. R.; Mahto, R. N.; Atlin. G. Characterization of the Effect of a QTL for Drought Resistance in Rice, qtl12.1, over a Range of Environments in the Philippines and Eastern India. *Euphytica* **2009b**, *166*, 207–217.

Bryan, G. T.; Wu, K.; Farrall, L.; Jia, Y.; Hershey, H. P.; McAdams, S. A.; Faulk, K. N.; Donaldson, G. K.; Tarchini, R.; Valent, B. A Single Amino Acid Difference Distinguishes Resistant and Susceptible Alleles of the Rice Blast Resistance Gene *Pi-ta. Plant Cell* **2000**, *12*, 2033–2045.

Causse, M. A.; Fulton, T. M.; Cho, Y. G.; Ahn, S. N.; Chunwongse, J.; Wu, K.; Xiao, J.; Yu, Z. H.; Ronald, P. C.; Harrington, S. E.; Second, G.; McCouch, S. R.; Tanksley, S. D. Saturated Molecular Map of the Rice Genome Based on an Interspecific Backcross Population. *Genetics* **1994**, *138*, 1251–1274.

Chauhan, R. S.; Farman, M. L.; Zhang, H. B.; Leong, S. A. Genetic and Physical Mapping of a Rice Blast Resistance Locus, Pi-CO39(t), That Corresponds to the Avirulence Gene AVR1-CO39 of *Magnaporthe grisea. Mol. Genet. Genomics* **2002**, *267*, 603–612.

Chen, D. H.; Dela Vina, M.; Inukai, T.; Mackill, D. J.; Ronald, P. C.; Nelson, R. J. Molecular Mapping of the Blast-resistance Gene, *Pi44(t)*, in a Line Derived from a Durably Resistant Rice Cultivar. *Theor. Appl. Genet.* **1999**, *98*, 1046–1053.

Chen, X. W.; Li, S. G.; Xu, J. C. Identification of Two Blast Resistance Genes in a Rice Variety, Digu. *J. Phytopathol.* **2004**, *152*, 77–85.

Chen, X. W.; Shang, J.; Chen, D.; Lei, C.; Zou, Y.; Zhai, W.; Liu, G.; Xu, J.; Ling, Z.; Cao, G.; Ma, B.; Wang, Y.; Zhao, X.; Li, S.; Zhu, L. A B-lectin Receptor Kinase Gene Conferring Rice Blast Resistance. *Plant J.* **2006**, *46*, 794–804.

Chin, J. H.; Gamuyao, R.; Dalid, C.; Bustamam, M.; Prasetiyono, J.; Moeljopawiro, S.; Wissuwa, M.; Heuer. S. Developing Rice with High Yield Under Phosphorus Deficiency: Pup1 Sequence to Application. *Plant Physiol.* **2011**, *156*, 1202–1216.

Das, A.; Soubam, D.; Singh, P. K.; Thakur, S.; Singh, N. K.; Sharma, T. R. A Novel Blast Resistance Gene, *Pi54rh* Cloned from Wild Species of Rice, *Oryza rhizomatis* Confers Broad Spectrum Resistance to *Magnaporthe oryzae. Funct. Integr. Genomics* **2012**. DOI: 10.1007/s10142-012-0284-1.

Deng, Y.; Zhu, X.; Shen, Y.; He, Z. Genetic Characterization and Fine Mapping of the Blast Resistance Locus *Pigm(t)* Tightly Linked to *Pi2* and *Pi9* in a Broad-spectrum Resistant Chinese Variety. *Theor. Appl. Genet.* **2006**, *113*, 705–713.

Dwinita, W. U.; Sugiono, M.; Hajrial, A.; Asep, S.; Ida, H. Blast Resistance Genes in Wild Rice (*Oryza rufipogon)* and Rice Cultivar IR64. *Indian J. Agric. Sci.* **2008**, *1*, 71–76.

FAO (Food and Agriculture Organization). *Rice Market Monitor.* Food and Agriculture Organization of the United Nations, Rome, 2014. http://www.fao.org/3/a-i4147e.pdf (accessed Oct 14, 2017).

Fan, C. et al. GS3, a Major QTL for Grain Length and Weight and Minor QTL for Grain Width and Thickness in Rice, Encodes a Putative Transmembrane Protein. *Theor. Appl. Genet.* **2006**, *112*, 1164–1171.

Fiyaz, R. A.; Gopalakrishnan, S.; Rajashekara, H.; Yadav, A. K.; Bashyal, B. M.; Bhowmick, P. K.; Singh, N. K.; Prabhu, K. V.; Singh, A. K. Development of High Throughput Screening Protocol and Identification of Novel Sources of Resistance Against Bakanae Disease in Rice (*Oryza sativa* L.). *Indian J. Genet.* **2014**, *74* (4), 414–422.

Fiyaz, R. A.; Yadav, A. K.; Gopala Krishnan, S.; Ellur, R. K. Bashyal, B. M.; Grover, N.; Bhowmick, P. K.; Nagarajan, M.; Vinod, K. K.; Singh, N. K.; Prabhu, K. V.; Singh, A.

K. Mapping Quantitative Trait Loci Responsible for Resistance to Bakanae Disease in Rice. Rice **2016,** *9* (45). DOI: 10.1186/s12284-016-0117-2.

Fjellstrom, R.; Conaway-Bormans, C. A.; McClung, A. M.; Marchetti, M. A.; Shank, A. R.; Park, W. D. Development of DNA Markers Suitable for Marker-assisted Selection of Three *Pi* Genes Conferring Resistance to Multiple *Pyricularia grisea* Pathotypes. *Crop Sci.* **2004,** *44,* 1790–1798.

Fu, C.; Wu, T.; Liu, W.; Wang, F.; Li, J.; Zhu, X.; Huang, H.; Liu, Z.; Liao, Y.; Zhu, M.; Chen, J.; Huang, Y. Genetic Improvement of Resistance to Blast and Bacterial Blight of the Elite Maintainer Line Rongfeng B in Hybrid Rice (*Oryza sativa* L.) by Using Marker-assisted Selection. *Afr. J. Biotech.* **2012,** *11,* 13104–13124.

Fujii, K.; Hayano-Saito, Y.; Shumiya, A.; Inoue, M. Genetical Mapping Based on the RFLP Analysis for the Panicle Blast Resistance Derived from a Rice Parental Line St. No.1. *Breed. Sci.* **1995,** *45,* 209.

Fujii, K.; Hayano-Saito, Y.; Sugiura, N.; Hayashi, N.; Saka, N.; Tooyama, T.; Izawa, T.; Shumiya, A. Gene Analysis of Panicle Blast Resistance in Rice Cultivars with Rice Stripe Resistance. *Breed. Res. 1,* **1999,** 203–210.

Fujino, K.; Sekiguchi, H. Origins of Functional Nucleotide Polymorphisms in a Major Quantitative Trait Locus, qLTG3-1, Controlling Low-temperature Germinability in Rice. *Plant Mol. Biol.* **2011,** *75,* 1–10.

Fukuoka, S.; Okuno, K. QTL Analysis and Mapping of pi21, a Recessive Gene for Field Resistance to Rice Blast in Japanese Upland Rice. *Theor. Appl. Genet.* **2001,** *103,* 185–190.

Goto, I. Genetic Studies on the Resistance of Rice Plant to the Blast Fungus I. Inheritance of Resistance in Crosses Sensho × H-79 and Imochishirazu × H-79. *Ann. Phytopathol. Soc. Japan* **1970,** *36,* 304–312.

Goto, I. Genetic Studies on Resistance of Rice Plant to Blast Fungus II. Difference in Resistance to the Blast Disease Between Fukunishiki and Its Parental Cultivar, Zenith. *Ann. Phytopathol. Soc. Japan* **1976,** *42,* 253–260.

Goto, I.; Jaw, Y. L.; Baluch, A. A. Genetic Studies on Resistance of Rice Plant to Blast Fungus IV. Linkage Analysis of Four Genes, *Pi-a, Pi-k, Pi-z* and *Pi-i. Ann. Phytopathol. Soc. Jpn.* **1981,** *47,* 252–254.

Goto, I. Genetic Studies on resistance of rice plant to blast fungus (VII). Blast resistance genes of Kuroka. *Ann. Phytopathol. Soc. Japan* **1988,** *54,* 460–465.

Gowda, M.; Barman-Roy, S.; Chatoo, B. B.; Molecular Mapping of a Novel Blast Resistance Gene Pi38 in Rice Using SSLP and AFLP Markers. *Plant Breed.* **2006,** *125,* 596–599.

Hamdani, A. R. In *Low Temperature Problems and Cold Tolerance Research Activities for Rice in India.* Report of a Rice Cold Tolerance Workshop; International Rice Research Institute: Los Baños, 1979; pp 45–60.

Hayashi, K.; Yoshida, H.; Ashikawa, I. Development of PCR-based Allele-specific and InDel Marker Sets for Nine Rice Blast Resistance Genes. *Theor. Appl. Genet.* **2006,** *113,* 251–260.

Hayashi, K.; Yasuda, N.; Fujita, Y.; Koizumi, S.; Yoshida, H. Identification of the Blast Resistance Gene *Pit* in Rice Cultivars Using Functional Markers. *Theor. Appl. Genet.* **2010a,** *121,* 1357–1367.

Hayashi, N.; Inoue, H.; Kato, T.; Funao, T.; Shirota, M.; Shimizu, T.; Kanamori, H.; Yamane, H.; Hayano-Saito, Y.; Matsumoto, T.; Yano, M.; Takatsuji, H. Durable

Panicle Blast-resistance Gene *Pb1* Encodes an Atypical CC-NBS-LRR Protein and was Generated by Acquiring a Promoter Through Local Genome Duplication. *Plant J.* **2010b**, *64*, 498–510.

Heffner, E. L.; Sorrells, M. E.; Jannink, J. L. Genomic Selection for Crop Improvement. *Crop Sci.* **2009**, *49*, 1–12.

Hittalmani, S.; Parco, A.; Mew, T. V.; Ziegler, R. S.; Huang, N. Fine Mapping and DNA Marker-assisted Pyramiding of the Three Major Genes for Blast Resistance in Rice. *Theor. Appl. Genet.* **2000**, *100*, 1121–1128.

Hua, L.; Wu, J. Chen, C.; Wu, W.; He, X.; Lin, F.; Wang, L.; Ashikawa, I, Pan, Q. The Isolation of Pi1, an Allele at the Pik Locus Which Confers Broad Spectrum Resistance to Rice Blast. *Theor. Appl. Genet.* **2012**, *125*, 1047–1055.

Huang, H.; Huang, L.; Feng, G.; Wang, S.; Wang, Y.; Liu, J.; Jiang, N.; Yan, W.; Xu, L.; Sun, P.; Li, Z.; Pan, S.; Liu, X.; Xiao, Y.; Liu, E.; Dai, L.; Wang, G. L. Molecular Mapping of the New Blast Resistance Genes *Pi47* and *Pi48* in the Durably Resistant Local Rice Cultivar Xiangzi 3150. *Phytopathology* **2011**, *101*, 620–626.

Imbe, T.; Oba, S.; Yanoria, M. J. T.; Tsunematsu, H. A New Gene for Blast Resistance in Rice Cultivar, IR24. *Rice Genet. Newslett.* **1997**, *14*, 60–62.

Inukai, T.; Nelson, R. J.; Zeigler, R. S.; Sarkarung, S.; Mackill, D. J.; Bonman, J. M.; Takamure, I.; Kinoshita, T. Allelism of Blast Resistance Genes in Near-isogenic Lines of Rice. *Phytopathology* **1994**, *84*, 1278–1283.

Inukai, T.; Nelson, R. J.; Zeigler, R. S.; Sarkarung, S.; Mackill, D. J.; Bonman, J. M.; Takamure, I.; Kinoshita, T. Genetic Analysis of Blast Resistance in Tropical Rice Cultivars Using Near-isogenic Lines. In *Rice Genetics III*. Khush, G. S. (Ed.), Proceeding of the Third International Rice Genetics Symposium, International Rice Research Institute, Manila, Philippines, 1996, pp 447–455.

Ise, K. Linkage Analysis of some Blast Resistance Gene in Rice, *Oryza sativa* L. *Jpn. J. Breed.* **1991**, *42*, 388–389.

Jannink, J. L.; Lorenz, A. J.; Iwata, H. Genomic Selection in Plant Breeding: From Theory to Practice. *Brief Funct. Genom.* **2010**, *9*, 166–177.

Jena, K. K.; Kim, S. M.; Suh, J. P.; Kim, Y. G. In *Development of Cold-tolerant Breeding Lines Using QTL Analysis in Rice*. Second Africa Rice Congress on Innovation and Partnerships to Realize Africa's Rice Potential Bamako, Mali, March 22–26, 2010.

Jeon, J. S.; Chen, D.; Yi, G. H.; Wang, G. L.; Ronald, P. C. Genetic and Physical Mapping of *Pi5*(t), a Locus Associated with Broad-spectrum Resistance to Rice Blast. *Mol. Genet. Genom.* **2003**, *269*, 280–289.

Jeung, J. U.; Kim, B. R.; Cho, Y. C.; Han, S. S.; Moon, H. P.; Lee, Y. T.; Jena, K. K. A Novel Gene, Pi40 (t), Linked to the DNA Markers Derived from NBS-LRR Motifs Confers Broad Spectrum of Blast Resistance in Rice. *Theor. Appl. Genet.* **2007**, *115*, 1163–1177.

Joshi, B. K.; Bimb, H. P. In *Classification of Nepalese Rice Cultivars Based on Isozymes*. Rice Research in Nepal, Proceedings of the 24th National Summer Crops Research Workshop, Khumaltar, Kathmandu, Nepal, June 28–30, 2004; Gautam, A. K., et al., Eds.; 2004; pp 175–187.

Kaushik, A.; Saini, N.; Jain, S.; Rana, P.; Singh, R. K.; Jain, R. K. Genetic Analysis of a CSR10 (indica) × Taraori Basmati F3 Population Segregating for Salt Tolerance Using ISSR Markers. *Euphytica* **2003**, *134*, 231–238.

Khan, G. H.; Gaur, A.; Shikari, A. B.; Najeeb, S.; Wani, S. A. Marker Assisted Development of Effective Fertility Restorers Suitable for use in Temperate Three-line Hybrids. *J. Pure App. Micro. Biol.* **2016**, *10* (1), 73–80.

Khanna, A.; Sharma, V.; Ellur, R. K.; Shikari, A. B.; Gopala Krishnan, S.; Singh, U. D.; Prakash, G.; Sharma, T. R.; Rathour, R.; Variar, M.; Prashanthi, S. K.; Nagarajan, M.; Vinod, K. K.; Bhowmick, P. K.; Singh, N. K.; Prabhu, K. V.; Singh B. D,. Singh, A. K. Development and Evaluation of Near-isogenic Lines for Major Blast Resistance Gene(s) in Basmati Rice. *Theor. Appl. Genet.* **2015**, *28*, 1243–1259.

Khush, G. S. Multiple Disease and Insect Resistance for Increased Yield Stability in Rice. In *Progress in Irrigated Rice Research;* International Rice Research Institute; Manila, Philippines, 1989; pp 79–92. (Record Number 19901144320).

Khush, G. S. Challenges for Meeting the Global Food and Nutrient Needs in the New Millennium. *Proc. Nutr. Soc.* **2001**, *60*, 15–26.

Kinoshita, T.; Kiyosawa, S. Some Considerations on Linkage Relationships Between *Pii* and *Piz* in the Blast Resistance of Rice. *Rice Genet. Newslett.* **1997**, *14*, 57–59.

Koizumi, S. Rice Blast Control with Multilines in Japan. In *Exploiting Biodiversity for Sustainable Pest Management*; International Rice Research Institute: Los Baños, The Philippines, 2001; pp 143–157.

Kottearachchi N. S.; Wijesekara U. A. D. S. L. Implementation of Pup-1 Gene Based Markers for Screening of Donor Varieties for Phosphorous Deficiency Tolerance in Rice. *Indian J. Plant Sci.* **2013**, *2* (4), 76–83.

Kumar, P.; Pathania, S.; Katoch, P.; Sharma, T. R.; Plaha, P.; Rathour, R. Genetic and Physical Mapping of Blast Resistance Gene *Pi-42(t)* on the Short Arm of Rice Chromosome 12. *Mol. Breed.* **2010**, *25*, 217–228.

Kwak, T. S.; Vergara, B. S.; Nanda, J. S.; Coffman, W. R. Inheritance of Seedling Cold Tolerance in Rice. *SABRAO J.* **1984**, *16*, 83–86.

Lei, C.; Huang, D.; Li, W.; Wang, J. L.; Liu, Z. L.; Wang, Z. L.; Wang, X. T.; Shi, K.; Cheng, Z. J.; Zhang, X.; Ling, Z.; Wan J. M. Molecular Mapping of a Blast Resistance Gene in an Indica Rice Cultivar Yanxian No 1. *J. Rice Genet. News* **2005**, *22*, 76–77.

Li, B.; Wang, J.; Wu, Y.; Hu, X.; Zhang, Z.; Zhang, Q.; Zhao, Q.; Feng, H.; Zhang, Z.; Wang, G. L.; Wang, G.; Lu, B.; Han, Z.; Wang, Z.; Zhou, B. The Magnaporthe Oryzae Avirulence Gene AvrPiz-t Encodes a Predicted Secreted Protein That Triggers the Immunity in Rice Mediated by the Blast Resistance Gene Piz-t. *Mol. Plant-Microbe Interact.* **2009**, *22*, 411–420.

Li, J. M.; Michael, T.; McCouch, S. R. Fine Mapping of a Grain-weight Quantitative Trait Locus in the Pericentromeric Region of Rice Chromosome 3. *Genetics* **2004**, *168*, 2187–95.

Li, J.; Xin, Y.; Yuan, L. In Hybrid Rice Technology Development: Ensuring China's Food Security. *Proven Successes in Agricultural Development: A Technical Compendium to Millions Fed*; Spielman, D. J., Pandya-Lorch, R., Eds.; International Food Policy Research Institute: Washington, DC, 2010.

Liu, B.; Zhang, S.; Zhu, X.; Yang, Q.; Wu, S.; Mei, M.; Mauleon, R.; Leach, J.; Mew, T.; Leung, H. Candidate Defense Genes as Predictors of Quantitative Blast Resistance in Rice. *MPMI* **2004**, *17*, 1146–1152.

Liu, X. Q.; Wang, L.; Chen, S.; Lin, F.; Pan, Q. H. Genetic and Physical Mapping of *Pi36(t)*, a Novel Rice Blast Resistance Gene Located on Rice Chromosome 8. *Mol. Genet. Genom.* **2005**, *274*, 394–401.

Liu, X.; Yang, Q.; Lin, F.; Hua, L.; Wang, C.; Wang, L.; Pan, Q. Identification and Fine Mapping of *Pi39(t)*, a Major Gene Conferring the Broad-spectrum Resistance to *Magnaporthe oryzae*. *Mol. Genet. Genom.* **2007**, *278*, 403–410.

Lone, Z. A.; Bhat, Z. A.; Najeeb, S.; Ahanger, M. A.; Shikari, A. B.; Parray, G. A.; Hussain, S. Screening of Rice Genotypes Against Bakanae Disease Caused by *Fusarium fujikuroi* Nirenberg. *Oryza* **2016**, *53* (1), 91–97.

Mackill, D. J.; Bonman, J. M. Inheritance of Blast Resistance in Near Isogenic Lines of Rice. *Phytopathology* **1992**, *82*, 746–749.

Mao, H. et al. Linking Differential Domain Functions of the GS3 Protein to Natural Variation of Grain Size in Rice. *Proc. Natl. Acad. Sci. U.S.A.* **2010**, *107*, 19579–19584.

McCouch, S. R.; Nelson, R. J.; Tohme, J.; Zeigler, R. S. In Mapping of Blast Resistance Genes in Rice. *Rice Blast Disease*; Zeigler, R. S., Leong, S. A., Teng, P. S., Eds.; C.A.B. International: Wallingford UK, 1994; pp 167–186.

Mew, T. W.; Leung, H.; Savary, S.; Vera Cruz C. M.; Leach J. E. Looking Ahead in Rice Disease Research and Management. *Crit. Rev. Plant Sci.* **2004**, *23* (2), 103–27.

Morsy, M. R.; Jouve, L.; Hausman, J. F.; Hoffmann, L.; Stewart, J. D. Alteration of Oxidative and Carbohydrate Metabolism Under Abiotic Stress in Two Rice (*Oryza sativa* L.) Genotypes Contrasting Chilling Tolerance. *J. Plant Physiol.* **2007**, *164*, 157–167.

Nagamine, T. Genetic Control of Tolerance to Chilling Injury at Seedling in Rice, *Oryza sativa* L. *Jpn. J. Breed.* **1991**, *41*, 35–40.

Nakamura, S.; Asakawa, S.; Ohmido, N.; Fukui, K.; Shimizu, N.; Kawasaki, S. Construction of an 800-kb Contig in the near Centromeric Region of the Rice Blast Resistance Gene Pi-ta2 Using a Highly Representative Rice BAC Library. *Mol. Genet. Genom.* **1997**, *254*, 611–620.

Naqvi, N. I.; Chattoo, B. B. Development of a Sequence Characterized Amplified Region (SCAR) Based Indirect Selection Method for a Dominant Blast-resistance Gene in Rice. *Genome* **1996**, *39*, 26–30.

Narayanan, N. N.; Baisakh, N.; Vera Cruz C. M.; Gnanamanickam, S. S.; Datta, K.; Datta S. K. Molecular Breeding for the Development of Blast and Bacterial Blight Resistance in Rice cv. IR50. *Crop Sci.* **2002**, *42* (6), 2072–2079.

Neeraja, C.; Maghirang-Rodriguez, R.; Pamplona, A.; Heuer, S.; Collard, B.; Septiningsih, E. et al. A Marker-assisted Backcross Approach for Developing Submergence-tolerant Rice Cultivars. *Theor. Appl. Genet.* **2007**, *115*, 767–776.

Nguyen, T. T. T.; Koizumi, S.; La, T. N.; Zenbayashi, K. S.; Ashizawa, T.; Yasuda, N.; Imazaki, I.; Miyasaka, A. *Pi35(t)*, a New Gene Conferring Partial Resistance to Leaf Blast in the Rice Cultivar Hokkai 188. *Theor. Appl. Genet.* **2006**, *113*, 697–704.

Pan, Q. H.; Tanisaka, T.; Ikehashi, H. Studies on the Genetics and Breeding of Blast Resistance in Rice IV Gene Analysis for the Blast Resistance of Indica Variety Kasalath. *Breed. Sci.* **1995**, *45*, 17094.

Pan, Q. H.; Wang, L.; Ikehashi, H.; Tanisaka, T. Identification of a New Blast Resistance Gene in the Indica Rice Cultivar Kasalath Using Japanese Differential Cultivars and Isozyme Markers. *Phytopathology* **1996**, *86*, 1071–1075.

Pan, Q. H.; Wang, L.; Ikehashi, H.; Yamagata, H.; Tanisaka, T. Identification of Two New Genes Conferring Resistance to Rice Blast in the Chinese Native Cultivar 'Maowangu'. *Plant Breed.* **1998**, *117*, 27–31.

Pan, X. B.; Zou, J. H.; Chen, Z. X.; Lu, J. F.; Yu, H. X.; Li, H. T.; Wang, Z. B.; Rush, M. C.; Zhu, L. H. Mapping the QTLs Responsible for Sheath Blight Resistance from Rice Cultivar Jasmine 85. *Chinese Sci. Bull.* **1999**, *44*, 1629–1635 (in Chinese).

Pandit, A.; Rai, V.; Bal, A.; Sinha, S.; Kumar, V.; Chauhan, M.; Gautam, R. K.; Singh, R.; Sharma, P. C.; Singh, A. K.; Gaikwad, K.; Sharma, T. R.; Mohapatra, T.; Singh, N. K. Combining QTL Mapping and Transcriptome Profiling of Bulked RILs for Identification of Functional Polymorphism for Salt Tolerance Genes in Rice (*Oryza sativa* L.). *Mol. Genet. Genom.* **2010**, *284*, 121–136.

Perata, P.; Voesenek, L. A. C. J. Submergence Tolerance in Rice Requires Sub1A, an Ethylene-response-factor-like Gene. *Trends Plant Sci.* **2007**, *120*, 43–46.

Parray, G. A.; Shikari, A. B. Conservation and Characterization of Indigenous Rice Germplasm Adapted to Temperate/Cooler Environments of Kashmir Valley. *Oryza* **2008**, *45* (3), 198–201.

Qian, Q.; Zeng, D. L.; He, P. QTL Analysis of the Rice Seedling Cold Tolerance in a Double Haploid Population Derived from Another Culture of a Hybrid Between Indica and Japonica Rice. *Chin. Sci. Bull.* **2000**, *145* (5), 448–453.

Qu, S.; Liu, G.; Zhou, B.; Bellizzi, M.; Zeng, L.; Dai, L.; Han, B.; Wang, G. The Broad-spectrum Blast Resistance Gene Pi9 Encodes an NBS-LRR Protein and Is a Member of a Multigene Family in Rice. *Genetics* **2006**, *172*, 1901–1914.

Sallaud, C.; Lorieux, M.; Roumen, E.; Tharreau, D.; Berruyer, R.; Svestasrani, P.; Garsmeur, O.; Ghesquiere, A.; Notteghem, J. L. Identification of Five New Blast Resistance Genes in the Highly Blast-resistant Rice Variety IR64 Using a QTL Mapping Strategy. *Theor. Appl. Genet.* **2003**, 106, 794–803.

Sarkar, R. K.; Reddy, J. N.; Sharma, S. G.; Ismail, A. M. Physiological Basis of Submergence Tolerance in Rice and Implications for Crop Improvement. *Curr. Sci.* **2006**, *91*, 899–906.

Septiningsih, E. M.; Pamplona, A. M.; Sanchez, D. L.; Neeraja, C. N.; Vergara, G. V.; Heuer, S.; Ismail, A. M.; Mackill, D. J. Development of Submergence-tolerant Rice Cultivars: the Sub1 Locus and Beyond. *Ann. Bot.* **2009**, *103*, 151–160.

Sharma, T. R.; Madhav, M. S.; Singh, B. K.; Shanker, P.; Jana, T. K.; Dalal, V.; Pandit, A.; Singh, A.; Gaikwad, K.; Upreti, H. C.; Singh. N. K. High-resolution Mapping, Cloning and Molecular Characterization of the *Pi-k*^{h} Gene of Rice, Which Confers Resistance to *Magnaporthe grisea. Mol. Genet. Genom.* **2005**, *274*, 569–578.

Shen, M. G.; and Lin, J. Y. In The Economic Impact of Rice Blast Disease in China. *Rice Blast Disease*; Zeigler, R. S., Leong, S. A., Teng, P. S., Eds.; CAB International/IRRI: Wallingford, U.K., 1994; pp 321–331.

Shikari, A. B.; Khanna, A.; Gopala Krishnan, S.; Singh, U. D.; Rathour, R.; Tonapi, V.; Sharma, T. R.; Nagarajan, M.; K. V.; Singh, A. K. Molecular Analysis and Phenotypic Validation of Blast Resistance Genes *Pita* and *Pita*^{2} in Landraces of Rice (*Oryza sativa* L.). *Indian J. Genet.* **2013**, *73* (2), 131–141.

Shikari, A. B.; Rajashekara, H.; Khanna, A.; Gopala Krishnan, S.; Rathour, R.; Singh, U. D.; Sharma, T. R.; Prabhu, K. V.; Singh, A. K. Identification and Validation of Rice Blast Resistance Genes in Indian Rice Germplasm. *Indian J. Genet.* **2014**, *74* (3), 286–299.

Shinoda, H.; Toriyama, K.; Yunoki, T.; Ezuka, A.; Sakurai, Y. Studies on the Varietal Resistance of Rice to Blast. VI. Linkage Relationship of Blast Resistance Genes. *Bull Chugoku Nat. Agric. Exp. Stn. Ser.* A **1971**, *20*, 1–25.

Singh, N.; Dang, T. T. M.; Vergara, G. V.; Pandey, D. M.; Sanchez, D.; Neeraja, C. N.; Septiningsih, E. M.; Mendioro, M.; Tecson- Mendoza, E. M.; Ismail, A. M.; Mackill, D. J.; Heuer, S. Molecular Marker Survey and Expression Analyses of the Rice Submergence-tolerance Gene SUB1A. *Theor. Appl. Genet.* **2010**, *121*, 1441–1453.

Singh, V. K.; Singh, A.; Singh, S. P.; Ellur, R. K.; Choudhary, V.; Sarkhel, S.; Singh, D.; Gopala Krishnan, S.; Nagarajan, M.; Vinod, K. K.; Singh, U. D.; Rathore, R.; Prasanthi, S. K.; Agrawal, P. K.; Bhatt, J. C.; Mohapatra, T.; Prabhu, K. V.; Singh, A. K. Incorporation of Blast Resistance into 'PRR78', an Elite Basmati Rice Restorer Line, Through Marker Assisted Backcross Breeding. *Field Crops Res.* **2012**, *128*, 8–16.

Skamnioti, P.; Gurr, S. J. Against the Grain: Safeguarding Rice from Rice Blast Disease. *Trends Biotech.* **2009**, *27*, 141–150.

Song, X. J. et al. A QTL for Rice Grain Width and Weight Encodes a Previously Unknown RING-type E3 Ubiquitin Ligase. *Nat. Genet.* **2007**, *39*, 623–630.

Sthapit, B. R.; J. R. Witcombe. Inheritance of Tolerance to Chilling Stress in Rice During Germination and Plumule Greening. *Crop Sci.* **1998**, *38*, 660–665.

Subbaiyan, G. K.; Waters, D. L. E.; Katiyar, S. K.; Sadananda, A. R.; Satyadev, V.; Henry, R. Genome-wide DNA Polymorphisms in Elite Indica Rice Inbreds Discovered by Whole-genome Sequencing. *Plant Biotechnol. J.* **2012**, *10*, 623–634.

Swamy, B. P. M, Vikram, P.; Dixit, S.; Ahmed, H. U, Kumar, A. Meta-analysis of Grain Yield QTL Identified During Agricultural Drought in Grasses Showed Consensus. *BMC Genom.* **2011**, *12*, 319.

Tabien, R. E.; Pinson, S. R. M.; Marchetti, M. A.; Li, Z.; Park, W. D.; Paterson, A. H.; Stansel, J. W. In *Blast Resistance Genes from Teqing and Lemont*. Rice genetics III; Proceedings of Third International Rice Genetics Symposium, International Rice Research Institute, Manila, The Philippines, Oct 16–20, 1996; Khush, G. S., Ed.; pp 451–455.

Tabien, R. E.; Li, Z.; Paterson, A. H.; Marchetti, M. A.; Stansel, J. W.; Pinson, S. R. M. Mapping of Four Rice Blast Resistance Genes from 'Lemont' and 'Teqing' and Evaluation of Their Combinatorial Effect for Field Resistance. *Theor. Appl. Genet.* **2000**, *101*, 1215–1225.

Talbot. On the Trail of a Cereal Killer: Exploring the Biology of *Magnaporthe grisea. Ann. Rev. Microbiol.* **2003**, *57*, 177–202.

Thomson, M. J.; Ocampo, M.; Egdane, J.; Rahman, M. A.; Sajise, A. G.; Adorada, D. L.; Tumimbang-Raiz E.; Blumwald, E.; Seraj, Z. I.; Singh, R. K.; Gregorio, G. B.; Ismail, A. M. Characterizing the Saltol Quantitative Trait Locus for Salinity Tolerance in Rice. *Rice* **2010**, *3*, 148–160.

Wang, Z. X.; Yano, M.; Yamanouchi, U.; Lwamoto, M.; Monna, L.; Hayasaka, H.; Katayose, Y.; Sasaki, T. The Pib Gene for Rice Blast Resistance Belongs to the Nucleotide Binding and Leucine-rich Repeat Class of Plant Disease Resistance Genes. *Plant J.* **1999**, *19*, 55–64.

Wang, L.; Xu, X. K.; Lin, F.; Pan, Q. H. Characterization of Rice Blast Resistance Genes in the *Pik* Cluster and Fine Mapping of the *Pik-p* Locus. *Phytopathology* **2009**, *99*, 900–905.

Wu, J. L.; Fan, Y. Y.; Li, D. B.; Zheng, K. L.; Leung, H.; Zhuang, J. Y. Genetic Control of Rice Blast Resistance in the Durably Resistant Cultivar Gumei 2 Against Multiple Isolates. *Theor. Appl. Genet.* **2005**, *111*, 50–56.

Xie, X. B.; Song, M. H.; Jin, F. X.; Ahn, S. N.; Suh, J. P. Fine Mapping of a Grain Weight Quantitative Trait Locus on Rice Chromosome 8 Using Near-isogenic Lines Derived from a Cross Between *Oryza sativa* and *Oryza rufipogon. Theor. Appl. Genet.* **2006,** *113,* 885–94.

Xie, X. B, Jin, F. X.; Song, M. H.; Suh, J. P.; Hwang, H. G. Fine Mapping of a Yield-enhancing QTL Cluster Associated with Transgressive Variation in an *Oryza sativa* × *O. rufipogon* Cross. *Theor. Appl. Genet.* **2008,** *116,* 613–22.

Xu, K.; Xia, X.; Fukao, T.; Canlas, P.; Maghirang-Rodriguez, R.; Heuer, S. et al. Sub1A is an Ethylene Response Factor-like Gene that Confers Submergence Tolerance to Rice. *Nature* **2006,** *442,* 705–708.

Xu, K.; Xu, X.; Ronald, P. C.; Mackill, D. J. A High-resolution Linkage Map in the Vicinity of the Rice Submergence Tolerance Locus Sub1. *Mol. Gen. Genet.* **2000,** *263,* 681–689.

Xu, X.; Liu, X.; Ge, S.; Jensen, J.; Hu, F.; Li, X.; Dong, Y.; Ryan, N.; Fang, L.; Huang L.; Li, J.; He, W.; Zhang, G.; Zheng, X.; Zhang, F.; Li, F.; Yu, C.; Kristiansen, K.; Zhang, X.; Wang, J.; Wright, M.; McCouch, S.; Nielsen, S.; Wang J.; Wang, W. Resequencing 50 Accessions of Cultivated and Wild Rice Yields Markers for Identifying Agronomically Important Genes. *Nat. Biotech.* **2012,** *30,* 105–111.

Yamaguchi, I. In Overview on the Chemical Control of Rice Blast Disease. *Rice Blast: Interaction with Rice and Control*; Kawasaki, S., Ed.; Kluwer Academic Press: Dordrecht, The Netherlands, 2004; pp 1–13.

Yang, Q.; Lin, F.; Wang, L.; Pan, Q. Identification and Mapping of Pi41, a Majorgene Conferring Resistance to Rice Blast in the *Oryza sativa* Subsp. Indica Reference Cultivar, 93–11. *Theor. App. Genet.* **2009,** *118,* 1027–1034.

Yu, Z. H.; Mackill, D. J.; Bonman, J. M.; Tanksley, S. D. Tagging Genes for Blast Resistance in Rice via Linkage to RFLP Markers. *Theor. Appl. Genet.* **1991,** *81,* 471–476.

Yu, S. B.; Xu, W. J.; Vijayakumar, C. H. M.; Ali, J.; Fu, B. Y.; Xu, J. L.; Jiang, Y. Z.; Marghirang, R.; Domingo, J.; Aquino, C.; Virmani, S. S.; Li, Z. K. Molecular Diversity and Multilocus Organization of the Parental Lines Used in the International Rice Molecular Breeding Program. *Theor. Appl. Genet.* **2003,** *108* (1), 131–140.

Zenbayashi-Sawata, K.; Fukuoka, S.; Katagiri, S.; Fujisawa, M.; Matsumoto, T.; Ashizawa, T.; Koizumi, S. Mapping of the QTL (Quantitative Trait Locus) Conferring Partial Resistance to Leaf Blast in Rice Cultivar Chubu 32. *Theor. Appl. Genet.* **2002,** *104,* 547–552.

Zhuang, J. Y.; Wu, J. L.; Fan, Y. Y.; Rao, Z. M.; Zheng, K. L. Genetic Drag Between a Blast Resistance Gene and QTL Conditioning Yield Trait Detected in a Recombinant Inbred Line Population in Rice. *RGN* **2001,** *18,* 67–69.

Zhang, Q.; Jiang, N.; Wang, G.; Hong, Y.; Wang, Z. Advances in Understanding Cold Sensing and the Cold-Responsive Network in Rice. *Adv. Crop Sci. Tech.* **2013,** *1.* DOI: org/ 10.4172.

Zhou, J. H.; Wang, J. L.; Xu, J. C.; Lei, C. L.; Ling, Z. Z. Identification and Mapping of a Rice Blast Resistance Gene Pi-g(t) in the Cultivar Guangchangzhan. *Plant Pathol.* **2004,** *53,* 191–196.

Zhu, Y.; Chen, H.; Fan, J.; Wang, Y.; Li, Y.; Chen, J.; Fan, J. X.; Yang, S.; Hu, L.; Leung, H.; Mew, T. W.; Teng, P. S.; Wang, Z.; Mundt, C. C. Genetic Diversity and Disease Control in Rice. *Nature* **2000,** *406,* 718–722.

CHAPTER 4

SYSTEM OF RICE INTENSIFICATION: RECENT ADVANCES AND TRENDS

ASHAQ HUSSAIN*, PARMEET SINGH, and AABID HUSSAIN LONE

Faculty of Agriculture, Sher-e-Kashmir University of Agricultural Sciences and Technology, Wadura, Sopore 193201, Jammu and Kashmir, India

Corresponding author. E-mail: ahshah71@gmail.com

ABSTRACT

Recently, there is a need to enhance the rice yield to meet the global food security. Several novel techniques such as rice intensification and integrated crop management are being practiced for efficient production with better soil health and yield. The characteristics of system of rice intensification (SRI) typically include the raised-bed nursery, transplanting younger seedlings with a wider spacing, keeping water to saturation level, weeding and incorporation of weeds in soil, and utilization of organic fertilizers. SRI aims at the reduction of quantity of irrigation water along with the reduced seed rate. The proper adoption of SRI improves the income per drop of water. The SRI is the good alternative under the intense water scarcity, climate change, and costly inputs conditions. This chapter provides the fundamentals of SRI, its principles, practices and environment impact, and recent trends in the dynamics of SRI.

4.1 INTRODUCTION

Rice is one of the most important crops of Asia being staple food for more than 90% of the population (Verma et al., 2012; Verma and Srivastav, 2017). Several innovations have been made in rice production technology systems in order to increase the yield and better meet the world food demands. The indiscriminate use of agrochemicals has resulted in soil and environmental degradation with declining yield trends. Rice has been cultivated under flooded conditions for centuries with the mindset that rice performs better under submerged conditions. Farmers are losing interest in rice cultivation as factor productivity is declining (Das et al., 2009) and its profitability is in question with the rise in input costs. Rice cultivation has evolved from direct-seeded to transplanted rice under properly puddled conditions. Nowadays, there is a thrust on the integrated crop and resource management. In this context, there are various new technologies among which, system of rice intensification (SRI) and integrated crop management (ICM) are most important and appear to have potential to save inputs, protect the environment, and improve soil health and productivity (Balasubramanian et al., 2007). The SRI was evolved by Fr. Henri de Laulinie during 1980s while working closely with the farmers of Malagasy. SRI has evolved as an empirical practice and has been credited with spectacular grain yields. Reports from more than 50 countries involving 10 million stake holders across Asian, Latin American, and African continents have shown that principles and methods of SRI have much potential for raising the yields and factor productivity under different circumstances, in environment friendly ways. SRI is communicated in terms of a set of practices reflecting some radical changes from the conventional methods which have been considered for experimentation and modification so as to make the system suitable under local conditions.

Traditionally, rice was grown by direct seeding (DS) method but with the advent of techniques of nursery raising, transplanting became more popular among the rice farmers. Farmers conventionally transplant the rice seedling at an age range that ranges from 30 to 60 days depending on cultivar and the agroclimatic situation. Research under SRI or otherwise on age of seedlings desirable for higher rice growth and productivity has proven beyond doubt that young seedlings of 8–12 days age are more productive than the older seedlings. In physiological terms, a two-leaf seedling shows higher growth, tillering, and productivity with minimum damage to roots. Transplanting the seedlings singly and at a wider spacing of 25 cm × 25 cm is recommended to exploit the maximum tillering potential of young rice seedling, allow the roots to proliferate, and promote the formation of a strong rice plant tolerant to diseases, pests, lodging, and moderate water stress. This is based on the principle of plastic response of rice to plant population and results in substantial seed saving, particularly more desirable in rice hybrids. Mechanical weeding is the most desirable weeding in agroecosystems for the purpose of weed removal, soil aeration, and being eco-friendly in nature. Since farmers plant randomly, use of machines has not picked up and they rely mostly on the chemicals followed by light manual weeding. Since SRI recommends intermittent drainage, the upland weeds grow at much faster rates and sole dependence on manual/mechanical weeding seems an uneconomical option in view of ever-escalating labor costs (Pandey, 2009). Presently, an integrated approach is more viable to reduce the dependence on herbicides partially. Among the crops, rice consumes around 1500–2500 mm of water for each crop depending on the agroclimatic situation. In the backdrop of climate change, increasing temperatures and water scarcity, increase in the demand of water from competing sectors, some water-saving technologies such as SRI, direct-seeded rice, aerobic rice, etc. have caught the attention of scientists, field functionaries, policy makers, and farmers. It has been widely reported that SRI has the advantage of 50% water savings, 80% seed saving coupled with yield advantages, and early maturity. Use of organic manures is recommended under the typical method of SRI developed in Madagascar. Use of organics is a desirable method of supplying plant nutrients and improving the overall soil health. With the advent of high-yielding and fertilizer-responsive varietal profile in almost all cereals, the application of fertilizers is inevitable in view of their huge nutrient uptake and limited availability of organic manures. Supplying as much organics as possible under a given situation is always desirable option, but socioeconomic factors are the final determinants of the adoption of a technology. Reports reveal that nutrient management under SRI method of rice cultivation shall be more pragmatically dealt by an integrated approach rather than solely depending on organics.

SRI is based on agroecological principles and interdisciplinary approach that help in exploring the growth and yield potential of rice. There is evidence that the concept of SRI and associated approaches can be extended to other major annual food crops such as pearl millet and sorghum. The beauty of SRI is that apparently, there still exists substantial scope to raise rice yields through relatively simple but profound adjustments in agronomic management practices. Further, SRI technique improves the seed quality parameters and can be used as seed production technique in rice. SRI helped in realizing higher yields, not by higher amounts of inputs but by suitable manipulations in cultural approaches that helped in productive use of the natural resources (land, water, seeds, and plant nutrients) and of labor, time, and space. Therefore, SRI approach offers interesting opportunities, across diverse and location-specific production systems particularly for small farmers who need to achieve both higher yields and increased net profits.

Crop establishment methods have a bearing on the important farm operations, including primary tillage, seedbed preparation, planting, weeding, and water management that in turn affect seedling establishment, rice growth, and rice canopy structure (Saha and Bharti, 2009). Due to increasing water scarcity as consequence of climate change, SRI will become more attractive to encourage efficient use of water. Since the droughts are becoming frequent and more serious, the SRI methods are inducing rice plants to grow much larger, and their deeper root systems give them more resistance to the impact of drought. SRI practices mark a paradigm shift from the conventional practices and therefore, its proponents are confronted with difficulties in promoting it. Rather, it gives small farmers additional opportunities to raise the productivity of their land, while trying to meet their staple food requirements. To analyze the subject, it is important to understand underlying principles that govern the practices in SRI method of rice cultivation.

4.2   PRINCIPLES AND PRACTICES OF SRI

4.2.1   PRINCIPLES OF SRI

4.2.1.1   YOUNG SEEDLINGS

Transplanting seedling at young age of 8–12 days, generally not more than 15 days, have proven advantageous over older seedlings. Rice seedlings display greater growth, tillering, and root growth if (1) transplanted at a young age and (2) transplanted very quickly to avoid desiccation and trauma of the plant and protect roots. There is drastic reduction in the tillering potential of the seedlings if transplanted beyond the age of 15 days. Rice seedlings produce maximum tillers if transplanted at fourth phyllochron of growth or two-leaf stage in conjunction with other SRI practices.

4.2.1.2   SINGLE SEEDLING/HILL

SRI advocates planting of single or at the most, two seedlings depending on the soil regime and control on the irrigation water. Planting one or two seedlings helps rice plants attaining their full root and shoot growth tillering potential and grain filling if spaced widely, rather than densely. Although yield is directly related to productive panicles per meter rather than per plant, more productive tillering has been recorded under SRI in a widely spaced crop. Avoiding the drying of roots by quick and careful transplanting, placing the roots in soil in an L shape rather than U helps in reducing seedling trauma and helps in quick establishment.

4.2.1.3   WIDER SPACING

A plant population of 9–16 plants/m^2 by practicing a wider spacing of 25 cm × 25 cm to 35 cm × 35 cm in a square pattern or even up to 50 cm × 50 cm under best soil conditions is proposed. In widely spaced plants, the potential for growth, tillering, and yield is fully realized.

4.2.1.4   INTERMITTENT IRRIGATION

Rice is adapted to flooded and hypoxic conditions but it is not necessary to do so. The soil has to be just moist, rather than flooded, following cycles of wetting and drying. Drying has to be allowed to the extent that hair-thin cracks develop in the soil. A thin layer of water has to be maintained during the flowering stage. Under continuous submergence, most of the rice roots degenerate at hypoxic conditions.

4.2.1.5   MECHANICAL WEEDING

Weeding is recommended 10–12 day after transplanting (DAT) using mechanical rotating hoe or a cono weeder that serves the dual purpose of removing the weeds and aerating the soil. Aeration prior to the panicle initiation results in better root growth and tillering of rice plants. Under flooded conditions, the hypoxic conditions result in early root degeneration and plant senescence, and on the other hand, aeration benefits the aerobic as well as anaerobic microbial populations.

4.2.1.6   USE OF ORGANIC MANURE

Application of any form of organic manure is recommended. Decomposition of organic matter under well-aerated conditions helps in release of plant nutrients resulting in vigorous root and shoot growth. The organic matter acts as substrate for the multiplication of a diversity of microbial flora that helps plant resist the damage from pests and diseases.

4.2.2 PRACTICES OF SRI

4.2.2.1 IN DIRECT-SEEDED RICE

Simplest form of SRI can be in the form of direct-seeded rice to reduce the transplanting shock and water and labor requirements. Among five established methods evaluated, namely, conventional transplanting (TP), seedling casting (SC), mechanical transplanting (MT), DS, and SRI, SRI produced significantly higher grain yield than TP and MT methods, whereas DS or SC methods did not produce higher yields. The seedling quality was higher in the case of DS and SC than TP and MT suggesting that robust seedlings with vigorous roots weaken the positive effect of SRI on rice yield. Although SRI produced a higher tillering rate than conventional method, the number of effective tillers was lower than SC and DS (Chen et al., 2013). In TP, the effect size differential and suppressive effect of standing water result in better weed control but in DSR; the risk of crop yield loss is due to competition from weeds that is always higher than for transplanted rice (Rao et al., 2007). SRI method of crop establishment produced significantly higher grain yield (6.95 t/ha) followed by DSR yield (6.2 t/ha) and the lowest yield was under transplanting method, that is, 4.18 t/ha. SRI produced 51% and DSR produce 48% higher grain yield than general TPR (Dhital, 2011). Direct-seeded rice is a viable option for saving labor and water but it is often difficult to optimize the plant population and control weeds, thus resulting in suboptimal yields. The popularity of DSR is increasing worldwide on account of reduced labor and water requirements. This has been possible due to development of new herbicide molecules suited for direct-seeded rice.

4.2.2.2 IN NURSERY MANAGEMENT

SRI advocates a raised-bed nursery and transplanting of the seedling at a time when the seed sac with sufficient amount of soil is still adhered to the seedling. Nursery is not regarded as miniature field to be kept flooded but rather it should be treated like a garden where the soil is kept only moist. Raised-bed nursery, 10–15 cm above ground level, is prepared close to the field to be transplanted. Application of organic @ 1 kg/m^2 is recommended and 100 m^2 area is sufficient for 1 ha. Sow 80–100 g of pregerminated seed per square meter. Nursery should not be flooded but irrigated with the help of a rose cane. Contrasting results have been reported in the literature. Superiority of conventional method of raising seedlings over dapog method in respect of grain yield has been reported. Although the conventional method was at par with polyethylene covered seed bed with regard to seedling quality. In comparison of short- and long-duration varieties, the long-duration variety BRRI Dhan29 yielded highest with SRI practices (Latif et al., 2005). Tillering behavior, phyllochron, and grain yield did not vary due to nursery techniques but crop spacing tillering behavior, phyllochron, and rate of leaf appearance was faster in widely planted rice cultivars compared to 25 cm × 25 cm spacing plantations. Light interception percentage was higher in CORH 3; closer spacing intercepted the more light than wider spacing (Baskar et al., 2013). Modified protected nursery is recommended for Kashmir valley conditions especially under aberrant weather and for early transplanting (Kamili et al., 2011). Seedlings raised by dapog and wet bed demonstrated higher seedling vigor (plant dry weight, specific leaf area, and N content) but among the methods, no significant difference was observed on the timing of tiller emergence and grain yield (Pasuquin et al., 2008). Soil media used for nursery raising influence the growth, vigor, and quality of seedlings produced. Among the nine treatment media studied, the maximum seedling height and root length of 17.06 and 10.75 cm was observed in farmyard manure (FYM) soil and vermin soil, respectively, prepared in 1:1 ratio (Dhananchezhiyan et al., 2013).

4.2.2.3   IN TRANSPLANTING OF YOUNG SEEDLINGS (8–12 DAYS OLD)

Before deeper insights on the use of younger seedlings are sought, an understanding of phyllochron concept needs to be introduced. Phyllochron is defined as the time interval of leaf emergence of two successive leaves. The factors that affect phyllochron are temperature, day length, nutrition, light intensity, planting density, and humidity (Nemoto et al., 1995). The concept of phyllochron was first developed by Katayama (1951). He worked out the rules for leaf emergence on the main stem and tillers of rice, wheat, and barley. De Laulanié used this model to explain the high tillering potential of young seedlings in SRI. From the Katayama model, the first tiller of the main stem appears at the fourth phyllochron. From his research, De Laulanié had discovered that if the rice seedling is transplanted later than the third phyllochron, there will be loss of all the forthcoming tillers from the first tiller which represents about 40% of the total tillers (ATS, 1992). Therefore, the concept of SRI advocates transplantation of the seedlings during the third phyllochron, which generally corresponds to two-leaf stage, in order to avoid reduction in subsequent tillering and root growth (Laulanié, 1993). Transplanting at young age synergizes with other practices of SRI that allows the realization of maximum tillering potential of rice plants (ATS, 1992). Transplanting rice seedlings at a younger age has been advocated by many other researchers (Horie et al., 2005). SRI capitalizes on the high tillering potential of young rice seedlings transplanted at 2–3 leaf (Katayama, 1951). Estimating the contribution individual component practices, it has been reported from Madagascar that the use of young seedlings is the single most important component practice of SRI (Uphoff and Fernandes, 2002).

Transplanting seedlings at an age of 10 days resulted in higher number of effective tillers/hill, grains/panicle, panicle length, test weight, and grain yield. Transplantation of either 6 or 14 days old seedling resulted in 18.7% and 25.0% reduction in grain yield over 10 days old seedlings, respectively. It has also been reported that for each day delay in transplanting beyond 10 days, yield was reduced to the extent of 4.5%/ha (Patra and Haque, 2011). Under Kashmir valley conditions, transplanting 1 or 2 seedlings/hill gave yield advantage of around 7% over 3 seedling/hill. Among the seedling ages, 14 and 21 days old seedlings resulted in significantly higher growth, yield attributes, and higher yield than 28 and 35 days old seedlings. Twenty one days old seedlings are more robust and convenient to transplant and produced grain yield at par with 14 days old seedlings but produced 5.8% and 13.9% more grain yield than 28 and 35 days olds seedlings, respectively (Hussain et al., 2012). Combining young seedling (14 days old), one seedling/hill, square planting with wider spacing (22.5 cm × 22.5 cm) and cono weeding four times at weekly interval starting from 15 DAT resulted in a significant increase in dry matter production, root growth, and nutrient uptake that finally got manifested in improvement the grain yield to the extent of 68.3% over the traditional practice (Sridevi and Chellamuthu, 2012). Combination of young seedling, one seedling, square planting and cono weeding recorded higher nutrient uptake, nitrogen harvest index (HI), soil available nitrogen, yield, and grain HI (Sridevi and Chellamuthu, 2015).

Young seedling with endosperm still attached results in reduction of stress during the uprooting and transplanting processes (Sakai and Yosida, 1957; Hoshikawa et al., 1995). In TP methods, uprooting of the seedlings results in root pruning to the extent of 40–60%. Pruning leads to shock and decrease in subsequent root and shoot growth and dry matter accumulation (Ros et al., 1998). The transplantation of younger seedlings (15 days old) yielded higher (6.59 t/ha) as compared to 29 days old seedlings (4.42 t/ha), whereas reduction in grain yield of 29 days seedlings was 32.9%. Similarly, highest grain yield from 25 cm × 25 cm (6.04 t/ha) and lowest grain yield (5.37 t/ha) from 30 cm × 30 cm was 11.2% (Karki, 2009). Growth, yield, quality parameters, and net monetary returns are significantly affected by age of seedlings. A yield advantage of 12.7%, 4.4%, and 17.5% was realized by transplanting 10 days old seedlings over 8, 12, and 14 days old seedlings, respectively (Ram et al., 2015). A difference in grain yields as high as 1 t/ha between 7 and 21 day transplanting and increments observed were consistent. Ten days old seedlings were also significantly better than those of others at the age of 12 and 14 days old seedlings in respect of yield attributes, grain yield, economics, nutrient uptake, and soil health (Singh et al., 2013). Significantly, lower grain and straw yield of rice were recorded under 25–28 days old seedling (Chaudhari et al., 2015). There are also contradictory reports that decrease in the yield by transplanting old age seedlings can be compensated by using higher number of seedlings per hill; therefore, transplanting younger seedlings under the SRI management may not necessarily enhance grain yields (Deb et al., 2012). It must be borne in mind that transplanting of 8–12 days old may be possible in tropical environments. A two-leaf stage may be attained in 15–21 days in cooler environments. Thus, physiological rather than chronological age is a more important and meaningful criterion.

4.2.2.4   IN TRANSPLANTING OF SINGLE SEEDLING PER HILL

Under normal sowing, dates recommended for specific region transplanting young seedlings means early transplanting. Early transplanting provides a longer vegetative growth period and single seedlings results in reduced competition and minimizes the shading of lower leaves. Rice compensates for decreased plant density by increasing yield per plant through more tillers and panicles. Higher photosynthetic rates coupled with enhanced supply of oxygen helped in higher root activity and growth (Tanaka, 1958; Horie et al., 2005). Since roots are the site of cytokinin synthesis, its higher concentration results delay in senescence of lower leaves. Higher photosynthetic activity is finally reflected in higher yield in case of single seedling/hill when compared with 3 seedlings/hill (San-oh et al., 2006). Single seedlings/hill had higher root length and density, and displayed a positive relationship with above canopy development and prolonged photosynthetic activity of older leaves (Mishra et al., 2006).

Planting of single seedlings reduces the competition and allows the rice to express its full growth potential. SRI (one seedling/hill) produced 10.17%

less yield than that of modified SRI (MSRI) (two seedlings/hill) but 21.7% more than in best management practices (BMP). Two seedlings is insurance for seedling mortality. Hybrid Ajay produced 9.1% more grains than cv. Tapaswini which could be possibly due to genetic potential of the former accompanied with lesser weed infestation. SRI had the highest weed biomass followed by MSRI and BMP. BMP had the highest percent of sedges, whereas MSRI and SRI had the highest percent of grasses and broad-leaved weeds, respectively (Dwibedi et al., 2016). Provision of optimum spacing to individual rice plants capitalizes upon the plastic response of rice to plant population. Due to reduced competition, single plants tiller profusely and develop a deep root system resulting in good growth and yield.

4.2.2.5   IN WIDE SPACING OF 25 cm × 25 cm OR MORE

With advent of dwarf high-yielding varieties, recommendations for rice cultivation have been to increase plant population. Keeping in view the limited land resources and a finite amount radiation, the research efforts have been reoriented to increase the rice productivity through (1) improving biomass production through more efficient use of radiation and (2) increasing the proportion of economic yield by increasing the HI. This philosophy has led to devising breeding strategies for higher number of grains/panicle but lesser tillers/plant (Khush, 1993). In contrast to this, the basic principles of SRI advocate for planting of young seedlings at wider spaces to harness the maximum tillering potential of rice (Laulanié, 1993).

Plants grown at wider spacing develop a strong root system which helps in uptake of larger amount of nutrients and thus, results in vigorous growth of above ground parts. Although at the initial stages the field looks sparsely populated, eventually, the tiller and leaf production increases manifold resulting in full ground cover and maximum interception of solar radiation. It has also been suggested that long-duration varieties perform better under SRI management (Stoop, 2005). In an experiment comparing rice varieties with varied spacing under SRI management, the grain yield of BRRI dhan29 was highest at a spacing of 25 cm × 25 cm and lowest at a spacing of 40 cm × 40 cm (Latif et al., 2005). Among the different spacing schedules (20 cm × 15 cm, 25 cm × 25 cm, 30 cm × 30 cm, 35 cm × 35 cm, and 40 cm × 40 cm) evaluated, maximum dry matter (156.2 g/hill), number of tiller (44/hill), number of effective tiller (36.3/hill), number of filled grains (101.5/panicle), grain yield (6.9 t/ha), straw yield (5.9 t/ha), biological yield (12.7 t/ha), and HI (54.5%) was obtained from 40 cm × 40 cm plant spacing while minimum was observed from 25 cm × 25 cm of plant spacing in this study (Chakrabortty et al., 2014). In contrast, yield under SRI management (planting 10 day old single seedlings at 25 cm × 25 cm + intermittent irrigation and cono weeding) was lower than that of recommended practices (20 day-old two seedlings at 20 × 15 cm + continuous irrigation and hand weeding) but was greater than that of farmers' practices (Anitha and Chellappan, 2011). Rice hybrid PHB 71 at a 25 cm × 25 cm under SRI management proved superior in respect of growth, yield attributes, yield (12.6%), HI, production efficiency, net monetary returns and benefit cost ratio over the rice variety NDR 359. PHB 71 also demonstrated a significant superiority in respect of quality parameters, namely, hulling, milling, head rice recovery, and protein NDR 359 (Ram et al., 2015). SRI practice for planting space of 25 cm × 25 cm to 30 cm × 30 cm, wetting and drying interval of three days, and younger seedling of 8–12 days are recommended as good combinations for SRI practices in Mkindo area, Morogoro region of Tanzania (Kahimba et al., 2014). At higher altitudes of 1200 m a.s.l. and at sea level in Madagascar, a spacing recommended under SRI varied from 25 cm × 25 cm to 40 cm × 40 cm, respectively. The yields varied from 2 to 8 or even 12 t/ha. The local varieties yielded at par with modern varieties under SRI management (Laulanié, 2011).

Breeding strategies have aimed at genotypes with best endowments for yield under suboptimal conditions and crowded plant populations. Under crowded conditions, shading of lower crop canopies results in suboptimal photosynthetic efficiency. Further, the flooded conditions result in fast root degeneration, thus foregoing the benefits of aerobic soil conditions. Chemicalized agriculture has adversely affected the soil microflora, soil health, and human health (Uphoff et al., 2015). Rice genotype PHB 71 was significantly superior over NDR 359 with respect to yield attributes, grain yield, economics, nutrient uptake, and soil health due to profuse root growth at wider row spacing of 30 cm × 30 cm (Singh et al., 2013). Reduction in the plant population is compensated by increase in per plant yield due to profuse tillering. Stronger root system, higher xylem exudation rates, erect and larger leaves, and higher number of tillers were recorded in SRI than recommended practice (Thakur et al., 2010).

Wider spacing enables rice plant to express its full growth potential of both above and below ground portions. However, the optimum spacing shall depend on the cultivar, soil fertility status, and other agroclimatic parameters. The spacing cannot be fixed for varied agroclimatic conditions and cultivars; however, SRI recommends wider spacing than the conventional methods of rice cultivation.

4.2.2.6 IN MECHANICAL WEEDING

Flooded conditions with controlled drainage in rice fields are used to control the weeds. However after prolonged flooding, the aquatic weeds start dominating, and the fields are drained to restrict the aquatic weed growth (Sahid and Hossain, 1995). For the management of terrestrial weeds, the field is again reflooded. This is the traditional way of controlling the weeds in rice paddies. Under SRI, weed infestation is higher (Haden et al., 2007) due to wider spacing of 25 cm × 25 cm (Singh et al., 2102) and alternate wetting and drying (Krupnik et al., 2012). Weed infestation can result in significant loss of yield under SRI management (Krupnik et al., 2012) even to the extent of 69% (Babar and Velayutham, 2012). Weeds under SRI have to be managed manually or mechanically which often becomes difficult due to high cost of labor. Integrated weed management practices such as hand weeding, mechanical weeding along with use of herbicides, and use of weed competitive cultivars can be used to keep the weeds at bay (Latif et al., 2005; Randriamiharisoa, 2002), although with different degrees of success. Weed management using rotary weeding at 15 DAT + hand weeding at 35 DAT produced the highest grain yield as compared to other weed management practices. Yield loss percent due to weed competition ranged from 59.5% to 74.06% and 49.97% to 73.4% in dry and wet season, respectively (Thuraa, 2010). Under SRI management, weed control using rotary push weeder relies on the early and frequent weeding generally from three to four times between the period of 10 and 40 DAT (ATS, 1992).

Rotary weeding at 12, 28, and 42 DAT served the dual purpose of aerating the soil and removing the weeds that resulted in a significant increase in number of effective tillers/m^2 (282.67), panicle weight (3.92 g), number of grains per panicle (184.54), lower sterility (7.36%), and higher grain yield (6.53 t/ha) (Pandey, 2009). Integrated approach produced lower weed density/m^2 and dry weight were recorded in SRI with butachlor 50 EC @ 1.5 kg a.i./ha at 3 DAT + cono weeding at 20 DAT, and the same was at par with ICM (Islam and Kalita 2016). Integrating the use of herbicides with single hand weeding resulted in highest yields as compared to other weed control methods (Latif et al., 2005). SRI registered a mean grain yield of 6.06 t/ha, which was significantly higher than conventional method of rice cultivation (5.42 t/ha). On an average across locations, SRI out yielded conventional method of rice cultivation by 11.1%. Four times, thrice, and twice rotary weedings resulted in yield increase to the extent of 24.1%, 15.4%, and 8.5%, respectively, over conventional weed control methods (Veeraputhiran et al., 2014).

Some other methods of weed management such as mulching have also been evaluated in SRI. Among the mulching treatments, weed density was significantly reduced with the rice straw mulching and commercial black plastic (Ramakrishna et al., 2006; Wayayoka et al., 2014). The dominant weed class associated with SRI was sedges, due to higher summed dominance ratio followed by grasses, and broad-leaved weeds, respectively. The effect of SRI mat mulch on weed control was 98.5%. The study further revealed that SRI mat was helpful in controlling the weeds up to rice canopy closure or 40 DAT as recommended in SRI, retaining the soil moisture and more number of tillers in SRI farming (Mohammad et al., 2015). The straight-spike floating weeder reduced weeding time by 32–49%, the twisted-spike floating weeder reduced weeding time by 32–56%, and the application of herbicide required 88–97% less time than hand weeding. Herbicide application provided the best weed control in two of the three seasons. No differences in weed control efficacy were observed between mechanical and hand weeding (Rodenburg et al., 215). Comparative evaluation of chemical methods and hand weedings revealed that application of fenoxaprop-p-ethyl 60 g/ha + ethoxysulfuron15 g/ha at 20 and 35 DAT was statistically at par with hand weeding (twice) at 20 and 40 DAT in respect of grain yield (Kumar and Singh, 2013). Among the mechanical weeders, the highest weeding efficiency (84.33%) was obtained with power as compared to cono weeder and rotary weeders (Alizadeh, 2011). Early weeding using a cono weeder revealed higher field capacity and better performance index and recorded a weeding efficiency of 72% and a damage factor of only 4% (Karhale et al., 2015). A 2-year study on the evaluation of manual hoeing at 20, 40, and 60 DAT revealed a superiority of 8.4% and 7.2% over orthosulfamuron and 61.0% and 64.9% over weedy check, respectively. Manual methods also registered superiority with regard to weed control efficiency to the extent of 87.89% and 82.32% during the year 2010 and 2011, respectively. Manual weeding at specified timings has proved as eco-friendly and efficient method of weeding to increase the grain yield of fine rice under SRI (Chadhar et al., 2014).

In a another study, hand weeding two times at 20 and 40 DAT was statically at par with postemergence application of fenoxaprop-p-ethyl 60 g/ha + ethoxysulfuron 15 g/ha at 20 and 35 DAT (Kumar and Singh, 2013). Crop establishment methods, namely, transplanting, SRI, and drum seeded and broadcast in main plots were evaluated along with four methods of weed control—pyrazosulfuron 0.02 kg/ha PE + mechanical weeding at 25 DAS or DAT, weeding by cono weeder at 25 DAS or 40 DAT, hand weeding at 25 and 40 DAS or DAT, and weedy check in subplots. The result revealed that transplanting and application of pyrazosulfuron 0.20 kg/ha + one mechanical weeding at 25 DAS or DAT out yielded the other treatment combinations. Application of pyrazosulfuron 0.20 kg/ha + one mechanical weeding at 25 DAS or DAT in transplanted or drum seeded or broadcasted rice was most effective in suppressing weed population and weed dry matter accumulation thereby producing higher rice grain yield compared to other weed control methods (Hassan and Upasani, 2015).

Although chemical methods of weed management in rice are more economical, mechanical methods have environmental safety, provide soil aeration, promote tillering and rooting, and play a synergistic role in combination with other SRI practices.

4.2.2.7 IN INTERMITTENT IRRIGATION

Intermittent irrigation practices are employed during vegetative growth stage. Water requirement of rice ranges from 1190 to 2650 ha-mm depending upon soil climatic conditions and varietal characteristics. Continuous flooding (CF) with 2–5 cm of water provides the opportunity for potential yield (Prasad, 1999). Intermittent irrigation could save 25–50% water without any adverse effect on rice yield (Ramamoorthy et al., 1993). It has also been reported by Boonjung and Fukai (1996) that limited water condition during vegetative growth did not result in any harm to rice growth. Rice plants can tolerate moderate water stress through osmotic adjustment (Steponkus et al., 1980). Exposing the rice plants to moderate water stress stimulates root growth and helps the plants to develop xeromorphic characteristics. Vigorous root development results in enhanced tillering and altered sink–source relationship. Reports from Tamil Nadu, Puducherry, and Tripura in India reveal that SRI management produced 30% higher yields with 40% water savings. India has enough potential to extend SRI to 20 m ha so that country could "produce more crop per drop" (Swaminathan and Kesavan, 2012).

Rice is essentially not an aquatic plant but tolerates waterlogged conditions through transport of oxygen to roots through aerenchymatous tissue. Under flooded conditions, most of roots remain restricted to the top soil layer and start degenerating before flowering. Intermittent drying results in sufficient aeration and stimulates root growth which is advantageous for enhanced nutrient uptake and crop vigor (Stoop et al., 2002). Upland cultivars are more prone to root degeneration under flooded conditions and high plant population. Shading results reduced photosynthetic activity of lower leaves and generation of roots to the extent of 78% before the flowering commences. High root activity at flowering is critical for achieving higher grain yield. Accumulation of CO_2 and other toxic gases near the rhizosphere results in the early senescence of roots (Kar et al., 1974). Minimal or even antagonistic interactions have also been reported between individual practices of SRI. Intermittent irrigation, compost application, and wider row spacing had a significant impact on grain yield. Synergistic effects of SRI practices were not observed in salt-affected soils (Menete et al., 2008). Alternate wetting and drying promoted soil aeration and resulted in 10.5% to 11.3% higher grain yield as compared to CF. The attribute that contributed more to the grain yield was number of grains/panicle. Addition of organic material in combination with intermittent irrigation resulted in increased redox potential (Eh) which is an index of degree of soil aeration (Lin et al., 2011). Significant increase in grain yield was achieved in Gambia at lower rates of inorganic fertilizers and lesser requirements of water using SRI practices. Water productivity increased from 0.10 to 0.76 g of rice per kg of water in CF and intermittent irrigation, respectively (Ceesay, 2011). SRI method resulted in water saving of about 34% and significantly less water was required to produce 1 kg of rice (Shahane et al., 2015).

Stimulation of higher root activity under SRI method resulted in 38% higher lateral root activity than conventional method of rice cultivation. As per the studies on vertical distribution of root activity, there was an increase of 48%, 34%, and 30% higher at 7.5, 15, and 22.5 cm vertical depth in SRI method over conventional method. The enhanced root activity under SRI was also reflected in 11% higher grain yield (Babu et al., 2014). Data from large-scale demonstrations have produced convincing evidence that water requirement was lesser in SRI (885 mm) as compared to conventional method (1180 mm). The grain yield and water use efficiency increased from 5284 kg/ha and 4.51 kg/ha-mm in conventional method to 6406 kg/ha and 7.31 kg/ha-mm in SRI, respectively. The water productivity in SRI was found to be 1398 L/kg as against 2274 L/kg in conventional irrigation (Pandian et al., 2014). To understand the mechanism of response, studies were conducted using four water treatments: (1) intermittent flooding through the vegetative stage (IF-V); (2) intermittent flooding extended into the productive stage (IF-R); (3) no standing water: maintaining soil at field capacity; and (4) CF condition at the Asian Institute of Technology in Thailand. It was observed that the senescence of lower leaf and flag leaf was delayed under IF-V compared to CF water condition. This delay was associated with higher root oxidizing activity rate (50% higher than CF), higher root length density (52% higher than CF), higher biomass production (14% higher than CF) along with higher grain yield (25% higher than CF) (Mishra, 2012). A 10-day cycle of wet days followed 10-day cycle of dry days resulted in 29% water saving without any significant reduction in grain yield. Water productivity index increased from 1.3 kg/m^3 in conventional water management to 1.73 kg/m^3 in SRI method (Chapagain et al., 2011). Water saving of 34.5–36.0% in SRI and 28.9–32.1% in aerobic rice was recorded as compared to CT rice (Singh, 2013). The reduction of ponded water depth sequentially from 80, 60, 40, to 20 mm could save irrigation water 17.3%, 28.2%, and 68.6%, respectively. These water regimes also resulted in reduction in total N and P load by 2.9%, 4.2%, and 10.9% and 4.5%, 6.0%, and 12.9%, respectively (Jung et al., 2013). Manipulations in water regimes affect the nutrient dynamics in rice soils due to its effect on soil pH and Eh. These parameters affect the solubility of nutrient-supplying minerals. Phosphorous and iron deficiencies in aerobic rice have been widely reported. Irrigation water saving 54% and 36% was recorded in aerobic rice and SRI, respectively. The corresponding figures for water productivity were 9.16 kg/ha-mm and 7.02 kg/ha-mm, respectively (Saha et al., 2015). Promising results were recorded in SRI for water use efficiency (WUE) that the little amount of water used during production was able to double the yield output compared with the conventional methods (Mgaya et al., 2016).

Irrigating rice crop at 1 or 3 days after the disappearance of ponded water (DADPW), being at par with each other recorded significantly higher net photosynthetic rate (NPR) of topmost fully expanded leaf than 5 DADPW. Moderate drying by irrigating at 1 and 3 DADPW resulted in an increase in NPR by 17.1% and 8.4% at tillering and 13.6% and 6.1% at flowering stage, respectively, compared with that of 5 DADPW. SRI improved grain yield by 16.9% and water productivity by 18.5% over TP (Das and Chandra, 2013). On the contrary, the tiller production increased even at 5 DADPW but the same did not translate into increased yield (Hazra and Chandra, 2014). Higher grain yield was registered under SRI with water saving of 16% and 7.8%, respectively, during *kharif* and *rabi* seasons, respectively (Bhuvaneswari et al., 2014). Even higher water savings of 31% and 37% have been reported in SRI during *kharif* and *rabi* seasons, respectively, over

BMP. Apart from water savings and yield increases, the added advantages of SRI were recorded in terms of increase on organic carbon (%) soil dehydrogenase activity, microbial biomass carbon, and total bacteria, fungi, and actinomycete count (Gopalakrishnan et al., 2014).

Planting young seedlings at wider spacing, in association with other component practices of SRI, results in a phenotypic change in rice plants. The plants are characterized by profuse tillering thrown at an angle rather than upright, with improved photosynthetic rates and xylem exudation. Most of the yield attributes such as panicle length, filled grains/panicle, and 1000-grain weight increased under SRI management. SRI management resulted in 49% higher grain yield, 14% less water use over TP. Water productivity recorded a sharp increase of 73% for SRI over TP (Thakur et al., 2013a). In Andhra Pradesh, increase in productivity and farmer's income was achieved through water savings and reduced labor and seed costs through use of SRI techniques. SRI has the potential to save water resource and power costs through systemic corrections in irrigation practices (Adusumilli and Laxmi, 2011). A clear advantage in water saving was recorded through the use descriptive statistical tests in SRI method over conventional method of rice cultivation. Water saving of 22% amounting to 3.3 million L/ha was recorded in SRI method. In another set of 17 studies, water saving of 35% amounting to 3.9 million L/ha (Jagannath et al., 2013) was recorded.

Among a large number of studies carried out on SRI under Indian conditions, majority have reported yield advantages and higher WUE. Use of SRI technique resulted in water saving to the extent of 25–50% or even more. In view of changing climate and increased frequency of droughts coupled with higher demand of water from competing sectors, the SRI will assume a significant importance in future. Policies need to be formulated to sensitize the farmers toward these kinds of techniques and promote their use at field level.

4.2.2.8   IN ADDITION OF ORGANIC MANURE

Application of organic manures improves the soil's physical, chemical, and biological properties, thereby increasing root length, density, active absorption area, root oxidation ability, and nutrient uptake (Yang et al., 2004). Application of organic manure under well-drained cognitions brings out synergistic effect on root growth and morphology, whereas waterlogged conditions results in early root degeneration (Sahrawat, 2000). Physical conditioning of soil and improved nutrient supply resulted in the most extensive root system of SRI plants as evidenced from results of many factorial trials (Uphoff, 2003). If sufficient organic manure is not available, SRI practices can be used in raising a rice crop either by fertilizers alone or integrated nutrient management (INM). Number of panicles/hill, dry matter accumulation, and test weight showed significant improvement due to impact of organic manure (Meena et al., 2014). Some of the researchers have attributed the early adoption of SRI to high N mineralization rates and supplying ability rather than synergistic effects of SRI components. Accumulation of soil organic carbon and high N supplying capacity were attributable to long-term and extensive organic manure applications. Changes in the soil hydrology under SRI management could be reason for higher N mineralization rates (Tsujimoto et al., 2009). However, substitution a part N of using organic manure did not result in increase in grain yield (Latif et al., 2005).

Rice plants harvested from SRI-managed plots exhibited considerably higher total biomass, root dry weight, seed fill, and more diverse AMF communities than those obtained from conventionally managed plots. The AMF community diversity was further enhanced due to the addition of compost to SRI plots (Watanarojanaporn et al., 2013). The changes in AMF diversity in response to the different cultivation systems and the application of compost might be due to several factors, such as pH, nutrient content, total soil C and N, and temperature, which are known to influence AMF distribution (Husband et al., 2002). The SRI increased grain yield (6.65 t/ha) by 18.0% and 25.8% over CT and DS, respectively, whereas the latter two treatments remained at par. INM registered higher grain yield by 11.9% and 19.2% over RDF and organic manure alone, respectively. SRI grown under INM recorded the highest productivity of 7.30 t/ha (Mohanty et al., 2015). Addition of organic matter acts in a synergistic way along with inorganic fertilizers and improves their use efficiency. Therefore, INM has emerged as the more viable alternative in intensive cropping systems. It improves the soil and plant health and makes the production systems more sustainable. The SRI is gaining popularity among paddy farmers in several states.

Nutrient management through integrated use of inorganic and organic fertilizers under SRI management resulted in significant increase in effective tillers, panicle length, dry matter, root dry weight, and root volume as compared to BMP. Increase in growth of above and below ground parts was finally translated into higher grain yield to the tune of 12–23% and 4–35% in *kharif* and *rabi* seasons, respectively (Gopalakrishnan et al., 2014). Significantly higher growth attributes, namely, plant height and leaf length, yield attributes, namely, number of effective tillers/hill, panicle length, and panicle weight in INM as compared to chemical fertilizers. INM also enhanced seed quality parameters, N uptake of soil organic carbon over chemical fertilizers alone (Singh et al., 2013). Some preliminary trials conducted temperate conditions of Kashmir valley at Mountain Research for Field Crops (SKUAST-K) have demonstrated that SRI (NPK + 10 t FYM) has yield superiority of 17% over the farmers practice (Hussain et al., 2009).

During *kharif*, grain yield was significantly higher in SRI (organic + inorganic fertilizers) than conventional method (inorganic fertilizers alone) and Eco-SRI (organics alone) by 10.3% and 33.4%, respectively. SRI and conventional method were on par and significantly superior to Eco-SRI with respect to N, P, and K uptake in both the seasons (Surekha et al., 2015). The results revealed MSRI (10 days aged seedlings + 100% nutrients through inorganic or 50% through organic + 50% through inorganic + irrigation as per SRI) gave 13.52% higher grain yield and 16.80% higher net income over recommended package of practices for hybrid rice (Verma et al., 2014).

Among the different nutrient management options, application of recommended dose of fertilizer along with *Azolla* and BGA in SRI resulted in significantly higher grain yield and HI (Jeyapandiyan and Lakshmanan, 2014). Biofertilizers including plant growth-promoting rhizobacteria enhanced root growth yield and quality of rice hybrids. There are also reports of enhancement in quality parameters of rice such as hulling, milling, head rice recovery, and protein content under the influence of *Azospirillum brasilense* and *Bacillus subtilis* significantly applied to rice seedlings over the control (Choudhary et al., 2013). Vigorous growth and tillering with higher leaf N content and delayed leaf senescence contributed to extension of photosynthetic processes, resulting in increased grain yield (Thakur et al., 2013b).

Addition of organic manures promotes soil's physical, biological, and chemical health, apart from acting as direct source of nutrients. Enhancement in soil microbial activity promotes the nutrient availability and plant growth. Efficient use of organic manure in conjunction with the inorganic fertilizers appears to be more practical keeping in view the dearth of on farm organic manures.

4.3 ENVIRONMENTAL IMPACT OF SRI

Flooding of rice paddies leads to anoxic conditions, conducive for anaerobic microbial activity, and reduction in Eh of rice soils. Anaerobic decomposition of organic matter produces CH_4, whereas N_2O emissions originate from synthetic nitrogenous fertilizers added to agricultural soils. Direct emission of N_2O from site of application occurs during the processes of nitrification and denitrification, whereas the indirect emissions result from the volatilization, leaching, and redeposition of N. Biochar being carbonaceous substance recalcitrant to decomposition released lesser amount of CH_4 (38%) as compared to FYM (230%) and straw compost (150%) averaged over different water regimes as compared to control. In alternate wetting and drying (AWD), FYM increased the release of N_2O by 30% straw compost and biochar. Even under permanent flooding, biochar released the least amount of N_2O which was at par with control. From this experiment, it was concluded that biochar has the potential to reduce the global warming potential of rice production (Pandey et al., 2015). SRI technology has the potential to reduce the GHG emissions both per hectare and per kg of rice produced. These advantages arise from reduced methane production in relatively aerobic soil conditions and reduced use of pumps used to lift the water for irrigation. The N_2O emissions per kg of rice produced are also lesser in SRI method (Gathorne-Hardy et al., 2013).

Intermittent irrigation in SRI results in aerobic conditions of rice soils. Application of soil organic manure under well-aerated conditions supports the growth of beneficial soil microorganisms. Higher enzymatic activity leads to higher N mineralization rates, more soil organic carbon, and larger nutrient pool for both soil microorganisms and plants. More exploratory research in this area can show the way of rice production while conserving the resources and making the activity cost effective (Anas et al., 2011).

Methane emissions increased progressively with the advance of crop growth until flowering and then declined thereafter. Notably, the levels of emission were consistently higher under conventional crop management compared to SRI methods of cultivation, by 27.8–42.6% during summer season, and by 33.0–43.1% in *kharif* season. Cumulative methane emissions for the entire cropping season were considerably lower in SRI (31.8 and 37.7 kg/ha) than conventional cultivation (44.6 and 55.5 kg/ha) in the summer and *kharif* seasons, respectively. The conclusion was that with SRI management, total methane emissions were reduced by 29% and 32% during the summer and *kharif* seasons, respectively. The data further indicated that cono weeding employed with SRI by itself contributes 19–63% of the reduction in methane emissions. The populations of methanogens which produce methane were significantly lower under SRI rather than conventional crop management regardless of stage, while the reverse was true with methanotrophs, microorganisms that consume methane. Overall, the data suggest that the SRI method of rice cultivation has a definite advantage of reducing methane emissions, which are closely associated with the practices of cono weeding and intermittent wetting and drying of paddies (Rajkishore et al., 2013). Developed ANN model can predict CH_4 and N_2O emissions accurately with determination coefficients of 0.93 and 0.70 for CH_4 and N_2O prediction, respectively. From the model, characteristics of those greenhouse gas emissions can be well identified. For the mitigation strategy, SRI treatments in which the water level was kept at nearly soil surface is the best strategy with highest yield production and lowest GHG emission (Arif et al., 2015).

Since WUE is quite high in SRI method as compared to conventional method, arsenic uptake is also less in SRI as compared to conventional method. So SRI method is very highly recommended for increased productivity, ecological security, and arsenic mitigation (Bokaria, 2015). The irrigation water requirements reduced by 55.6% in SRI plots. Loss of water due to runoff was also reduced by 5–15%. SRI recorded concentrations of SS 89.4 mg/L, COD_{Cr} 26.1 mg/ L, COD_{Mn} 7.5 mg/L, BOD 2.0 mg/L, TN 4.2 mg/L, and TP 0.4 mg/L in runoff water. Except for COD_{Cr} and TN, these concentrations were significantly lower, by 15.8–44.1%, than those from CT plots. Although the rice grain yield in SRI plots ranged between 76% and 92% in SRI of that of CT plots due to reduced plant population, it has the potential to reduce the nonpoint source (NPS) pollution in Korean conditions (Choi et al., 2013). Reduction of 61.1% and 64% in methane emissions in SRI (12 days old seedling) and MSRI (18 days old seedling) was recorded as compared to conventional transplanted rice (TPR). However, a corresponding increase of 22.5% increase in N_2O was observed in SRI and MSRI. Overall, SRI resulted in 36% water saving and 28–30% reductions in global warming potential (Jain et al., 2014). Global warming potential was highest in CT (807.4 kg/CO_2/ha) and lowest in SRI (498.25 kg/CO_2/ha) (Suryavanshi et al., 2013).

The sole or integrated use of organic manures results in improvement in soil health and nutrient supply. From a number studies conducted on the effect of SRI methods on GHG emission, it can be concluded that SRI results in a significant reduction methane emission despite an increase in N_2O emission. The net effect is a reduction in GHG emissions and global warming potential.

4.4   ADOPTION DYNAMICS OF SRI

SRI has been tried now in more than 50 countries across the world as it has the potential to improve the rice productivity with less capital, water, and labor in socially acceptable, economically viable, and environmentally sustainable ways. In India, Tamil Nadu Agriculture University (TNAU) initiated experiments on SRI in a collaborative research project on growing rice with less water. The results revealed that grain yield increased by 1.5 t/ha with a concomitant 8% reduction in labor. Survey on the adoption of SRI farmers in Sri Lanka reported a yield increase of 44% but used the technology on only a small portion of their farms. However, latter many farmers disadopted SRI largely on account of heavy labor requirements in transplanting, weeding, and organic fertilizer collection. Nonadopters reported the labor in transplanting and weeding requirement as the major impediment in adoption. The main determinants of adoption were labor availability, years of schooling, access to training programs, farm or field location, and the poverty status of the household (Namara et al., 2003).

Adoption of SRI in Tamil Nadu showed an increasing trend from 28.3% in 2007–2008 to 32.4% in 2010–2011. Increase in grain yield by 40–50% and water saving of 25% was the main factor for higher level of adoption (Pandian et al., 2014). SRI technology has emerged as a prominent tool to achieve increased rice production and thus productivity in Kerala. Factors such as experience in farming, income from off-farm and nonfarm sources, size of landholding and contact with agriculture extension officers are significant determinants of adoption of SRI technology. On the other hand, factors such as difficulty in water management and early transplantation, nonavailability of skilled labor and difficulty in using the cono weeder cause the disadoption of SRI. In addition to promoting SRI technology, it is also imperative to develop effective irrigation facilities and promote participatory irrigation management to produce more crop per drop of water. Many farmers who discontinued SRI technology cited the lack of institutional support and capacity building as the factors influencing their decision. Therefore, the state government needs to focus on these aspects so that not only disadoption can be minimized significantly but more farmers can also be encouraged to adopt the technology in the coming years (Durga and Kumar, 2016). Higher grain and straw yield, water and seed saving, and less cost of cultivation aroused interest among the farmers and put SRI technology at takeoff stage. Some of the farmers who got disinterested in SRI quoted labor and institutional constraints as the main impediments (Johnson, 2011). In spite of higher grain yield with a corresponding reduction in cost and increase in benefit to cost (B:C) ratio in SRI method, psychological fear and shortage of labor does exist among the farmers (Ghosh and Chakma, 2015). In northern coastal region of Andhra Pradesh, SRI practices resulted in 31% increase in grain yield and higher B:C ratio but the most important constraint was identified as nursery management (Rao et al., 2013).

In 10 villages of Kurud block in the Dhamtari district of Chhattisgarh, India, a total of 126 selected farmers were treated as respondents. The study revealed the shortage of agriculture labor (93.65%) as a major constraint in adoption of SRI technology, followed by 67.46% of the respondents who faced nonavailability of cono weeder and marked nonavailability of FYM, 36.50% of them faced labor demanded high labor cost, 34.12% of them faced sometimes seedling died in early stage, 30.1% of respondents faced more infestation of weed due to wider spacing and sometimes seedling died in early stage, and the respondents (61.1%) suggested that the trained labor should be available on low wage and the amount of subsidies on seed and fertilizers should be increased, 57.9% of the respondents suggested that government should provide cono weeder and marker, 47.6% of them suggested that the price of hybrid rice should be low (Narbaria et al., 2015). This innovation has now become an important strategy by the Sichuan (China) government and its relevant departments to increase and sustain rice yield. In 2007, demonstrations were established in more than 60 counties in the hilly areas of Sichuan. By 2009, the application area had reached 66,700 ha. From the yields of different demonstration sites, it was seen that in both drought and wet years, this technology combining no-till raised beds with plastic-cover technology caused significantly increase in yield by SRI crop management. Average yield increase has been 2383 kg/ha, a 31% increase compared with farmers' usual practice. This integrated technology is giving higher yields with less water, fertilizer, seeds, and labor, with reduced soil and water pollution, and better environmental conditions. It can serve as a good technique for ensuring local and national food security in the future (Lu et al., 2013). More faith in the traditional practices, involvement of more efforts, and lack of knowledge on scientific water management are the main constraints that come in way of adoption of SRI (Islam et al., 2014)

The relationship of variables, namely, age, education, farming experience, SRI experience, trainings attended, extension orientation, economic motivation, risk orientation, market perception, innovativeness, and attitude were found to be positively significant at 1% level of probability with their extent of adoption of SRI technology. The most important constraint in SRI cultivation has been identified as usage of cono weeder (58.2%) followed by nursery management (56.6%). SRI adoption has increased productivity of rice and emerged as an alternative method of rice cultivation (Ravichandran and Prakash, 2015). SRI demonstrations that Indonesia generated significant yield gains but with reallocation of family labor from nonfarming to farming activities. Although there was a reallocation of labor but there was no evidence of its effect on child labor (Takahashi and Barrett, 2013). Moreover, SRI is labor-intensive methodology and family composition is significant determinant of its adoption and continuing use (Takahashi, 2013).

Although work is underway in Tanzania and Malawi for the promotion of SRI but there are many constrains in the smooth adoption of the technology such as poor irrigation projects, lack of quality seeds and manures, hand-pushed rotary weeds and capacity building on SRI. Further, lack of field leveling, poor market, and fluctuating rice prices are other constraints in SRI adoption (Aune et al., 2014). Yield advantage of 33%, seed saving of 87%, and water saving of 28% has been reported from Mwea, Kenya. Although there was an increase in labor requirement by 9% in some units, at other places, there was a decrease in labor requirement by 13–15% with the introduction of rotary weeders. Benefit–cost ratio increased to 1.8 was realized under in second season as compared to 1.35 in farmers practice. Upscaling of SRI technology has the potential to ensure food security in Mwea (Ndiiri et al., 2013). Study on the relative economics of SRI has revealed that SRI gave significantly higher B:C ratio of 2.27 as compared to 1.44 in traditional methods in Assam state of India. Main constraints in adoption of the method were lack of skilled labor, awareness, and training programs on SRI (Ishani et al., 2013).

Education, training undergone, social participation, extension contact, economic motivation, scientific orientation, management orientation, achievement motivation, innovativeness, mass media exposure, and risk orientation were found to be positively correlated ($p < 0.01$) with their extent of adoption of SRI technology. Age and farming experience showed significant negative relationship, whereas land holding had a nonsignificant relationship with their extent of adoption of SRI technology (Kumar et al., 2014).

Extrapolation of SRI methods to other crops have been tried with success in some other crops such as finger millet, wheat, sugarcane, tef, oilseeds such as mustard, legumes such as soya and kidney beans, and various vegetables in some Asian an African countries. The system is referred to as system of crop intensification. Application of SRI principles in other food crops has shown increased productivity with reduced costs and higher resilience due to effects of climate change (Abraham et al., 2014).

Higher biomass production in SRI is expected to mine more nutrients from the soil but experiments have shown that higher soil nutrient status after the harvest of rice crop. The increase of 10%, 42%, and 13% in the level of N, P, and K, respectively, in available soil was observed in SRI plots. Vigorous crop growth and favorable changes in rhizosphere in SRI method has been observed to reduce the pest and relative abundance of plant parasitic nematodes (Kumar et al., 2013b).

SRI practices are a radical shift from conventional practices and therefore, farmers are reluctant to adopt the new method. The handling of the small seedling with markings in field in a square pattern is relatively difficult and perfect water level maintenance is essential for successful establishment. To change the mindset of farmers, it is imperative to hold large number of field demonstrations to motivate them toward this resource-saving technology. Off course capacity building particularly with regard to nursery raising, transplanting, water management, and mechanical weeding is essential to spread its application over extensive areas.

4.5 ADVANTAGES OF SRI

When SRI practices are used together and as recommended, the following results are common: (1) Grain yields are usually increased by 50–100%, or sometimes more, while water applications are reduced by 30–50% since there is no CF, straw yields usually also increase, which is an additional benefit to many farmers; (2) The need to use agrochemicals for crop protection is reduced because SRI plants are naturally more resistant to pest and disease damage; (3) With reduced costs of production, including often reduced labor requirements, farmers' net income is greatly increased with the higher yields; (4) SRI plants are better suited to withstand the effects of climate change, having greater resistance as a rule to most biotic and abiotic stresses; and (5) SRI paddy usually gives higher milling outturn, about 15%, because when milled, there is less chaff (fewer unfilled grains) and less breaking of grains. These qualities are probably attributable to the effects of better root systems which can more effectively take up micronutrients from lower soil horizons. Currently, SRI practices have been introduced in many countries with modifications and adaptation to local conditions (Iswandi et al., 2011).

Although organic, inorganic, and INM did not affect nematode infestation, it is significantly lower in SRI and conventional approach than aerobic method of rice cultivation (Singh, 2013). SRI method has a unique characteristic of maintaining a balance between the pests and beneficial insects (predators and parasitoids) with any appreciable yield loss. SRI has shown the way for conservation of biodiversity crucial for food security (Norela, 2013).

Conventional management recorded higher values for *Rhizobium* population but nitrogen fixers, phosphate solubilizers, and cyanobacteria exhibited a 10-fold increase in the plots under SRI management. There was substantial decrease in methane production (50–80%) in SRI plots inoculated with cyanobacteria and recommended fertilizers. The grain N and P content in SRI method was higher due to microbial treatments. Cyanobacteria have the potential of increasing the rice productivity and decreasing N requirement by 60 kg/ha (Prasanna et al., 2015). The grain yield ranged from 3.68 t/ha (Loktantra under conventional transplanting) to 7.38 million t/ha (Radha-4 under SRI) among the combination of different method of crop establishment and varieties. SRI method of crop establishment produced significantly higher grain yield (6.95 t/ha) followed by DSR yield (6.2 t/ha) and the lowest yield was under transplanting (TPR) method, that is, 4.18 t/ha. SRI produced 51% and DSR produce 48% higher grain yield than general TPR (Dhital, 2011).

Incidence of pests such as blue beetle, case worm, leaf folder, and gundhi bug/m^2 were lower in SRI as compared to conventional system. This was attributed to prevalence of natural enemies such as dragon flies and wolf spiders which was higher in SRI. The incidence of fungal diseases such as blast, sheath blight, brown and false smut spot was about 7–8% in SRI plots (Mahesh et al., 2012). The interaction of various components of SRI resulted in diversification of predator population and a reduction in stem borer dead heart incidence by 8–9%. A 3% reduction in incidence of gall midge incidence was also observed in SRI plots (Visalakshmi et al., 2014).

Metadata to analyze the impact of SRI on rice production scenario in China revealed that there was a 10% increase in the grain yield with associated advantages of reduced water and seed. Reduction in cost and increase in yield has resulted in enhancement of net income of the rice farmers. Greater diseases resistance, higher photosynthetic efficiency, N use efficiency, and improved physiological traits have been reported widely (Wu et al., 2015).

On-farm demonstrations in Madagascar were conducted to evaluate the performance of SRI over standard cultural practices and asses as to how the management practices affect genetic potential of rice plant. Harnessing the synergistic effects of different SRI practices, farmers could get yield 2.63 times than that of standard practices. There was significant change in the phenotype particularly in root growth as assessed by root pulling resistance (RPR). SRI plants offered eight times more resistance than conventionally grown plants. To uproot a single SRI plant, a force of 55.2 kg was required, whereas in the case of conventionally grown plants, it was 6.2 kg/plant. On the basis of direct root measurements, the higher RPR was attributed to more prolific and deep root system. Although N and K recorded a nonsignificant increase in internal efficiency (IE) in SRI over conventional practice but there was a significant improvement in IE of P. It was concluded that use of SRI practices could bring favorable changes in the structure and functioning of rice plants particularly its roots that got reflected in higher grain yield (Barison and Uphoff, 2011). Higher growth during vegetative and reproductive phases under SRI management attributed higher photosynthetic rates and chlorophyll content and higher N and P uptake. On the other hand, no differences were recorded in transpiration rate and leaf temperature. Higher rates growth and photosynthesis resulted in 24% increase in grain yield in SRI over conventional method (Hidayati et al., 2016). Results comparison trials 12,133 in number spread over an area of 9429 ha revealed an increase of 78% in grain yield, 40% water saving, 50% fertilizer saving, and 20% cost reduction. These results supported the earlier findings that reported a reduction in inputs, enhanced productivity of land, labor, water, and capital (Sato and Uphoff, 2007). Superiority of SRI practices has been reported from Godavari delta of Andhra Pradesh in respect of 3–4 days earlier maturity, 83% productive tillers, 83% spikelet fertility, 23% water saving, and significantly higher grain and straw yield (Rao et al., 2013). All India coordinated research at 25 locations for 4 years across the country revealed a 7–20% higher grain yield over conventional practices. Varieties with higher tillering ability and hybrids recorded higher grain over HYVs with moderate tillering capacity. Higher dry root biomass, root volume, and dehydrogenase activity was recorded under SRI practice. The additional benefits from SRI were 80% seed saving, 29% water saving, earlier maturity by 8–12 days, and therefore higher productivity/day. Identification of cultivars suitable for SRI, modification of SRI practices according to local agroclimatic conditions, and harnessing of synergistic effects of SRI practices are needed for wider adoption of SRI technology (Kumar et al., 2013).

In a number of studies, SRI has yielded higher, saved water and seed, and proved more profitable than conventional practices. Apart from this, SRI plants grow stronger deeper root system that makes them tolerant to drought and lodging. There are reports of reduced pest and disease incidence because of the presence of healthier plants coupled with lesser use of pesticides that promotes the biodiversity and a balance in pest and predator population. There are also reports of seed quality improvement and therefore can be used as a method of seed production in rice.

4.6   CRITICISM AND LIMITATIONS OF SRI

Rice ecosystems widely vary in respect of soil and climatic conditions and so do the farmer's practices and yield obtained, particularly water management practices. In SRI, we deal with a set of empirical practices and the principles underlying them are still being explored. The practices need skilful hands and meticulous planning and implementation of SRI operations (Stoop, 2003). Management of small and young seedlings in soils with impeded drainage becomes often difficult and therefore, a patchy performance. Reports of extraordinary high yields in China and Madagascar have been attributed to error and therefore, SRI has no inherent yield advantage (Sheehy et al., 2004). Widely spaced single plants gives bare look of field for initial 6 weeks and the methodology appears risky. It is for the farmers to test different ages of seedlings, seedlings/hill, and spacing depending on the local agroclimatic conditions (Sheehy et al., 2004; McDonald et al., 2006). The estimation of the effects of SRI practices on the rice yield hinges on the scientific rigor and accuracy of measurements (Glover, 2011). Variations in yield under SRI method can result from varying degrees of adherence to SRI practices (Takahashi and Barrett, 2014; Sinha and Talati, 2007). Prolific root system is developed under low input and intermittent irrigation but the same is not required for achieving high yield under well-irrigated, high input intensive cropping systems. SRI is reported to have little potential in improving rice productivity in intensively irrigated systems but may help improve productivity of resource-poor farmers in areas with poor soils. (Dobermann, 2004). Only a modest increase in grain yield was associated with the adoption of SRI practices but there was a concurrent increase in labor and fertilizer use. The SRI has often been tested on fertile plots with no firm evidence of improvement in factor productivity. Partial factor productivities have also shown mixed results. Diversity of SRI practices also adds complications in assessing its impact on the yield (Berkhout et al., 2014). SRI is being criticized from scientific circles as just being unconfirmed field observations (Sinclair and Cassman, 2004).

Experiments on the use of SRI practices in Bangladesh revealed that SRI had little effect on rice yield but increased the labor demand and decreased the profitability, making it unattractive for farmers (Latifa et al., 2004). To confront these criticisms, Uphoff (2011) pointed out that initially, little agronomic research had been done to support some of the claims of SRI and few published articles in peer-reviewed journals. Since green revolution technologies based on high-yielding agrochemical responsive varieties have dominated the modern farming, SRI is still facing resistance from the scientific communities.

4.7 SUMMARY AND CONCLUSIONS

SRI was initially demonstrated on poor soils of resource-poor farmers of Madagascar; however, latter it was demonstrated in highly productive soils of China, India, Indonesia, Japan, Thailand, etc. There has been ever-increasing number of evidences of yield advantages of SRI over conventional methods of rice cultivation both in the old varieties and new hybrids having high tillering potential. The method is more productive when all the practices are followed so that the positive interaction among them is harnessed by exploiting the innate genetic potential of the rice plant. Apart from yield advantage, SRI saves substantial amounts of water and seed, improves soil health, and reduces the dependence on chemical inputs. Growth of strong root and shoot system bestows the plant resistance against the biotic and abiotic stresses. It is worthwhile to popularize the technology where yield or other resource-saving advantages has been reported. These practices are not fixed set of rules but need to be evolved and modified according to the agroclimatic situation. The method needs to be evaluated in the regions where it has not been introduced so far. The SRI shall go long in enhancing the grain and straw yield with reduced dependence on water, seeds, and other agri-inputs. In view of climate change, increasing scarcity of water, changing pest scenario, and ever-increasing cost of inputs, SRI appears to be a viable alternative.

KEYWORDS

- aerobic rice
- continuous flooding
- direct-seeded rice
- dry matter accumulation
- hand weeding
- herbicides
- inorganic fertilizers

REFERENCES

Abraham, B.; Araya, H.; Tareke, B.; Edwards, S.; Biksham, G.; Khadka, R. B.; Koma, Y. S; Adusumilli, R.; Laxmi, S. B. Potential of the System of Rice Intensification for Systemic Improvement in Rice Production and Water Use: The Case of Andhra Pradesh, India. *Paddy Water Environ.* **2011**, *9*, 89–97.

Abraham, B.; Araya, H.; Berhe, T.; Edwards, S.; Gujja, B.; Khadka, R. B.; Koma, Y. S.; Sen, D., Sharif, A.; Styger, E.; Uphoff, N.; Verma, A. The System of Crop Intensification: Reports from the Field on Improving Agricultural Production, Food Security, and Resilience to Climate Change for Multiple Crops. *Agric. Food Secur.* **2014**, *3* (4), 2–12.

Alizadeh, M. R. Field Performance Evaluation of Mechanical Weeders in the Paddy Field. *Sci. Res. Essays* **2011**, *6* (25), 5427–5434.

Anas, I.; Rupela, O. P.; Thiyagarajan, T. M.; Uphoff, N. A Review of Studies on SRI Effects on Beneficial Organisms in Rice Soil Rhizospheres. *Paddy Water Environ.* **2011**, *9*, 53–64.

Anitha, S.; Chellappan, M. Comparison of the System of Rice Intensification (SRI), Recommended Practices, and Farmers' Methods of Rice (*Oryza sativa* L.) Production in the Humid Tropics of Kerala, India. *J. Trop. Agric.* **2011**, *49* (1–2), 64–71.

Arif, C.; Setiawan, B. I.; Widodo, S.; Rudiyanto.; Hasanah, N. A. I.; Mizoguchi, M. Development of Artificial Neural Network to Predict Greenhouse Gas Emissions from Rice Fields with Different Water Regimes. *J. Irigasi* **2015**, *10* (1), 1–10.

ATS (Association Tefy Saina). *The System of Rice Intensification*; 1992. http://www.tefy-saina.org (accessed Nov 15, 2016).

Aune, J. B.; Sekhar, N. U.; Esser, K.; Tesfai, M. *Opportunities for Support to System of Rice Intensification in Tanzania, Zambia and Malawi. Report Commissioned by NORAD Under the NMBU–Norad Frame Agreement.* Noragric Report No. 71. Department of International Environment and Development Studies, Noragric, Faculty of Social Sciences, Norwegian University of Life Sciences, 2014; pp 1–45.

Babar, S. R.; Velayutham, A. Weed Management Practices on Nutrient Uptake, Yield Attributes and Yield of Rice Under System of Rice Intensification. *Madras Agric. J.* **2012**, *99* (1–3), 51–54.

Babu, P. S.; Madhavi, A.; Reddy, P. V. Root Activity of Rice Crop Under Normal (Flooded) and SRI Method of Cultivation. *J. Res. ANGRAU* **2014**, *42* (2), 1–3.

Balasubramanian, V.; Sie, M. R.; Hijmans J.; Otsuka, K. Increasing Rice Production Sub-Saharan Africa: Challenges and Opportunities. *Adv. Agron.* **2007**, *94*, 55–126.

Barison, J.; Uphoff, N. Rice Yield and its Relation to Root Growth and Nutrient-use Efficiency Under SRI and Conventional Cultivation: An Evaluation in Madagascar. *Paddy Water Environ.* **2011**, *9* (1), 65–78.

Baskar, P.; Siddeswaran, K.; Thavaprakaash, N. Tiller Dynamics, Light Interception Percentage and Yield of Rice Cultivars Under System of Rice Intensification (SRI) as Influenced by Nursery Techniques and Spacing. *Madras Agric. J.* **2013**, *100* (1–3), 131–134.

Berkhout, E.; Glover, D.; Kuyvenhoven, A. On-farm Impact of the System of Rice Intensification (SRI): Evidence and Knowledge Gaps. *Paddy Water Environ.* **2014**, *12*, 193–202.

Bhuvaneswari, K.; Geethalakshmi, V,; Lakshmanan, A.; Anbhazhagan, R.; Nagothu, D.; Sekhar, U.; Climate Change Impact Assessment and Developing Adaptation Strategies for Rice Crop in Western Zone of Tamil Nadu. *J. Agrometeorol.* **2014**, *16* (1), 38–43.

Bokaria, K. Importance of System of Rice Intensification Method for Mitigation of Arsenic in Rice. *Int. J. Adv. Res.* **2015**, *3* (5), 1398–1409.

Boonjung, H.; Fukai, S. Effect of Soil Water Deficit at Different Growth Stages on Rice Growth and Yield Under Upland Conditions. *Field Crops Res.* **1996**, *48*, 37–45.

Ceesay, M. An Opportunity for Increasing Factor Productivity for Rice Cultivation in Gambia Through SRI. *Paddy Water Environ.* **2011,** *9* (1), 129–135.

Chadhar, A. R.; Nadeem, M. A.; Tanveer, A.; Yaseen, M. Weed Management Boosts Yield in Fine Rice Under System of Rice Intensification. *Planta Daninha, Viçosa-MG* **2014,** *32* (2), 291–299.

Chakrabortty, S.; Biswas, P. K.; Roy, T. S.; Mahmud, M. A. A.; Mehraj, H.; Jamal-Uddin, A. F. M. Growth and Yield of Boro Rice (BRRI Dhan 50) as Affected by Planting Geometry Under System of Rice Intensification. *J. Biosci. Agric. Res.* **2014,** *2* (1), 36–43.

Chapagain, T.; Riseman, A.; Yamaji, E. Achieving More with Less Water: Alternate Wet and Dry Irrigation (AWDI) as an Alternative to the Conventional Water Management Practices in Rice Farming. *J. Agric. Sci.* **2011,** *3* (3), 1–11.

Chaudhari, P. R.; Patel, A. P.; Patel, V. P.; Desai, L. J.; Patel, J. V.; Chaudhari, D. R.; Tandel, D. H. Effect of Age of Seedlings and Fertilizer Management on Yield, Nutrient Content and Uptake of Rice (*Oryza sativa* L.). *The Bioscan* **2015,** *10* (1), 351–353.

Chen, S.; Zheng, X.; Wang, D.; Xu, C.; Zhang, X. Influence of the Improved System of Rice Intensification (SRI) on Rice Yield, Yield Components and Tillering Characteristics Under Different Rice Establishment Methods. *Plant Prod. Sci.* **2013,** *16* (2), 191–198.

Choi, J. D.; Park, W. J.; Park, K. W.; Lim, K. J. Feasibility of SRI Methods for Reduction of Irrigation and NPS Pollution in Korea. *Paddy Water Environ.* **2013,** *11* (7), 241–248.

Choudhary, R. L.; Dinesh, K.; Shivay, Y. S.; Anjali, A.; Lata, N. Yield and Quality of Rice (*Oryza sativa*) Hybrids Grown by SRI Method with and Without Plant Growth Promoting Rhizobacteria. *Indian J. Agron.* **2013,** *58* (3), 430–433.

Das, A.; Tomar, J. M. S.; Ramesh, T.; Munda, G. C.; Ghosh, P. K.; Patel, D. P. Productivity and Economics of Low Land Rice as Influenced by N-fixing Tree Leaves Under Mid-altitude Subtropical Meghalaya. *Nutr. Cycl. Agroecosys.* **2009,** *87* (1):9–19.

Das, A.; Chandra, S. Irrigation, Spacing and Cultivar Effects on Net Photosynthetic Rate, Dry Matter Partitioning and Productivity of Rice Under System of Rice Intensification in Mollisols of Northern India. *Exp. Agr.* **2013,** *49* (4), 504–523.

Deb, D.; Lässi, G. J.; Kloft, M. A Critical Assessment of the Importance of Seedling Age in the System of Rice Intensification (SRI) in Eastern India. *Exp. Agr.* **2012,** *48,* 326–346.

Dhananchezhiyan, P.; Durairaj, C. D.; Parveen, S. Development of Nursery Raising Technique for "System of Rice Intensification" Machine Transplanting. *Afr. J. Agr. Res.* **2013,** *8* (29), 3873–3882.

Dhital, K. Study on System of Rice Intensification in Transplanted and Direct-seeded Versions Compared with Standard Farmer Practice in Chitwan, Nepal. M.Sc. Thesis, Institute of Agriculture and Animal Science, Rampur, Tribhuvan University, Chitwan, Nepal, 2011.

Dobermann, A. A Critical Assessment of the System of Rice Intensification (SRI). *Agric. Syst.* **2004,** *79,* 261–281.

Durga, A. R.; Kumar, S. D. More Crop per Drop of Water: Adoption and Dis-adoption Dynamics of System of Rice Intensification. *IIM Kozhikode Soc. Manag. Rev.* **2016,** *5* (1) 74–82.

Dwibedi, S. K.; De, G. C.; Dhua, S. R. Weed Dynamics and Grain Yield as Influenced by Sowing Time and System of Cultivation of Rice Genotypes. *J. Crop Weed* **2016,** *12* (1), 107–111.

Gathorne-Hardy, A.; Reddy, D. N.; Venkatanarayana, M.; Harriss-White, B. A Life Cycle Assessment (LCA) of Greenhouse Gas Emissions from SRI and Flooded Rice Production in SE India. *Taiwan Water Conserv.* **2013**, *61* (4), 110–125.

Ghosh, B.; Chakma. N. Impacts of Rice Intensification System on Two C. D. Blocks of Barddhaman District, West Bengal. *Curr. Sci.* **2015**, *109* (2), 342–346.

Glover, D. The System of Rice Intensification: Time for an Empirical Turn, *NJAS–Wageningen J. Life Sci.* **2011**, *57* (3–4), 217–224.

Gopalakrishnan, S.; Mahender, K, P.; Humayun, P.; Srinivas, V.; Ratna, K. B.; Vijayabharathi, R.; Singh, A.; Surekha, K.; Padmavathi, Ch. Somashekar, N.; Rao, P. R.; Latha, P. C.; Subba Rao, L. V.; Babu, V. R.; Viraktamath, B. C.; Goud, V. V.; Loganandhan, N.; Gujja, B.; Rupela, O. M. Assessment of Different Methods of Rice (*Oryza sativa.* L.) Cultivation Affecting Growth Parameters, Soil Chemical, Biological, and Microbiological Properties, Water Saving, and Grain Yield in Rice–Rice System. *Paddy Water Environ.* **2014**, *12* (1), 79–87.

Haden, V. R.; Duxbury, J. M.; Tommaso, A. D.; Losey, J. E. Weed Community Dynamics in the System of Rice Intensification (SRI) and the Efficacy of Mechanical Cultivation and Competitive Rice Cultivars for Weed Control in Indonesia. *J. Sustain. Agric.* **2007**, *30* (4), 5–26.

Hassan, D.; Upasani, R. R. Effect of Crop Establishment and Weed Control Methods on Productivity of Rice (*Oryza sativa* L.). *J. Crop Weed* **2015**, *11*, 228–230.

Hazra, K. K.; Chandra, S. Mild to Prolonged Stress Increased Rice Tillering and Source-to-sink Nutrient Translocation Under SRI Management. *Paddy Water Environ.* **2014**, *12* (1), 245–250.

Hidayati, N.; Triadiati and Anas, I. Photosynthesis and Transpiration Rates of Rice Cultivated Under the System of Rice Intensification and the Effects on Growth and Yield. *HAYATI J. Biosci.* **2016**, *23*, 67–72.

Horie, T.; Shiraiwa, K.; Homma, K.; Katsura, Y.; Maeda and Yoshida, H. Can Yields of Low Land Resume the Increases that They Showed in the 1980s? *Plant Prod. Sci.* **2005**, *8*, 251–272.

Hoshikawa, K.; Sasaki, R.; Hasebe, K. Development and Rooting Capacity of Rice Nursery Seedlings Grown Under Different Raising Conditions. *Jpn. J. Crop Sci.* **1995**, *64*, 328–332.

Husband, R.; Herre, E. A.; Turner, S. L.; Gallery, R.; Young, J. P. W. Molecular Diversity of Arbuscular Mycorrhizal Fungi and Patterns of Host Association Over Time and Space in a Tropical Forest. *Mol. Ecol.* **2002**, *11*, 2669–2678.

Hussain, A.; Bhat, M. A.; Ganai, M. A.; Hussain, T. Comparative Performance of System of Rice Intensification (SRI) and Conventional Methods of Rice Cultivation Under Kashmir Valley Conditions. *Environ. Ecol.* **2009**, *27* (1A), 399–402.

Hussain, A.; Bhat, M. A.; Ganie, M. A. Effect of Number and Age of Seedlings on Growth, Yield, Nutrient Uptake and Economics of Rice (*Oryza sativa* L.) Under System of Rice Intensification in Temperate Conditions. *Indian J. Agron.* **2012**, *57* (2), 33–37.

Ishani, P.; Hazarika J. P.; Nivedita, D. A Comparative Economic Analysis of SRI and Traditional Methods of Rice Cultivation in Assam. *Agric. Econom. Res. Rev.* **2013**, *26*, 236–236.

Islam, M.; Kalita, D. C. Effect of Rice (*Oryza sativa* L.) Establishment Methods and Integrated Weed Management on Productivity and Soil Fertility in Eastern Himalayas, India. *J. Agric. Nat. Resour. Sci.* **2014**, *1* (2), 87–103.

Islam, M. M.; Kalita, D. C. Weed Dynamics and Productivity of Wetland Rice as Influenced by Establishment Methods and Integrated Weed Management. *Bangl. J. Bot.* **2016,** *45* (1), 9–16.

Islam, M.; Nath, L. K.; Pate, D. P.; Das, A.; Munda, G. C.; Samajdar, T.; Ngachan, S. V. Productivity and Socio-economic Impact of System of Rice Intensification and Integrated Crop Management over Conventional Methods of Rice Establishment in Eastern Himalayas, India. *Paddy Water Environ.* **2014,** *12* (1), 193–202.

Iswandi, A .; Barison, J.; Kassam, A.; Mishra, A.; Rupela, O. P.; Thakur, A. K.; Thiyagarajan, T. M.; Uphoff, N. The System of Rice Intensification (SRI) as a Beneficial Human Intervention into Root and Soil Interaction. *J. Tanah Lingk* **2011,** *13* (2), 72–88.

Jagannath, P.; Pullabhotla, H.; Uphoff, N. Meta-analysis Evaluating Water Use, Water Saving, and Water Productivity in Irrigated Production of Rice with SRI vs. Standard Management Methods. *Taiwan Water Conserv.* **2013,** *61* (4), 14–49.

Jain, N.; Dubey, R.; Dubey, D. S.; Singh, J.; Khanna M.; Pathak. H.; Bhatia, A. Mitigation of Greenhouse Gas Emission with System of Rice Intensification in the Indo-Gangetic Plains. *Paddy Water Environ.* **2014,** *12* (3), 355–363.

Jeyapandiyan, N.; Lakshmanan, A. Yield Comparison of Rice in Different Cultivation Systems. *Trends Biosci.* **2014,** *7,* 1635–1637.

Johnson, B.; Vijayaragavan, K. Diffusion of System of Rice Intensification (SRI) Across Tamil Nadu and Andhra Pradesh in India. *Indian Res. J. Ext. Educ.* **2011,** *11* (3), 72–79.

Jung, C. G.; Park, J. Y.; Kim, S. J.; Park, G. A. The SRI (System of Rice Intensification) Water Management Evaluation by SWAPP (SWAT–APEX Program) Modeling in an Agricultural Watershed of South Korea. *Paddy Water Environ.* **2013,** *12* (1), 251–261.

Kahimba, F. C.; Kombe, E. E.; Mahoo, H. F. The Potential of System of Rice Intensification (SRI) to Increase Rice Water Productivity: A Case of Mkindo Irrigation Scheme in Morogoro Region, Tanzania. *Tanzania J. Agric. Sci.* **2014,** *12* (2), 10–19.

Kamili, A. S.; Shafiq A. Wani.; Singh K. N.; Khan N. A.; Wani, R. A.; Ram, D., Khan, I. M.; Kirmani.; N. A.; Dar, Z. A.; Gul. A., Eds. *Cereal Crops (Kharif and Rabi) Package of Practices Manual,* Directorate of Extension, SKUAST-Kashmir, 2011; Vol. III p 2.

Kar, S.; Varade, S. B.; Subramanyam, T. K.; Ghildyal. B. P. Nature and Growth Pattern of Rice Root System Under Submerged and Unsaturated Conditions. *RISO (Italy)* **1974,** *23,* 173–179.

Karhale, S. S.; Lambe, S. P.; Neharkar, P. S. Mechanical Weed Control by Conoweeder in SRI Method of Paddy Cultivation, *Int. J. Adv. Res. Sci. Eng.* **2015,** *4* (2), 744–752.

Karki, K. B. *Productivity and Economic Viability of Rice Under Different Planting Pattern and Age of Seedlings Through System of Rice Intensification (SRI)*; Institute of Agriculture and Animal Science, Tribhuvan University: Chitwan, Nepal, 2009.

Katayama, T. *Inemugi no bungetsukenkyu (Studies on Tillering in Rice, Wheat and Barley)*; Yokendo Publishing: Tokyo, Japan, 1951.

Khush, G. S. Breeding Rice for Sustainable Agricultural Systems. *Int. Crop Sci.* **1993,** I, 189–199.

Krupnik, T. J.; Rodenburg, J.; Haden, V. R.; Mbaye, D.; Shennan, C. Genotypic Trade-offs Between Water Productivity and Weed Competition Under the System of Rice Intensification in the Sahel. *Agric. Water Manag.* **2012,** *115,* 156–166.

Kumar, R. M. System of Rice Intensification (SRI)-present Status and Future Prospects. http://www. rkmp.co.in (accessed June 6, 2016).

Kumar, D. D.; Singh, A. P. Effect of Integrated Weed Management on Weed Flora, Distribution, Weed Dynamics and Performance of Rice (*Oryza sativa* L.) Under System of Rice Intensification (SRI) in Chhattisgarh. *J. Progress. Agric.* **2013**, *4* (1), 96–101.

Kumar, M. R.; Rao R. P.; Somasekhar, N.; Surekha, K.; Padmavathi, C. H.; Prasad, M. S.; Ravindra-Babu, V.; Subba-Rao, L. V.; Latha, P. C.; Sreedevi, B.; Ravichandran, S.; Ramprasad, A. S.; Muthuraman, P.; Gopalakrishnan, S.; Vinod-Goud, V.; Viraktamath, B. C. SRI: A Method for Sustainable Intensification of Rice Production with Enhanced Water Productivity. *Agrotechnology* **2013a**, *11*, 1–6.

Kumar, M. R.; Subba Rao, L. V.; Babu, V. R.; Gopalakrishnan, S.; Surekha, K.; Padmavathi, Ch.; Somashekar, N.; Raghuveer-Rao, P.; Sreenivas, P. B.; Latha, P. C.; Nirmala, B.; Muthuraman, P.; Ravichandran, S.; Vinod, G. V.; Viraktamath, B. C. *System of Rice Intensification: Its Present Status, Future Prospects and Role in Seed Production in India. Mukhapatra-Annu. Tech. Issue* **2013b**, *17*, 22–43.

Kumar, R. M.; Surekha, K.; Padmavathi, C.; Subba Rao, L. V.; Latha, P. C.; Prasad, M. S.; Babu, R. V.; Ramprasad, A. S.; Rupela, O. P.; Goud, V.; Raman, P. M.; Somashekar, N.; Ravichandran, S., Singh, S. P.; Viraktamath, B. C. Research Experiences on System of Rice Intensification and Future Directions. *J. Rice Res.* **2013c**, *2* (2), 61–71.

Kumar, G. A.; Sailaja, V.; Satyagopal, P. V.; Prasad, S. V. Evaluation of Profile Characteristics of SRI Cultivation Farmers in Relation to Their Extent of Adoption of Technologies. *Curr. Biotica* **2014**, *8* (1), 36–41.

Latif, M. A.; Islam, M. R.; Ali, M. Y.; Saleque, M. A. Validation of the System of Rice Intensification (SRI) in Bangladesh. *Field Crops Res.* **2005**, *93* (2–3), 281–292.

Laulanié, H. Le Système de Riziculture Intensive Malgache. *Tropicultura (Brussels)* **1993**, *11*, 110–114.

Laulanié, H. D. Intensive Rice Farming in Madagascar. *Tropicultura* **2011**, *29* (3), 183–187.

Lin, X.; Zhu, D.; Lin, X. Effects of Water Management and Organic Fertilization with SRI Crop Practices on Hybrid Rice Performance and Rhizosphere Dynamics. *Paddy and Water Environ.* **2011**, *9* (1), 33–39.

Lu, S. H.; Dong, Y. J.; Yuan, J.; Lee, H.; Padilla, H. A. High-yielding, Water-saving Innovation Combining SRI with Plastic Cover on No-till Raised Beds in Sichuan, China. *Taiwan Water Conserv.* **2013**, *61* (4), 94–109.

Mahesh, P.; Shakywar, R. C.; Dinesh; Shyam, S. S. Prevalence of Insect Pests, Natural Enemies and Diseases in SRI (System of Rice Intensification) of Rice Cultivation in North East Region. *Ann. Plant Protect. Sci.* **2012**, 20 (2), 375–379.

McDonald, A. J.; Hobbs, P. R.; Riha, S. J. Does the System of Rice Intensification Out Perform Conventional Best Management? A Synopsis of the Empirical Record. *Field Crops Res.* **2006**, *96*, 31–36.

Meena, M.; Patel, M. V.; Das, T.; Verma, H. P. Effect of Organic Sources and Nitrogen Levels on Growth and Yield of *Kharif* Rice (*Oryza sativa.*) Under SRI Technique. *Agric. Sustain. Dev.* **2014**, *2* (1), 39–42.

Menete, M. Z. L.; Van, H. M.; Brito, R. M. L.; DeGloria, S. D.; Famba, S. Evaluation of System of Rice Intensification (SRI) Component Practices and Their Synergies on Salt-affected Soils. *Field Crops Res.* **2008**, *109*, 34–44.

Mgaya, A. M.; Thobunluepop, P.; Sreewongchai, T.; Sarobol, E.; Onwimol, D. Integral Effect of Seed Treatments and Production Systems for Sustainability of Rice Production Under Acid Soil. *J. Agron.* **2016,** *15* (3), 122–129.

Mishra, A. Intermittent Irrigation Enhances Morphological and Physiological Efficiency of Rice Plants. *Agriculture (Pol'nohospodárstvo)* **2012,** *58* (4), 121–130.

Mishra, A. M.; Whitten, J.; Ketelaar, W.; Salokhe. V. M. The System of Rice Intensification (SRI): A Challenge for Science, and an Opportunity for Farmer Empowerment Towards Sustainable Agriculture. *Int. J. Agric. Syst.* **2006,** *4,* 193–212.

Mohammad, U.; Wayayoka, A.; Sooma, M. A. M.; Abdan, K. Performance of Umar-Srimat on Soil Water Conservation and Weed Control in System of Rice Intensification. *J. Teknol.* **2015,** 76, 83–88.

Mohammed, U.; Aimrun, W. Amin, M. S.; M. Khalina, A.; Zubairu, U. B. Influence of Soil Cover on Moisture Content and Weed Suppression Under System of Rice Intensification (SRI). *Paddy Water Environ.* **2016,** *14,* 159–167.

Mohanty, T. R.; Maity, S. K.; Roul, P. K. Response of Rice to Establishment Methods and Nutrient Management Practices in Medium Land. *ORYZA Int. J. Rice* **2015,** *51* (2), 136–142.

Namara, R. E.; Weligamage, P.; Barker, R. *Prospects for Adopting System of Rice Intensification in Sri Lanka: A Socioeconomic Assessment.* Research Report 75; International Water Management Institute Colombo: Sri Lanka, 2003.

Narbaria, S.; Sharma, M. L.; Khan, M. A.; Dhruv, Y.; Painkra, V. K. System of Rice Intensification Technology: An Analysis of Constraints Perceived by the Farmers in Adoption of SRI Technology. *Plant Arch.* **2015,** *15* (1), 159–162.

Ndiiri, J. A.; Mati, B. M.; Home, P. G.; Odongo, B.; Uphoff, N. Adoption, Constraints and Economic Returns of Paddy Rice Under the System of Rice Intensification in Mwea, Kenya. *Agric. Water Manag.* **2013,** *129,* 44–55.

Nemoto, K. S.; Morita, T. B. Shoot and Root Development in Rice Related to the Phyllochron. *Crop Sci.* **1995,** *35,* 24–29.

Norela, S.; Anizan, I.; Ismail, B. S.; Maimon, A. Diversity of Pest and Non-pest Insects in an Organic Paddy Field Cultivated Under the System of Rice Intensification (SRI): A Case Study in Lubok China, Melaka, Malaysia. *J. Food Agric. Environ.* **2013,** *11* (3–4), 2861–2865.

Pandey, S. Effect of Weed Control Methods on Rice Cultivars Under the System of Rice Intensification (SRI). M.Sc. Thesis, Institute of Agriculture and Animal Science, Tribhuvan University, Rampur, Chitwan, Nepal, 2009.

Pandey, A.; Mai, V. T.; Vu, D. Q.; Bui, T. P. L.; Mai, T. L. A.; Jensen, L. S.; Neergaard, A. Organic Matter and Water Management Strategies to Reduce Methane and Nitrous Oxide Emissions from Rice Paddies in Vietnam.*Agr. Ecosyst. Environ.* **2015,** *196,* 15,137–146..

Pandian, B. J.; Sampathkumar, T.; Chandrasekaran, R. System of Rice Intensification (SRI): Packages of Technologies Sustaining the Production and Increased the Rice Yield in Tamil Nadu, India. *Irrig. Drain. Syst. Eng.* **2014,** *3* (1), 1–6.

Pasuquin, E.; Lafarge, T.; Tubana, B. Transplanting Young Seedlings in Irrigated Rice Fields: Early and High Tiller Production Enhanced Grain Yield. *Field Crops Res.* 105, of Rice

(*Oryza sativa* L.) Under System of Rice Intensification. *ARPN J. Agric. Biol. Sci.* **2008**, *6* (11), 33–35.

Patra, P. S.; Haque, S. Effect of Seedling Age on Tillering Pattern and Yield of Rice (*Oryza sativa* L.) Under System of Rice Intensification. *ARPN J. Agric. Biol. Sci.* **2011**, *6* (11), 33–35.

Prasad R. *Rice Agronomy*; Jain Brothers: New Delhi, 1999; p 131.

Prasanna, R.; Adak, A. Verma, S., Bidyarani, N., Babu, S., Pal, M., Shivay, Y. S. and Nain, L. Cyanobacterial Inoculation in Rice Grown Under Flooded and SRI Modes of Cultivation Elicits Differential Effects on Plant Growth and Nutrient Dynamics. *Ecol. Eng.* **2015**, *84*, 532–541.

Rajkishore, S. K.; Doraisamy, P.; Subramanian K. S.; Maheswari, M. Methane Emission Patterns and Their Associated Soil Microflora with SRI and Conventional Systems of Rice Cultivation in Tamil Nadu, India. *Taiwan Water Conserv.* **2013**, *61* (4), 126–134.

Ram, H.; Singh, J. P.; Bohra, J. S.; Yadav, A. S.; Sutaliya, J. M. Assessment of Productivity, Profitability and Quality of Rice (*Oryza sativa*) Under System of Rice Intensification in Eastern Uttar Pradesh. *Indian J. Agric. Sci.* **2015**, *85* (1), 38–42.

Ramakrishna, A.; Tam, H. M.; Wani, S. P.; Long, T. D. Effect of Mulch on Soil Temperature, Moisture, Weed Infestation and Yield of Groundnut in Northern Vietnam. *Field Crops Res.* **2006**, *95* (2), 115–125.

Ramamoorthy, K.; Selvarao K. V.; Chinnaswami. K. N. Varietal Response of Rice to Different Water Regimes. *Indian J. Agron.* **1993**, *38*, 468–469.

Randriamiharisoa, R. In *Research Results on Biological Nitrogen Fixation with the System of Rice Intensification. Assessments of the System for rice intensification (SRI)*; Proceedings of an International Conference, Sanya, China, Cornell International Institute for Food, Agriculture and Development: New York, 2002.

Rao, A. N.; Johnson, D. E.; Sivaprasad, B.; Ladha, J. K.; Mortimer, A. M. Weed Management in Direct Seeded Rice. *Adv. Agron.* **2007**, *93*, 153–255.

Rao, A. U.; Ramana A. V.; Sridhar T. V. Performance of System of Rice Intensification (SRI) in Godavari Delta of Andhra Pradesh. *Ann. Agric. Res.* **2013**, *34* (2), 118–121.

Ravichandran, V. K.; Prakash, K. C. Socio-economic Impact of System of Rice Intensification (SRI) and Traditional Rice Cultivation in Villupuram District of Tamil Nadu: Experiences from TN-IAMWARM Project. *Int. J. Agric. Sci.* **2015**, *11* (1), 166–171.

Rodenburg, J.; Saito, K.; Irakiza, R.; Derek,W.; Enos A.; Onyuka, M.; Senthilkumar, K. Labor-saving Weed Technologies for Lowland Rice Farmers in Sub-Saharan Africa. *Weed Technol.* **2015**, *29*, 751–757.

Ros, C.; White, P. F.; Bell, R. W. Field Survey on Nursery and Main Field Fertilizer Management. *Cambodian J. Agric.* **1998**, *1*, 22–33.

Saha, A.; Bharti, V.; Effect of Different Crop Establishment Methods on Growth, Yield and Economics of Rice. *Environ. Ecol.* **2009**, *28*, 23–29.

Saha, S.; Singh, Y. V.; Gaind, S.; Kumar, D. Water Productivity and Nutrient Status of Rice Soil in Response to Cultivation Techniques and Nitrogen Fertilization. *Paddy Water Environ.* **2015**, *13* (4), 443–453.

Sahid, I. B.; Hossain. M. S. The Effect of Flooding and Sowing Depth on the Survival and Growth of Five Rice-weeds Species. *Plant Prot. Q.* **1995**, *10*, 139–142.

Sahrawat, K. L. Elemental Composition of the Rice Plant as Affected by Iron Toxicity Under Field Conditions. *Commun. Soil Sci. Plant Anal.* **2000**, *132*, 2819–2827.

Sakai, H.; T. Yoshida. Studies on Conditions Inducing Murenae Incidence of Rice Plants.1α-Naphtylamine Oxidizing Power of Roots. *Bull. Hokkaido Nat. Agric. Exp. Stn.* **1957**, *72*, 82–91.

San-oh, Y.; Sugiyama, T.; Yoshhita, D.; Ookawa, T.; Hirasawa, T. The Effect of Planting Pattern on the Rate of Photosynthesis and Related Process During Ripening in Rice Plants. *Field Crops Res.* **2006**, *96* (1), 113–124.

Sato, S.; Uphoff, N. A Review of On-farm Evaluations of System of Rice Intensification Methods in Eastern Indonesia Reviews: Perspectives in Agriculture, Veterinary Science. *Nutr. Nat. Resour.* **2007**, *2* (54), 1–12.

Shahane, A. A.; Singh, Y. V.; Kumar, D.; Prasanna, R.; Chakraborty, D. Effect of Planting Methods and Cyanobacterial Inoculants on Yield, Water Productivity and Economics of Rice Cultivation. *J. Agric. Rural Develop. Trop. Subtrop.* **2015**, *116* (2), 107–121.

Sheehy, J. E.; Peng, S.; Dobermann, A., Mitchell, P. L.; Ferrer, A.; Yang, J.; Zou, Y.; Zhong, X.; Huang, J. Fantastic Yields in the System of Rice Intensification: Fact or Fallacy? *Field Crops Res.* **2004**, *88*, 1–8.

Sinclair, T. R.; Cassman, K. G. Agronomic UFOs? *Field Crops Res.* **2004**, *88*, 9–10.

Singh, Y. V. Crop and Water Productivity as Influenced by Rice Cultivation Methods Under Organic and Inorganic Sources of Nutrient Supply. *Paddy Water Environ.* **2013**, *11* (1–4), 321–329.

Singh, N.; Kumar, D.; Thenua, O. V. S.; Tyagi, V. K. Influence of Spacing and Weed Management on Rice (*Oryza sativa*) Varieties Under System of Rice Intensification. *Indian J. Agron.* **2012**, *57* (2), 138–142.

Singh, Y. V.; Singh, K. K.; Sharma S. K. Influence of Crop Nutrition on Grain Yield, Seed Quality and Water Productivity Under Two Rice Cultivation Systems. *Rice Sci.* **2013**, *20* (2), 129–138.

Singh, K.; Singh, S. R.; Singh, J. K.; Rathore, R. S.; Singh, S. P.; Roy, R. Effect of Age of Seedling and Spacing on Yield, Economics, Soil Health and Digestibility of Rice (*Oryza sativa*) Genotypes Under System of Rice Intensification. *Indian J. Agric. Sci.* **2013**, *83* (5), 479–83.

Sinha, S. K.; Talati. J. Productivity Impacts of the System of Rice Intensification (SRI): A Case Study in West Bengal, India. *Agric. Water Manag.* **2007**, *87* (1), 55–60.

Sridevi V.; Chellamuthu, V. Advantages of SRI Cultivation in the Tail End of Cauvery Delta. The System of Crop Intensification: Reports from the Field on Improving Agricultural Production, Food Security, and Resilience to Climate Change for Multiple Crops. *J. Crop Weed* **2012**, *8* (2), 40–44.

Sridevi, V; Chellamuthu, V. Nutrient Harvest Index and Soil Available Nutrients of Rice (*Sativa*) as Influenced by System of Rice Intensification (SRI) Practices. *Int. J. Sci. Res.* **2015**, *4* (6), 2277–8179.

Steponkus, P. L.; Culture J. C.; Toole. J. C. O. Adaptation to Water Deficit in Rice. In *Adaptation of Plants to Water and Temperature Stress*, Turner N. C., Kramer, Eds.; Wiley Interscience: NY, USA, 1980; pp 401–418.

Stoop, W. A. The System of Rice Intensification (SRI) from Madagascar Myth or Missed Opportunity? Report on a Study Visit to the "HautsPlateaux" Region of Madagascar.

R and D for Tropical Agriculture. Akkerweg 13A 3972 AA, Driebergen–R., The Netherlands, 2003; pp 1–16.

Stoop, W. A. The System of Rice Intensification (SRI): Results from Exploratory Field Research in Ivory Coast-research Needs and Prospects for Adaptation to Diverse Production Systems of Resource-poor Farmers. 2005. http: //ciffad.cornell.edu/sri/ (accessed July 05, 2013).

Stoop, W. A.; Uphoff, N.; Kassam. A. A Review of Agricultural Research Issue Raised by the System of Rice Intensification (SRI) from Madagascar: Opportunities for Improving System for Resource Poor Farmers. *Agric. Syst.* **2002,** *71,* 249–274.

Surekha, K.; Kumar, M. R.; Padmavathi, Ch. Evaluation of Crop Establishment Methods for Their Productivity, Nutrient Uptake and Use Efficiency Under Rice-rice System. *J. Rice Res.* **2015,** *8* (1), 41–47.

Suryavanshi, P.; Singh, Y. V.; Prasanna, R. Bhatia, A.; Shivay, Y. S. Pattern of Methane Emission and Water Productivity Under Different Methods of Rice Crop Establishment. *Paddy Water Environ.* **2013,** *1,* 321–329.

Swaminathan, M. S.; Kesavan. P. C. Agricultural Research in an Era of Climate Change. *Agric. Res.* **2012,** *1* (1), 3–11.

Takahashi, K. The Roles of Risk and Ambiguity in the Adoption of the System of Rice Intensification (SRI): Evidence from Indonesia. *Food Secur.* **2013,** *5,* 513–524.

Takahashi, K.; Barrett C. B. The System of Rice Intensification and Its Impacts on Household Income and Child Schooling: Evidence from Rural Indonesia. *Paddy Water Environ.* **2013,** *11* (1), 241–248.

Takahashi, K.; Barrett. C. B. The System of Rice Intensification and its Impacts on Household Income and Child Schooling: Evidence from Rural Indonesia. *Am. J. Agric. Econ.* **2014,** *96* (1), 269–289.

Tanaka, A. Studies on the Physiological Function of the Leaf at a Definite Position on a Stem of the Rice Plant. *J. Sci. Soil Manure* **1958,** *29,* 291–294.

Thakur, A. K.; Uphoff, N.; Antonyan, E. Assessment of Physiological Effects of System of Rice Intensification (SRI) Practices Compared with Recommended Rice Cultivation Practices in India. *Exp. Agric.* **2010,** *46* (1), 77–98.

Thakur, A. K.; Mohanty, R. K.; Patil, D. U.; Kumar, A. Impact of Water Management on Yield and Water Productivity with System of Rice Intensification (SRI) and Conventional Transplanting System in Rice. *Paddy Water Environ.* **2013a,** *11* (1–4), 297–308.

Thakur, A. K.; Rath, S.; Mandal, K. G. Differential Responses of System of Rice Intensification (SRI) and Conventional Flooded-rice Management Methods to Applications of Nitrogen Fertilizer. *Plant Soil* **2013b,** *370* (1), 59–71.

Thuraa, S. Evaluation of Weed Management Practices in the System of Rice Intensification (SRI). M.Sc. Thesis (Agronomy), Department of Agronomy, Yezin Agricultural University, Myanmar, 2010.

Tsujimoto, Y.; Horie, T.; Randriamihary, H.; Shiraiwa, T.; Homma. K. Soil K Management: The Key Factors for Higher Productivity in the Fields Utilizing the System of Rice Intensification (SRI) in the Central Highland of Madagascar. *Agric. Syst.* **2009,** *100,* 61–71.

Uphoff, N. Higher Yields with Fewer External Inputs? The System of Rice Intensification and Potential Contributions to Agricultural Sustainability. *Int. J. Agric. Sustain.* **2003,** *1* (1), 38–50.

Uphoff, N. Comment to 'the System of Rice Intensification: Time for an Empirical Turn.' *NJAS-Wageningen J. Life Sci.* **2011**, *57*, 217–224.

Uphoff, N.; Fernandes, E. System of Rice Intensification Gain Momentum. *LEISA* **2002**, *18*, 24–29.

Uphoff, N.; Fasoula, V.; Iswandi, A.; Kassam, A.; Thakur, A. K. Improving the Phenotypic Expression of Rice Genotypes: Rethinking "Intensification" for Production Systems and Selection Practices for Rice Breeding. *Crop J.* **2015**, *3*, 174–189.

Veeraputhiran, R.; Balasubramanian, R.; Pandian, B. J. Effect of Mechanical Weeding in System of Rice Intensification and Its Adoption. *Indian J. Weed Sci.* **2014**, *46* (4), 383–385.

Verma, D. K.; Srivastav, P. P. Proximate Composition, Mineral Content and Fatty Acids Analyses of Aromatic and Non-Aromatic Indian Rice. *Rice Sci.* **2017**, *24* (1), 21–31.

Verma, D. K.; Mohan, M.; Yadav, V. K.; Asthir, B.; Soni, S. K. Inquisition of Some Physico-chemical Characteristics of Newly Evolved Basmati Rice. *Environ. Ecol.* **2012**, *30* (1), 114–117.

Verma, A. K.; Pandey, N.; Shrivastava, G. K. Production Potential and Economics of Hybrid Rice Under System of rice Intensification and its Manipulation. *SAARC J. Agric.* **2014**, *12* (2), 71–78.

Visalakshmi, V.; Rao, P. R. M.; Satyanarayana, N. H. Impact of Paddy Cultivation Systems on Insect Pest Incidence. *J. Crop Weed* **2014**, *10* (1), 139–142.

Watanarojanaporn, N.; Boonkerd, N.; Tittabutr, P.; Longtonglang, A.; Young, P. W. J.; Teaumroong, N. Effect of Rice Cultivation Systems on Indigenous Arbuscular Mycorrhizal Fungal Community Structure. *Microbes Environ.* **2013**, *28* (3), 316–324.

Wayayoka, A.; Amin, M.; Sooma, M.; Abdana, K.; Mohammed, U. Impact of Mulch on Weed Infestation in System of Rice Intensification (SRI) Farming. *Agric. Agric. Sci. Procedia* **2014**, *2*, 353–360.

Wu, W.; Ma, B.; Uphoff, N. A Review of the System of Rice Intensification in China. *Plant Soil* **2015**, *393* (1), 361–381.

Yang, C.; Yang, L.; Yang, Y.; Ouyang, Z. Rice Root Growth and Nutrient Uptake as Influenced by Organic Manure in Continuously and Alternately Flooded Paddy Soils. *Agric. Water Manag.* **2004**, *70*, 67–81.

PART 2

Nutrient Management for Rice Production and Quality Improvement

CHAPTER 5

EFFECT OF ZINC (Zn) ON GROWTH, YIELD, AND QUALITY ATTRIBUTES OF RICE (*Oryza sativa* L.) FOR IMPROVED RICE PRODUCTION

V. K. VERMA[1]*, DEEPAK KUMAR VERMA[2], and RAM KUMAR SINGH[1]

[1]Department of Agronomy, Institute of Agricultural Sciences, Banaras Hindu University, Varanasi 221005, Uttar Pradesh, India

[2]Department of Agricultural and Food Engineering, Indian Institute of Technology, Kharagpur 721302, West Bengal, India

*Corresponding author. E-mail: vermaagribhu@gmail.com

ABSTRACT

In an intensive rice production system, the climatic changes, nutritional disorders, moisture stress, insect-pest infestation, etc. are the major challenges which produce the underdeveloped rice with low yield. Further, the excessive application of NPK fertilizers has reduced the other nutrients significantly resulting in their deficiencies. Zinc is the most important micronutrient for the growth and yield of rice, and its deficiency in the soil leads to poor quality rice. The application of zinc in soil is not readily transported to the shoot and likewise, its foliar use is inefficiently transported to roots. Therefore, zinc when applied either in soils or on foliage can enhance the productivity only under well-defined agroecosystem. Also, zinc fertilization is the most economic way to alleviate its deficiency in rice and improves the productivity and grain Zn concentration, thus contributing to the higher nutritional value of rice. Keeping this in view, this chapter highlights the several rice attributes, such as growth, yield, and quality indices, being affected by zinc to improve the rice production and sort out the Zn deficiency problems.

5.1   INTRODUCTION

Rice (*Oryza sativa*, L.) is one of the most important cereal crops of India among all which feed as staple food by more than 70% of the populations living in the Asian continent where more than 90% of rice is produced and consumed (Verma et al., 2012, 2013, 2015). It is extensively grown in tropical and subtropical regions of the world (Verma, 2011). It is grown in 112 countries in the world, covering every continent and is consumed by 2500 million people in developing countries (Datta and Khush, 2002). Rice is globally cultivated under varied agroclimatic conditions. Consequently it is faced with limitations of various kinds. However, the rice is subjected to face more challenges as it is grown in a more risky environment, notably, nutritional disorders, moisture stress, infestation by insect-pest and diseases along with weeds, etc., which are responsible for its poor growth, development, and consequently low yield. Exploitation of the genetic yield potential of any crop depends entirely on appropriate crop management practices, particularly better crop establishment, adequate supply of water, and mineral nutrition along with effective plant protection measures. In recent years, intensive cultivation of high-yielding varieties of rice with excessive use of macronutrients (NPK) fertilizers have accelerated the depletion of certain secondary micronutrients from the soil leading to emergence of their deficiencies in many parts of country.

Rice is already discussed in above paragraph as most important crop in India and may also the hub of food security of the global population (Verma, 2011; Verma et al., 2012, 2013, 2015). Among different production factors, planting methods produce significant effects not only on paddy yield and grain quality of rice but also on soil health and productivity. To meet the challenges of food security for rapidly increasing population, intensive cropping pattern were adopted which resulted in declining nutrient status of soil and therefore the use of chemical fertilizer are increasing day by day. Micronutrient deficiencies affecting crop yields are difficult to be corrected due to large temporal and special variation in availability, quick conversion and fixation into unavailable form and poor movement of applied micronutrients in soil profile. Soil application of micronutrients is not readily transported to the shoot; similarly foliar application of micronutrients is also inefficiently transported to roots. Therefore, application of micronutrients applied either in soils or on foliage may be successful in enhancing grain yield only under well-defined agroecosystem. Hence, extrapolating results to a new situation is unlikely to be successful without a specific adjustment in the fertilizer schedule. Keeping pace with the above-discussed facts in view, the present chapter focused on various attributes of rice, namely, growth, yield, and quality parameters, how they are affected by Zn which will help to improve rice production and sort out the Zn-deficiency problems.

5.2   IMPORTANCE OF ZINC (Zn) AS MICRONUTRIENTS

Zinc (Zn) is one of the 17 essential elements necessary for the normal growth and development of all humans, animals (Broadley et al., 2007), and plants (Prasad 2008). In one side, Zn is vital for the proper functioning of the immune system and crucial for healthy growth, physical and mental development of children while another side, considered as one among eight micronutrients which have key role in plants involved with enzymes and proteins in carbohydrate metabolism, protein synthesis, gene expression, auxin (growth regulator) metabolism, pollen formation, maintenance of biological membranes, protection against photooxidative damage and heat stress, and resistance to infection by certain pathogens (Alloway, 2008). In plant's biochemical and physiological processes, Zn plays multiple roles even slight deficiency of Zn may causes a decrease in growth, yield, and Zn content of edible plant parts.

Zn content in food is very important for human health as the artificial supplementation of food with essential minerals is often difficult to achieve, particularly in developing countries. Therefore, it has been suggested that increased levels of Zn in staple food, for example, rice may play a role in reducing Zn deficiency (Graham et al., 1990; Ruel and Bouis, 1998; Welch and Graham, 1999).

Zn deficiency is a serious agricultural problem around one-half of the cereal-growing soils in the world which contain low Zn in the soil (Graham and Welch, 1996); Cakmak et al., 1999). The solubility and availability of Zn is determined by various factors like high $CaCO_3$, high pH, high clay soils, low organic matter, low soil moisture, and high iron (Fe) and aluminum oxides (Cakmak, 2008). Zn deficiency causes decrease in plant growth and yield. In addition, Zn is critical to the control of gene transcription and the coordination of other biological processes regulated by proteins containing DNA-binding Zn-finger motifs (Rhodes and Klug, 1993).

In India, 26% population is suffering from Zn deficiency. Children are the most affected part of this and about 54% children suffering from Zn deficiency (Sinha, 2004). Effective Zn nutrition in a person require 15 mg Zn/day (https://ods.od.nih.gov/factsheets/Zinc-HealthProfessional/) but our food grains contain only 15–35 mg Zn/kg and out of which only 13–35%

are bioavailable (Cakmak et al., 2004). So there is a big gap between daily requirement—daily intakes, and to fulfill the gap, our food grains should contain 40–60 mg Zn/kg (Pfeiffer and McClafferty, 2007). Zn deficiency leads to the hidden hunger or malnutrition. Experts estimate that 2 billion people, mostly in poorer countries are suffering from the micronutrient malnutrition (Verma, 2015).

Zn is now recognized as the fifth leading risk factor in developing Asian countries (IFA, 2004) and efforts are underway for encouraging Zn fertilization not only from the view point of getting higher yields and quality produce, but also for increasing Zn concentration in grain and straw to improve human and animal Zn nutrition (Singh et al., 2016).

5.3   IMPROVING SUPPLY AND ACQUISITION OF ZINC (Zn)

Zn-deficient soil and Zn deficiency in rice is reported worldwide (Yoshida, 1972; Silanpaa, 1990; Fageria et al., 2002; Norman et al., 2003; Alloway, 2004; Gao et al., 2006; Cakmak, 2009). Zn deficiency in rice was first reported by Nene (Neue et al., 1998) from G. B. Pant University of Agriculture and Technology, India and is characterized by the appearance of dark brown spots on the leaves, which in severe deficiency may coalesce to give dark brown color to the entire plant. This deficiency was given the name *kharia* disease due to the brown color of the extract of *Acacia catechu*. In India, about 49% soils are deficient in Zn (Behera et al., 2009a, 2009b).

Apart from the incorporation of fertilizers, adequate supply of organic matter is known to increase availability of almost all the nutrients including the Zn. High-molecular-weight organic carbon exudates released into the rhizosphere, activates microorganism activities around roots, and may indirectly affect the solubility and availability of micronutrients (Curl and True-love, 1986; Marschner, 1995). Microorganisms present in the rhizosphere are known to produce better plant growth by enhancing nutrient availability.

In addition to specific Zn fertilizers, organic manures such as farmyard manure (FYM), compost, vermicompost, biogas slurry, and crop residues are also known as the good sources of Zn and other micronutrients. Their continue application help to maintain micronutrient deficiencies away or at least to reduce their occurrence. References pertaining to effect of applied organic matter, that is, green manure (GM), chelated materials are well documented.

5.3.1 SOIL APPLICATION OF ZINC (Zn)

In calcareous soil, the soil application of Zn resulted in marked increase in grain yield (Serry et al., 1974). Increased production of number of effective tillers, plant height, length of panicle, number of grains per panicle, test weight and grain, and straw yield with increasing levels of $ZnSO_4$ applied as soil proved superior to its foliar spray (Panda and Nayak, 1974). Zn application together with gypsum showed much effective growth (Takkar and Singh, 1978). Since rice is prone to Zn deficiency under cold weather conditions, soil application of Zn proved superior to foliar spray. Incorporation of 60 kg $ZnSO_4$ ha^{-1} as soil application gave highest grain and straw yield associated with maximum number of ear-bearing tillers per m^2, tallest plant, longest panicle, maximum number of grains $panicle^{-1}$ and maximum weight of 1000 grains, whereas foliar application showed no advantage over soil application of Zn. Krishnaswamy et al. (1994) and Sharma et al. (1982) concluded that soil application proved to be advantageous in building up soil Zn status and left residual effects to the succeeding crops (Prasad and Umar, 1993). Biswas and Dravid (2001) recommended the application of 25 kg $ZnSO_4$ ha^{-1} at puddling under normal soil conditions.

Methods of application of Zn also depend upon the use of Zn fertilizer. Soluble Zn fertilizers such as $ZnSO_4$, incorporated as band placement proved much superior to its broadcast application, whereas relatively less soluble fertilizers such as ZnO and Zn frits also showed better response as broadcast application (Katyal et al., 2004; Prasad, 2008). For low land rice, broadcast application of 1 kg Zn-EDTA at puddling proved effective. The highest mean Zn uptake by rice grain and straw recorded to be 209.2 and 133.8 g ha^{-1}, respectively (Naik and Das, 2008). Among the different Zn fertilizer sources, residual effect of Zn-EDTA was found to be the best with respect to growth, yield attributes, grain, straw, Zn content and uptake, and economics (Singh and Shivay, 2013).

5.3.2 FOLIAR APPLICATION OF ZINC (Zn)

Foliar application of Zn is often used for quick correction of severe Zn deficiencies (Martens and Westernmann, 1991). Foliar application of Zn is usually applied under emergencies to save the crops if symptoms of Zn deficiencies appear. However, single foliar application may not be adequate for correcting moderate to severe Zn deficiency. Foliar application of 0.5% aqueous solution of $ZnSO_4$ twice at 20 and 30 days after transplanting (DAT) proved superior to its soil application of 40 kg $ZnSO_4$ ha^{-1} in sodic soils (Dargan et al., 1982). They further reported marked increase in the number of effective tillers and plant height, with increasing levels of $ZnSO_4$ applied either as soil or foliar. Application of 25 kg $ZnSO_4$ ha^{-1} significantly increased numbers of panicle/m^2, length of panicles, test weight, effective grains $panicle^{-1}$, and grain yield of rice over control, but remained at par with seeding roots dipped in 2% $ZnSO_4$ suspension and spraying of 0.5% $ZnSO_4$ solution at 3 and 5 week after transplanting (Kumar and Singh, 1996). The application of Zn @ 2.53 kg ha^{-1} as foliar spray significantly produced higher grain yield over soil application and control (Singh and Jain, 1964). While evaluating different methods of Zn application, Zn applied at the rate of 25 kg $ZnSO_4$ ha^{-1} on transplanting (Zn_1), 0.5% as foliar spray at 30 DAT (Zn_2), 0.5% as foliar spray at 30 and 45 DAT (Zn_3), and root dipping in 2.5% solution for 24 h before transplanting (Zn_4), it was observed that soil application produced significantly higher grain and straw yields (Yadav et al., 2016).

5.4 IMPORTANT PARAMETERS FOR IMPROVED RICE PRODUCTION AND ZINC (Zn) EFFECT

5.4.1 PLANT GROWTH PARAMETER

All the plant growth parameters are affected by various treatments of Zn at different stages of growth, namely, 30 DAT, 60 DAT, 90 DAT, and at harvest. The work of many researchers revealed marked variation on growth attributes, namely, plant height, tiller number $hill^{-1}$, fresh weight, dry weight, number of leaves $hill^{-1}$, leaf area $hill^{-1}$, leaf area index (LAI), and chlorophyll content are also affected due to different levels of Zn.

5.4.1.1 EFFECT OF ZINC (Zn) ON PLANT GROWTH AND GROWTH ATTRIBUTES OF RICE: CASE STUDY

Zn is the most important trace element which is absorbed by rice in enough quantity to complete their life cycle (Sommer and Lipman, 1926; Arnon and Stout, 1939). The role of Zn has long been recognized to have a stimulating effect on rice causing significant increase the plant growth parameters such as number of productive tillers, dry matter production, LAI, and crop growth rate with every increase in the levels of Zn sulfate (Grewal et al., 1997).

Increase in number of tillers due to micronutrient application can be discussed in light of the fact that more photosynthates produced was made available for initiation of tiller and their growth under these treatments were duly supported by more number of green leaves (Verma and Neue, 1984). Zn nutrition is known to increase tillering in rice, which caused a significant increase in dry matter production (Ghatak et al., 2005). Slaton et al. (2005) also observed an increase in dry matter accumulation of rice by application of Zn. Application of Zn, from Zn-coated urea and other sources is reported to have beneficial effect on LAI (Grewal et al., 1997). Similar findings were also reported by Kulandaivel et al. (2004).

Jat et al. (2011) reported that the highest agronomic efficiency (AE) was obtained with the application of 2% Zn-enriched urea (ZEU), namely, $ZnSO_4.7H_2O$, Physiological efficiency (PE) and zinc harvest index (ZnHI) of applied Zn were also significantly influenced due to dual purpose summer legumes and applied in aromatic hybrid rice. Treatments like FYM (10 t ha^{-1}) + ESR + 33% extra fertilizers (EF) + FSS; FYM (9 t ha^{-1}) + RSR + RF + FSS; and FYM (8 t ha^{-1}) + recommended seed rate (RSR) + adjusted fertilizers + ferrous sulfate spray (FSS) recorded the highest values for leaf area, dry matter accumulation, plant height, number of functional leaves and number of grains per panicle. The three highest grain (1273, 1230, and 1212 kg ha^{-1}) and straw (2049, 1979, and 1939 kg ha^{-1}) yields were also obtained from the above-mentioned treatments (Asewar et al., 2000).

Low Zn content in grain (Cakmak et al., 1999; Graham and Welch, 1996) can be attributed to dilution of Zn in plant tissue because of enhanced dry matter production (Rengel and Graham, 1995). Zn application @ 25–30 kg ha^{-1} was sufficient to correct Zn deficiency for 4–7 years and the residual effect was best with soil application on long range basis. It has also been observed that after a certain limit in micronutrient fertilization, further increase in fertilizer application caused not only reduction in yield but also decreased micronutrient grain density (Nambiar and Motiramani, 1983).

5.4.2 YIELD PARAMETER

The maximum grain and straw yield is due to marked improvement in dry matter accumulation. The grain yield of rice can only be improved to a degree by increasing grain size. This is because of the growth, restricted by the size of hulls (Yilmez et al., 1997). Among the different sources of Zn, Zn-EDTA @ 1 kg ha^{-1} registered for maximum yield attributes such as number of panicles/m^2, panicle weight, panicle length, number of spikelets (grains)/panicle, grain filling percent, and test weight. The response of the yield attributes to different Zn sources was in the order of $Z_2 > Z_1 > Z_3 > Z_0$. The yield attributing characters with slight improvement might be due to the crop genotype which does not change with the above said agronomic management practices (Kulandaivel et al., 2004).

5.4.2.1   EFFECT OF ZINC ON YIELD AND YIELD ATTRIBUTES OF RICE: CASE STUDY

Zn application invariably enhanced translocation of photosynthates which was responsible for realization of higher grain and straw yields (Kulandaivel et al., 2004). Zn application is generally irrespective of methods of its application (Yasuhir et al., 2011) which significantly increased the grain yield. This was attributed to dilution of Zn in plant tissue because of enhanced dry matter production (Rengel and Graham, 1995). Zn application @ 25–30 kg ha^{-1} was sufficient to correct Zn deficiency for 4–7 years and the residual effect was best with soil application on long range basis. It has also been observed that after a certain limit in micronutrient fertilization, further increase in fertilizer application caused not only reduction in yield but also decreased micronutrient grain density as reported by Nambiar and Motiramani (1983). Prashad et al. (1995) reported that rice transplanted with 10 t ha^{-1} compost and 100 kg N + 80 kg P_2O_5 + 40 kg K_2O and 0, 1.25, or 2.6 ppm Zn produced higher grain yield of 4.37, 4.37, 4.00, and 4.07 t ha^{-1} with NPK and compost as compared to control (2.48 t ha^{-1}) and NPK alone (3.65 t ha^{-1}).

Singh et al. (1987) advocated that rice grain and straw yields increased progressively with the increase in the level from 10 to 30 kg $ZnSO_4$ ha^{-1}. The linear and significant response of rice grain and straw yield up to 40 kg $ZnSO_4$ ha^{-1} was also observed by Saraswat and Bansal (1991) and Singh et al. (1992). The results were confirmed further by Zhang and Barrow (1993), Prasad and Umar (1993), Singh and Ghose (1996), and Grewal et al. (1997). Thus, it is well established that application of zinc sulfate ($ZnSO_4$) had significant influence on rice plant growth characters, yield attributing parameters, and grain and straw yields. However, the degree of influence varied depending upon the soil type, weather condition, duration of the crop, method of rice culture, and mode of application of $ZnSO_4$. Dravid and Goswami (1987) reported significant increase in the number of panicles/m^2, more number of grains per panicle, and increased panicle weight with the increase in the levels of $ZnSO_4$ which were further confirmed by Ahmad et al. (2012), Mali and Shaikh (1994) and Devarajan and Krishnasamy (1996). Rengel and Graham

(1995) also reported an appreciable improvement in the rice yield attributes such as panicle length, fertility percentage, and test weight due to increase in the levels of $ZnSO_4$. Tripathi and Tripathi (2004) found that application of 30 kg $ZnSO_4$ ha^{-1} resulted in significantly higher grain and straw yields, harvest index, seed weight per panicle, and 1000-seed weight. However, contrary to this, Agarwal et al. (1987) found no significant difference in panicle length and fertility percentage of rice due to the application of Zn.

Yield of rice was reported to be significantly increased with the application of Zn. There was a correlation established between Zn application and Fe availability/fertilization rate. While Fe application brought about a significant improvement in available soil and plant Fe and Mn, it decreased significantly Zn content of the crop. After crop harvest, recovery of added Fe was 20% and Zn 12%. Results suggested that benefits of Fe application to rice in sodic soils could only be realized if it was applied along with Zn (Swarup, 1993). Tripathi (1993) at Varanasi in India observed significantly more dry matter yield of rice with combined soil application of 10 mg kg^{-1} Zn and 10 mg kg^{-1} Fe. Fe with Mn application increased rice yield on silicon-treated soil.

Chelated Zn showed high solubility and stability of Zn and increased the movement of Zn ions in to the plants to increase the grain yield (Sachdev and Deb, 1990). The favorable influence of applied Zn on yield may be due to its catalytic or stimulatory effect on most of the physiological and metabolic process of plants (Mandal et al., 2009). Singh et al. (1987) advocated that rice grain and straw yields could be increased progressively with the increase in the level from 10 to 30 kg $ZnSO_4$ ha^{-1}.

Yao-Zheng et al. (1998) and Kumar et al. (1999) reported that $ZnSO_4$ level could not influence panicle weight, fertility percentage, and test weight of rice. However, Rengel and Graham (1995) reported an appreciable improvement in the rice yield attributes such as panicle length, fertility percentage, and test weight due to increase in the levels of $ZnSO_4$. Application of GM with *Azospirillum* recorded significantly shorter period for 50% flowering, highest number of productive tillers m^{-2}, filled grains per panicle, panicle length, and grain yield (5282 and 5218 kg ha^{-1}). The highest level of N (187.5 kg ha^{-1}) with Zn application recorded significantly higher yield attributes and yield (5516 and 5376 kg ha^{-1}) than 150 kg N ha^{-1} with Zn (Shanmugam and Veeraputhran, 2001).

The frequency of Zn application based on the study of 10 cropping systems reported by Prasad et al. (2002) in which @ 25 kg Zn $SO_4.7H_2O$ ha^{-1} used as soil application after a two-crop interval was optimal. The rate of increase in yields of rice and wheat was 52.4 and 21.0 kg per kg of $ZnSO_4$, respectively, and the percent increase in yield of rice was 46.6 and in wheat 38.1. The rice and wheat yield in the cropping system was significantly correlated with Zn removal. Kumar et al. (1999) obtained best results with application of 25 kg $ZnSO_4$ to transplanted plants, spraying with 0.5% $ZnSO_4$ solution 3 weeks after transplanting or dipping seedling roots in 2% ZnO suspension. Zn application in the nursery was effective in correcting Zn deficiency and improving yield even when Zn was not applied after transplanting.

Biswas and Dravid (2001) recommended application of 25 kg $ZnSO_4$ ha^{-1} at puddling under normal soil condition mixed with equal amount of dry soil. Supplementing Zn with 0.52 mol Zn L^{-1} resulted in the largest root–shoot ratio and highest yield of biomass and rice grain (Gang et al., 2003). Significant increase in yield and Zn uptake were recorded due to application of 30 kg ha^{-1} (Rao, 2003). Patil and Meisheri (2003) studied the direct and residual effect of applied Zn in rice–rice cropping system and reported that direct effect of 25 kg $ZnSO_4$ ha^{-1} or 10 kg Zn (chelated) ha^{-1} though were at par but significantly increased paddy yield over N, P, and K. However, when combined with 10 t FYM ha^{-1} it produced significantly higher yield over other treatments. Uptake of Zn in straw, grain as well as its residual content also increased due to applied Zn along with FYM. Similarly, application of Zn with or without FYM brought about significant increase in grain and straw yield of subsequent rice crop.

A significant increase in yield and yield components was recorded for increasing levels of Zn. The application of 1.0% Zn solution appeared to be an optimum level for rice crops in these soil series. Tikken soil series gave the highest paddy and straw yields, while the Ramak series gave the lowest. The interaction between the soil series and Zn levels was significant for all parameters (Khan et al., 2005). Tripathi and Tripathi (2004) found that application of 30 kg $ZnSO_4$ ha^{-1} resulted in significantly higher grain and straw yields, harvest index, seed weight per panicle, and 1000-seed weight. The levels and mode of application of Zn and Fe considerably increased the growth, yield attributes, and yield of rice–wheat cropping system and also improved the soil contents of these micronutrients. Application of 30 kg $ZnSO_4$ + 5 kg $FeSO_4$ ha^{-1} through chelating with FYM was found to be the best combination for rice (Kulandaivel et al., 2004).

Karak et al. (2006) revealed that the use of different sources of Zn (Zn-EDTA, and $ZnSO_4.7H_2O$) gave significantly increase in both rice grain (4.56 t ha^{-1}) and straw (6.88 t ha^{-1}) yield during 2-year experiment where 1.0 kg Zn applied as Zn-EDTA. Suri and Kumar (2006) reported equal affectivity of Zn-coated urea with $ZnSO_4$. Application of 0.5 kg Zn ha^{-1} as

Zn-EDTA, and 1.0 kg ha^{-1} as Zn-EDTA in two equal split revealed that split application of chelated Zn proved much effective and produced highest rice yield of 4.56 t ha^{-1} (Karak et al., 2005).

Alamdari et al. (2007) observed maximum Zn content in the leaves due to NPK, S, and $ZnSO_4$ application which proved superior to the control. However, Zn and Cu contents in the grain markedly increased when NPK, S, Zn, Cu, and $MnSO_4$ were applied together. Pandey et al. (2007) and Mandal et al. (2009) reported that the application of FYM at the rate of N_{150}, P_{75}, K_{60} or N_{100}, P_{60}, K_{40}, or organic fertilizer level of N_{100}, P_{60}, K_{40} + Zn remained at par but produced significantly higher panicle/m^2 and test weight over control. Similarly, treatment with N_{150}, P_{75}, K_{46} + FYM and N_{100}, P_{60}, K_{40} + FYM significantly increased the yield and nutrient uptake. It could be due to slow release of nutrients for longer period after decomposition of FYM, which favored better plant growth and improved the yield component of hybrid rice.

Shivay et al. (2008) observed increased uptake of Zn due to ZEU. Field experiment during rainy (Rice) and winter (Wheat) seasons of 2004–2006 on sandy loam soil were done for 2 consecutive years at research farm of Indian Agricultural Research Institute, New Delhi, India. While studying the response of prilled urea (PU) and ZEU results revealed that the ZEU produced significant growth yield attributes and yields of aromatic rice. Highest values for all these attributes and yields were recorded with highest enriched (3.5%) of the PU with Zn. However, 1% ZEU proved the most economic source for aromatic rice and wheat cropping system and recommended for rice–wheat cropping system. Naik and Das (2008) examined the relative performance of chelated Zn-EDTA and $ZnSO_4$. Initial incorporation of chelated Zn showed pronounced effect on growth over single basal application of $ZnSO_4$. The highest rice grain and straw yield were recorded due to application of 1 kg Zn ha^{-1} as Zn-EDTA as basal with highest mean filled grain percentage, thousands grain weight, and number of panicle/m^2. Varshney et al. (2008) observed that application of 10 kg $ZnSO_4$ to rice in the first year followed by 5 kg Zn ha^{-1} in second year under hybrid rice wheat sequence in Tarai region, significantly increased the grain yield of both crops along with increased Zn uptake. Hybrid rice showed marked efficiency in uptake of Zn over wheat.

Mandal et al. (2009) observed the effect of Zn in combination with FYM and recommended their application to transplanted rice. Application of 10 kg Zn in three splits combined with recommended dose (RD) of NPK recorded significantly higher growth, yield attributes, grain and straw yield, and showed additional net return. Jana et al. (2009) conducted a field experiment to study the effect of Zn application on transplanted rice grown on farmer's field in red and laterite soil and reported that Zn application produced significantly greater yield attributes, higher grain and straw yields of rice. Application of 30–40 kg $ZnSO_4$ ha^{-1} gave significantly higher value of plant height, number of effective tillers, panicle length, grain number per panicle, grain and straw yields, and higher uptake of NPK and Zn in grain and straw of rice. Also, 30–40 kg $ZnSO_4$ gave greater yield response and higher net return.

Sridevi et al. (2010) reported positive effect of Zn-enriched organic manures. The results of the field experiment revealed that RD of NPK + 200 kg FYM enriched with 5.0 kg Zn increased the grain and straw yields due to increased availability of Zn. Uptake of nitrogen (N), phosphorous (P), and potassium (K), and Zn also showed uniform increase in uptake. Chandrapala et al. (2010) observed that application of S, Zn, and FYM significantly improved the productivity of rice and succeeding maize crop. Among nutrient management practices, NPK + Zn + S recorded higher grain yield and yield attributes of rice cultivation. NPK + FYM application to rice crop recorded significantly highest quantity of available soil N, P, and K content after crop harvest. Kumar and Kumar (2010) reported on effects of rates and methods of Zn fertilizers applications in rice under flood-prone conditions. There was a significant increase in yield and yield attributes of rice up to 45 kg $ZnSO_4$ ha^{-1}. The content and uptake of Zn also increased significantly with increasing levels of $ZnSO_4$. The application of Zn in soil proved superior as compared to its foliar application.

Jat et al. (2011) found that application of 2% ZEU ($ZnSO_4.7H_2O$) was the best in terms of all the yield attributes and grain yield of aromatic hybrid rice. In respect to Zn concentrations in grain and straw of aromatic hybrid rice, 2% ZEU ($ZnSO_4.7H_2O$) recorded higher Zn concentrations in grain as well as in straw of aromatic hybrid rice. Overall grain yield was increased by all treatments compared with the control and it was highest with 11 kg Zn as zinc frits followed by $ZnSO_4$. Leaf Zn content ranged from 5.5 ppm in the control to 13.4 ppm with $ZnSO_4$ application. Grain yield was significantly correlated with leaf Zn content (Subbaiah et al., 1994). Results suggested that benefits of Fe application to rice in sodic soils could only be realized if it was applied along with Zn (Swarup, 1993).

5.4.3   QUALITY PARAMETER

The quality parameters depend upon a number of factors such as variety, shape, size, and chalkiness of endosperm, nature of postharvest handling and inherent strength of grain to resist breakage and damage. Adequate supply of nutrients along with micronutrients might have resulted in strengthening of grain to resist breakage (Govindaswami, 1967; Govindaswami and Murthy 1966). The researchers shown that application of Zn have maximum important values in various quality parameters, namely, hulling, milling, head rice recovery (HRR), kernel length, kernel breadth, kernel length after cooking (KLaC), volume expansion ratio (VER), water uptake (WU), amylose content (AC), alkali spreading value (ASV), protein content, and protein yield (Verma et al., 2012; 2013; 2015), which remained significantly superior to other doses. The response of quality parameters to different Zn sources was in the order of $Z_2 > Z_1 > Z_3 > Z_0$. Application of Zn might be due to increased vigor, photosynthetic accumulation and better translocation of photosynthates to the sink (Kanda and Dixit, 1995).

5.4.3.1 EFFECT OF ZINC (Zn) ON QUALITY AND QUALITY ATTRIBUTES OF RICE: CASE STUDY

In food, Zn content has been recognized as a vital element for human health. The artificial augmentation of foods with essential minerals is often difficult to achieve, particularly in developing countries. Therefore, it has been registered that increased levels of Zn in staple foods, for example, rice may play a role in reducing Zn deficiency in human being (Ruel and Bouis, 1998; Graham et al., 1999; Welch and Graham, 1999). Zn deficiency is a serious nutritional problem and is affecting estimated one-third of the world's population (Sandstead, 1991) with the contribution of 0.8 million deaths per year.

Various researches further revealed that hulling (%), milling (%), HRR (%), kernel length (mm), kernel breadth (mm), L/B ratio, KLaC, VER, WU, AC, ASV, protein content, and protein yield are considered as important quality parameters of rice (Verma et al., 2012; 2013; 2015) were significantly increased by the application of different sources of Zn. In addition, Zn is also known for its critical role in control of gene transcription because of the proteins that regulates control of gene transcription and the coordination of other biological processes by DNA-binding Zn-finger motifs (Rhodes and Klug, 1993). Jat et al. (2011) observed that cooking quality parameters could be significantly improved by legume residue incorporation along with application of ZEU. The N and Zn fertilization is directly responsible for higher 1000-grain weight because N increased the crude protein content in grains (Naik and Das, 2008).

Singh et al. (2005) investigated the response of varying rates and mode of Zn application on rice and concluded that the highest amount of Zn acquisition in grain and straw was recorded due to 3 split application of N (135.55 g ha^{-1}) with two foliar spray and root dipping in Zn solution (127.15g ha^{-1}) in calcareous soil, Rahman et al. (2007) reported a positive correlation with dry matter yield and plant Zn content. Mollah et al. (2009) failed to observe any marked variation in yield of rice due to different brands of Zn fertilizers. However, the nutrient content in grain and straw significantly increased.

5.4.4 NUTRIENT UPTAKE PARAMETER

Raising the micronutrient content in grain is an effective approach toward nutritional security. Food biofortification either by plant breeding (genetic biofortification) or by the use of micronutrient fertilizers (agronomic biofortification) can both contribute to this goal. It is important to note that biofortification for health purposes require higher levels of micronutrients in grain than is necessary to achieve optimum crop yields (IFA, 2004). Biofortification focuses on enhancing the Zn nutritional qualities of crops at source, which encompasses process that increase both Zn levels and their bioavailability in the edible part of the staple crops. Biofortification is of two types, namely, agronomic and genetic. Agronomic approaches for Zn enrichment in food grains adequate Zn fertilization, type, time, and method of Zn application. With the increase in the level of Zn fertilizers the Zn concentration in grains of aromatic rice also increases (Shivay et al., 2010) and $ZnSO_4$-enriched urea is better in enriching the Zn in rice grains than the zinc oxide (ZnO)-enriched urea (Prasad et al., 2013).

Soil chemical analysis showed that available soil nitrogen (N), phosphorus (P) and potassium (K) contents were affected significantly due to seeding methods and tillage, significantly greater available soil N, P, and K were recorded under direct seeding of rice followed by manual and mechanical transplanting. Conventional tillage recorded significantly lower available soil N and higher value of soil P and K; whereas, zero tillage recorded higher value of available soil N and lower value of available soil P and K during the 3 years of study (Gangwar et al., 2004).

5.4.4.1 EFFECT OF ZINC (Zn) ON NUTRIENT AND NUTRIENT UPTAKE OF RICE: CASE STUDY

The uptake of nutrient by a crop depends upon the total biomass production and nutrient concentration in plant parts which in turn is influenced by soil climate, cultural practices, level of nutrients applied, and age of the plant (Gupta and Toole, 1986). Favorable increase in Zn uptake due to increased supply of Zn has been reported by Gurmani et al. (1988), Chhibba et al. (1989), and Sachdev and Deb (1990). From the studies of Indulkar and Malewar (1991), Singh et al. (1992), and Kumar et al. (1999) noticed a suggestive phenomenon to see the correlation between $ZnSO_4.7H_2O$ application and Zn uptake. Rice genotypes were found to be greatly differing with respect to Zn use efficiency and thereby different levels of Zn content in grains. Significant increase in Zn uptake with increasing $ZnSO_4.7H_2O$ application but subsequently on higher application, it showed saturation effect and this aspect has been investigated by various researchers (Neue et al., 1998; Graham et al., 1999; Wissuwa et al., 2006; Wissuwa et al., 2008; Refuerzo et al., 2009; Nagarathna et al., 2010). In rice grain, Zn concentrations ranged from 15.9 to 58.4 mg kg^{-1} depending on soil Zn status (Graham et al., 1999; Wissuwa et al., 2006; Wissuwa et al., 2008). The seed, Fe, and Zn concentrations did not increase in plants grown under Fe-deficient conditions in calcareous paddy fields or under Fe-sufficient conditions, if driven by a suitable promoter, it can result in a significant increase in grain Fe (Suzuki et al., 2008; Mollah et al. 2009).

Giordano et al. (1974) studied the absorption of Zn by 14-day-old intact rice seedling in short term uptake experiments in solution culture. They suggested that high solution concentrations of reduced Fe and manganese which is developed in paddy culture may be related to Zn nutrition of rice. Prasad (2016) reported that critical values of DTPA extractable Zn varied from 0.38 to 1.34 mg kg^{-1} (with a general value 0.6) depending upon soil and crop. The lower critical limit of Zn in plant tissue is generally accepted as 20 mg kg^{-1} dry matter. Ugurluoglu and Kacar (1996) reported different sources of Zn, namely, ZnO, $ZnSO_4.7H_2O$, and Zn-EDTA for rice; he found that application of Zn @ 8 mg kg^{-1} as Zn-EDTA was found most effective in the enhancement of Zn content in rice plant.

Crops as well as crop cultivars vary in respect of micronutrient density. Quality also appears to be dependent on historical, socioeconomic, and cultivar adaptation. In fact, it is due to high Fe concentration in the inner skin of grains of red rice. However, Zn concentration is higher inside the polished grain. The aromatic cultivars have constantly higher Fe concentration in grain and also have more Zn than nonaromatic types. It was observed that traditional cultivars had better Zn and Fe concentration in grain than high yielding new cultivars. Jalmagna had almost double the amount of Fe present in IR-36 and IR-64 and Zn concentration was about 40% higher than IR-64. Mobilization of micronutrients from vegetative tissues to grain can form a significant source of micronutrients but little is known about mechanism governing such mobilization (Pearson and Rengel, 1994). Red rice genotype had higher Zn concentration than white genotypes. Unpolished red rice had higher Fe and Mn than unpolished white rice grain (Qui et al., 1993).

Polished rice seeds with IDS3 inserts had up to 1.40 and 1.35 times higher Fe and Zn concentrations, respectively, compared to nontransgenic rice seeds. Enhanced mugineic acid family phytosiderophores production due to the introduced barley genes is suggested to be effective for increasing Fe and Zn concentrations in rice grains (Masuda et al., 2008). Combined application of Zn fertilizers and organic matter (FYM) or incorporation of GM resulted in significant increment in Zn concentration in food grains than Zn fertilizers alone. Combined application of Zn-treated seeds along with soil and foliar Zn application contributed more Zn to grain than untreated seeds (Pooniya and Shivay, 2013; Singh and Shivay, 2013).

5.5   SUMMARY AND CONCLUSION

This chapter focused on different important parameters of rice, namely, plant growth attributes, yield attributes, quality attributes, and nutrient uptake attributes for improved production, the importance of Zn and its effect on said parameters for improved rice production and also to focused on the work, findings, and results of many researchers what they confirmed in their experiments and what they suggested for improving rice production.

KEYWORDS

- agronomic efficiency
- augmentation
- crop growth rate
- farmyard manure
- leaf area index
- organic matter
- physiological efficiency

REFERENCES

Agarwal, S. K.; Suraj, B.; Bhan, S. Effect of Levels of Zinc Sulphate Application on the Yield in Rice–Wheat Cropping Sequence. *Indian J. Agric. Res.* **1987,** *31* (3), 174–178.

Ahmad, H. R.; Aziz, T.; Hussain, S.; Akraam, M.; Sabir, M.; Kashif, S. R.; Hanafi, M. M. Zinc Enriched Farm Yard Manure Improves Grain Yield and Grain Zn Concentration in Rice Grown on a Saline-sodic Soil. *Int. J. Agric. Biol.* **2012,** *14* (5), 787–792.

Alamdari, M. G.; Rajurkar, N. S.; Patwardhan, A. M.; Mabasser, H. R. The Effect of Sulphur and Sulphate Fertilizers on Zn and Cu Uptake by the Rice Plant (*Oryza sativa* L.), *Asian J. Plant Sci.* **2007,** *60* (2), 407–410.

Alloway, B. J. *Zinc in Soils and Crop Nutrition*, International Zinc Association: Brussels, Belgium, 2004.

Alloway, B. J. *Zinc in Soils and Crop Nutrition*, IFA: Paris, France and IZA: Brussels, Belgium, 2008.

Arnon, D. I.; Stout, P. R. *Agronomy Terminology*, Indian Society of Agronomy, Division of Agronomy, IARI: New Delhi, 1939, pp 307–312.

Asewar, B. V.; Dahiphale, V. V.; Chavan, G. V.; Katare, N. B.; Sontakke, J. S. Effect of Ferrous Sulphate on Grain Yield of Upland Basmati Rice. *J. Maharashtra Agric. Univ.* **2000,** *25* (2), 209–210.

Behera, S. K.; Singh, D.; Dwivedi, B. S. Change in Fractions of Iron, Manganese, Copper and Zinc in Soil Under Continuous Cropping for More than Three Decades. *Commun. Soil Sci. Plant Anal.* **2009a,** *40* (9–10), 1380–1407.

Behera, S. K.; Singh, M. V.; Lakaria, B. L. Micronutrient Deficiencies in Indian Soil and Their Amelioration Through Fertilization. *Indian Farm.* **2009b,** *59* (2), 28–31.

Biswas, D. R.; Dravid, M. S. P and Zn Interaction as Influenced by Mg on Yield and Nutrient Uptake by Rice Under Varying Moisture Regimes. *Ann. Agric. Res. New Ser.* **2001,** *22* (3), 329–334.

Broadley, M. R.; White, P. J.; Hammond, J. P.; et al. Zinc in Plants. *Newphilol* **2007,** *173*, 677–702.

Cakmak, I. Enrichment of Cereal Grains with Zinc: Agronomic or Genetic Biofortification? *Plant Soil* **2008,** *302*, 1–17.

Cakmak, I. Enrichment of Fertilizers with Zinc: An Excellent Investment for Humanity and Crop Production in India. *J. Trace Elem. Med. Biol.* **2009,** *23*, 981–989.

Cakmak, I.; Kalayci, M.; Ekiz, H.; Braun, H. J.; Yilmaz, A. Zinc Deficiency as Anactual Problem in Plant and Human Nutrition in Turkey: A NATO Science for Stability Project. *Field Crops Res.* **1999,** *60*, 175–188.

Cakmak, I.; Torun, A.; Millet, E.; Feldman, M.; Fahima, T.; Korol, A.; Nevo, E.; Braun, H. J.; Ozkan, H. *Triticum dicoccoides*: An Important Genetic Resource for Increasing Zn and Fe Concentration in Modern Cultivated Wheat. *Soil Sci. Plant Nutr.* **2004,** *50*, 1047–1054.

Chandrapala, A. G.; Yakadri, M.; Kumar, R. M.; Raj, G. B. Establishment, Zn and S Application in Rice. *Indian J. Agron.* **2010,** *55* (3), 171–176.

Chhibba, I. M.; Nayyar, V. K.; Takkar, P. N. Direct and Residual Effect of Some Zn Carriers in Rice–Wheat Rotation. *J. Indian Soc. Soil Sci.* **1989,** *37* (3), 585–587.

Curl, E. A.; Truelove, B. *The Rhizosphere*. Springer: Verlag, NY, 1986.

Dargan, K. S.; Singh, O. P.; Gupta, I. C. *Crop Production in Salt Affected Soils*. Oxford and IBH Publishing Co. Pvt. Ltd.: New Delhi, 1982; pp 28–29.

Datta, S. K.; Khush, G. S. Improving Rice to Meet Food and Nutritional Needs: Biotechnological Approaches. *J. Crop Product.* **2002,** *6* (1–2), 224–229.

Devarajan, R.; Krishnasamy, R. Effect of Zn-enriched Manures on Rice Yield. *Madras Agric. J.* **1996,** *83* (5), 280–283.

Dravid, M. S.; Goswami, N. N. Effect of Farm Yard Manure P and Zn on Dry Matter Yield and Uptake of P, K, Zn and Fe by Rice Under Non-saline and Saline Conditions. *J. Nucl. Agric. Biol.* **1987,** *16*, 201–205.

Fageria, N. K.; Bajigar, V. C.; Clark, R. B. Micronutrients in Crop Production. *Adv. Agron.* **2002,** *77,* 187–266.

Gang, L. Z.; Zheng, Y.; Qian, F. Y. Y.; Xia, Y. Effect of Zinc on Plant Growth and Zinc Partitioning in Rice Plant. *Chinese J. Rice Sci.* **2003,** *17* (1), 61–66.

Gangwar, F. S.; Singh, K. K.; Sharma, S. K. Effect of Tillage on Growth, Yield and Nutrient Uptake in Wheat After Rice in Indogangetic Plain of India. *J. Agric. Sci.* **2004,** *142* (4), 453–459.

Gao, X.; Zou, C.; Fan, X.; Zhang, F.; Hoffland, E. Form Flooded to Aerobic Conditions in Rice Cultivation: Consequences from Zn Uptake. *Plant Soil* **2006,** *280*, 41–47.

Ghatak, R.; Jana, P. K.; Sounda, G.; Ghosh, R. K.; Bandopadhyay, P. Responses of Transplanted Rice to Zn Fertilization at Farmer Field on Red and Laterite Soil of West Bengal. *J. Interacadem.* **2005,** *9* (2), 231–234.

Giordano, P. M.; Noggle, J. C.; Mortvedt, J. J. Zinc Uptake by Rice as Affected by Metabolic Inhibitions and Competing Cations. *Plant and Soil* **1974,** *41*, 637–646.

Govindaswami, S. In *Increasing Rice Production by Better Breeding and Technology*, Proceedings of the Symposium on "Science and India's Food Problem" Sponsored by the National Institute of Science, India, 1967, pp 336–346.

Govindaswami, S.; Murthy, P. S. N. *Effect of Manorial and Cultural Practices on the Quality Characterizes*. CRRI: Cuttack, Orissa, 1966.

Graham, R. D.; Welch, R. M. Breeding for Staple Food Crops with High Micronutrient Density. In *Worker Papers on Agricultural Strategies for Micronutrient*; Bouis, H. D. Ed.; International Food Policy Research Institute: Washington, D.C., 1996; pp 1–72.

Graham, R. D.; Senadhira, D.; Beebe, S.; Iglesias, C.; Monasterio, I. Breeding for Micronutrient Density in Edible Portions of Staple Food Crops: Conventional Approaches. *Field Crops Res.* **1999,** *60*, 57–80.

Grewal, H. S.; Zand, L.; Graham, R. D. Influence of Subsoil Zinc on Dry Matter Production, Seed Yield and Distribution of Zinc in Oilseed Rape Genotype Differing in Zinc Efficiency. *Plant and Soil* **1997,** *192*, 81–189.

Gupta, A.; Toole, O. Zinc Fertilization on Flooded Rice in Vertisols of Hyderabad. *J. Nucl. Agric. Biol.* **1986,** *12* (4) 96–99.

Gurmani, A. H.; Yousaf, M.; Khattak, J. K. *Effect of Zinc, Copper, Iron and Manganese on Yield of Rice*. National Seminar on Micronutrient in Soils and Crops in Pakistan (Peshawar), Dec 13–15, 1988, pp 170–178.

IFA. How Can the Fertilizer Industry Contribute to the Nutrition Security Challenge. International Fertilizer Association (IFA) International Symposium on Micronutrients, Feb 23–24, New Delhi, India, 2004.

Indulkar, B. S.; Malewar, G. U. Response of Rice to Different Zinc Sources and Their Residual Effect on Succeeding Chickpea. *Indian J. Agron.* **1991,** *36*, 5–6.

Jana, P. K.; Ghatak, R.; Sounda, G.; Ghosh, R. K.; Bandhyopadhyay, P. Effect of Zinc Fertilization on Yield N, P, K and Zn Uptake Transplanted Rice at Farmers Field of Red and Lateritic Soils of West Bengal. *Indian Agric.* **2009,** *53* (3–4), 129–132.

Jat, S. L.; Shivay, Y. S.; Parihar, C. M. Dual Purpose Summer Legumes and Zinc Fertilization for Improving Productivity and Zinc Utilization in Aromatic Hybrid Rice (*Oryza sativa*). *Indian J. Agron.* **2011,** *56* (4), 328–333.

Kanda, C.; Dixit, L. Effect of Zinc and Nitrogen Fertilization on Summer Rice. *Indian J. Agron.* **1995,** *40* (2), 695–697.

Karak, T. Singh, U. K.; Das, D. K.; Kuzyakov, Y. Comparative Efficacy of ZnSO$_4$ and Zn-EDTA Application for Fertilization of Rice (*Oryza sativa* L). *Arch. Agron. Soil Sci.* **2005,** *51* (3), 253–264.

Karak, T.; Das, D. K.; Maiti, D. Yield and Zn Uptake in Rice (*Oryza sativa* L.) as Influenced by Sources and Times of Zn Application. *Indian J. Agric. Sci.* **2006,** *76* (6), 346–348.

Katyal, J. C.; Rattan, R. K.; Datta, S. P. Management of Zn and Bo for Sustainable Food Production. *Fertilizer News* **2004,** *94* (12), 83–89.

Khan, M. U.; Qasim, M.; Jamil, M. Effect of Different Levels of Zn on the Yield and Yield Components of Rice in Different Soils of D. I. Khan, Pakistan. *Sarhad J. Agric.* **2005,** *21* (1), 63–69.

Krishnaswamy, R. S.; Poongothai, S. V.; Mani, K.; Vanagamudi, S. M.; Savithri, P.; Devrajan, R. *Twentyfive Years of Micronutrient Research in Soils and Crops of Tamil Nadu*, Dept. Of Soil Science and Agricultural Chemistry: Tamil Nadu Agricultural University Coimbatore, India, 1994.

Kulandaivel, S.; Mishra, B. N.; Gangiah, B.; Mishra, P. K. Effect on Levels of Zinc and Iron and Their Chelation on Yield and Soil Micronutrient Status in Hybrid Rice (*Oryza sativa*)— wheat (*Triticum aestivum*), Cropping System. *Indian J. Agron.* **2004,** *49* (2), 80–83.

Kumar, B.; Singh, S. P. Zinc Management in Nursery and Transplanted Rice (*Oryza sativa*, L.). *Indian J. Agron.* **1996,** *41* (1), 153–154.

Kumar, D.; Chauhan, R. P. S.; Singh, B. B.; Singh, V. Response of Rice (*Oryza sativa*) to Zinc Sulphate Incubated and Blended with Organic Materials in Sodic Soil. *Indian J. Agric. Sci.* **1999,** *69* (6), 402–405.

Kumar, T.; Kumar, V. Effect of Rates and Methods of Zinc Application on Yields Economics and Uptake of Zinc by Rice Crop in Flood Prone Situation. *Asian J. Agron. Soil Sci.* **2010,** *4* (1), 96–98.

Mali, C. V.; Shaikh, A. R. Management of Zinc Sources in Rice-gram Cropping System. *J. Maharashtra Agric. Univ.* **1994,** *19*, 4–7.

Mandal, L.; Maiti, D.; Bandyopathyay, P. Response of Zinc in Transplanted Rice Under Integrated Nutrient Management in New Alluvial Zone of West Bengal. *Oryza* **2009,** *46* (2), 113–115.

Marschner, H. *Mineral Nutrition of Higher Plants*, 2nd ed.; Academic Press: London, 1995; pp.889.

Martens, D. C.; Westernmann, D. T. Fertilizer Application for Correcting Micronutrient Deficiency. In *Micronutrients in Agricultural,* 2nd ed.; Mortvedt, J. J., Cox, F. R., Shuman, L. M., Welch, R. M., Eds.; Soil Science Society of America, Inc. Madison: Wisconsin, USA, 1991; pp 549–592.

Masuda, H. S.; Morikawa, K. C.; Kobayashi, T. N.; Takahashi, H.; Saigusa, M. M.; Mori, S.; Nishizawa, N. K. Increase in Iron and Zinc Concentrations in Rice Grains via the Introduction of Barley Genes Involved in Phytosiderophore Synthesis. *Rice* **2008,** *1*, 100–108.

Mollah, M. Z. I.; Talukdar, N. M.; Islam, M. N.; Rahman, M. A.; Ferdous, Z. Effects of Nutrients Content in Rice as Influenced by Zn Fertilization. *World Appl. Sci. J.* **2009**, *6* (8), 1082–1088.

Nagarathna, T. K.; Shankar, A. G.; Udayakumar, M Assessment of Genetic Variation in Zinc Acquisition and Transport to Seed in Diversified Germplasm Lines of Rice (*Oryza sativa* L.). *J. Agric. Technol.* **2010**, *6*, 171–178.

Naik, S. K.; Das, D. K. Relative Performance of Chelated Zinc and Zinc Sulphate for Lowland Rice (*Oryza sativa* L.). *Nutr. Cycl. Agroecosyst.* **2008**, *81* (3), 219–227.

Nambiar, K. K. M.; Motiramani, D. P. Zinc Nutrition on the Yield and Nutrient Composition of Chickpea as Related to Four Soil Types. *J. Crop Sci.* **1983**, *152*, 165–172.

Neue, H. U.; Quijano, C. Senadhira, D.; Setter, T. Strategies for Dealing with Micronutrient Disorders and Salinity in Lowland Rice Systems. *Field Crops Res.* **1998**, *56*, 139–55.

Norman, R. J.; Wilson, C. E. J.; Slaton, N. A. Soil Fertilization and Mineral Nutrition in US Mechanized Rice Culture. In *Rice: Origin, History and Production*; Smith, C. W., Dilday, R. H., Eds.; John Wiley: NJ, USA, 2003; pp 31–412.

Panda, S. C.; Nayak, R. C. Effect of Zinc on Growth and Yield of Rice. *Indian J. Agron.* **1974**, *19*, 9–13.

Pandey, N.; Verma, A.; Anurag, K.; Tripathi, R. S, Integrated Nutrient Management in Transplanted Hybrid Rice (*Oryza sativa* L.). *Indian J. Agron.* **2007**, *52* (1), 40–42.

Patil, K. D.; Meisheri, M. B. Direct, Residual Effect of Applied Zinc Along with FYM on Rice in Soil of Konkan Region of Maharashtra. *Ann. Agric. Res.* **2003**, *24* (4), 927–933.

Pearson, J. N.; Rengel, Z. Distribution and Remobilization of Zn and Mn During Grain Development in Wheat. *J. Exp. Bot.* **1994**, *45*, 1829–1835.

Pfeiffer, W. H.; McClafferty, B. Biofortification: Breeding Micronutrient Dense Crops. In *Breeding Major Food Staples*. Manjit, S. K., Priyadarshan, P. M., Eds.; Blackwell Publishing Ltd.: UK, 2007; pp 61–92.

Pooniya, V.; Shivay, Y. S. Enrichment of Basmati Rice Grain and Straw with Zinc and Nitrogen Through Fertilization and Summer Green Manuring Under Indogangetic Plains of India. *J. Plant Nutr.* **2013**, *36* (1), 91–117.

Prasad, A. S. Zinc in Human Health: Effect of Zinc on Immune Cells. *Mol. Med.* **2008**, *14* (5–6), 353–357.

Prasad, B.; Umar, S. M. Direct and Residual Effect of Soil Application of Zincsulphate on Yield and Zinc Uptake in a Rice-wheat Rotation. *J. Indian Soc. Soil Sci.* **1993**, *41* (1), 192–194.

Prasad, B.; Sharma, M. M.; Sinha, S. K. Evaluating Zinc Fertilizer Requirement on Typic Haplaquent in the Rice–Wheat Cropping System. *J. Sustain. Agric.* **2002**, *19* (3), 39–49.

Prasad, R. Zinc in Soils and in Plant, Human & Animal Nutrition. (Special Issue: Seminar Special). *Indian J. Fertil.* **2006**, *2* (9), 103–119.

Prasad, R.; Shivay, Y. S.; Kumar, D. Zinc Fertilization of Cereals for Increased Production and Alleviation of Zinc Malnutrition in India. *Agric. Res.* **2013**, *2* (2), 111–118.

Prashad, B.; Prashad, J.; Prashad, R. Nutrient Management for Sustainable Rice and Wheat Production in Calcareous Soil Amended with Green Manuring, Organic Manures and Zinc. *Fertil. News* **1995**, *40*, 39–45.

Qui, L. C.; Pan, J.; Daun, B. W. The Mineral Nutrient Component and Characteristics of Coloured and White Brown Rice. *Chinese J. Rice Sci.* **1993**, *7* (2), 95–100.

Rahman, M. A.; Jahiruddin, M.; Islam, M. R. Critical Limit of Zinc for Rice in Calcareous Soils. *J. Agric. Rural Dev. (Gazipur)* **2007**, *11* (1–2), 43–47.

Rao, C. P. Nutrient Utilization Efficiency of Rice Influenced by Different Sources and Levels of Phosphorus and Rates of Zinc. *Ann. Agric. Res.* **2003**, *24* (1), 7–11.

Refuerzo, L.; Mercado, E. F.; Arceta, M.; Sajese, A. G.; Gregorio, G.; Singh, R. K. QTL Mapping for Zinc Deficiency Tolerance in Rice (*Oryza sativa* L.). International Rice Research Institute, Manila, Philippines, 2009. http://agris.fao.org/agris-search/search.do?recordID=PH2009001755 (accessed April 10, 2016).

Rengel, Z.; Graham, R. D. Importance of Seed Zn Content for Rice Growth on Zn-deficient Soil. *Plant Soil* **1995**, *173* (2), 259–266.

Rhodes, D.; Klug, A. Zinc Fingers. *Sci. Am.* **1993**, *268*, 56–65.

Ruel, M. T.; Bouis, H. E. Plant Breeding: A Long-term Strategy for the Control of Zn Deficiency in Vulnerable Populations. *Am. J. Clin. Nutraceutical Suppl.* **1998**, *68*, 488–494.

Sachdev, P.; Deb, D. L. Influence of Gypsum and Farmyard Manure on Fertilizer Zinc Uptake by Wheat and Its Residual Effect on Succeeding Rice and Wheat Crops in a Sodic Soil. *J. Nucl. Agric. Biol.* **1990**, *19* (3) 173–178.

Sandstead, H. Zinc Deficiency: A Public Health Problem? *Am. J. Dis. Child.* **1991**, *145*, 853–859.

Saraswat, V. K.; Bansal, K. N. Methods of Zinc Application and Its Effect on Yield and Zinc Content of Rice (*Oryza sativa*) and Wheat (*Triticum vulgare*). *Madras Agric. J.* **1991**, *78* (5–8), 174–177.

Serry, A.; Mawardi, A.; Awad, S.; Aziz, I. A. Effect of Zinc and Manganese on Wheat Production. I. FAO/SIDA Seminar for Plant Scientists from Africa and Near East, FAO Rome, 1974, pp.404–409.

Shanmugam, P. M.; Veeraputhran, R. Effect of Organic Manure, Biofertilizers, Inorganic Nitrogen and Zinc on Growth and Yield of Rabi Rice (*Oryza sativa* L.). *Madras Agric. J.* **2001**, *87* (1–3), 90–93.

Sharma, C. P.; Sharma, P. N.; Bisht, S. S.; Nautiyal, B. D. In *Zinc Defiency Induced Changes in Cabbages in Cabbage*, Aug 22–27, 1982; Proceedings of 9th Plant Nutrition, Scaife, C. A., Awick, W., Eds.; Commonwealth Agriculture Buro (CAB), Farnham House: Sough, U.K, 1982, pp.601–606.

Shivay, Y. S.; Kumar, D.; Prasad, R. Effect of Zn Enriched Urea on Productivity, Zinc Uptake and Efficiency of an Aromatic Rice–Wheat Cropping Systems. *Nutr. Cycl. Agroecocyst* **2008**, *81*, 229–243.

Shivay, Y. S.; Prasd, R.; Rahal, A. Genotypic Variation for Productivity, Zn Utilization Effiencies and Kernel Quality in Aromatic Rice Under Low Available Zinc Conditions. *J. Plant Nutr.* **2010**, *33* (12), 1835–1848.

Silanpaa, M. *Micronutrient Assessment at the Country Level: An International Study*; Food and Agriculture Organization (FAO) of the United Nations, Rome, 1990.

Singh, R. M.; Jain, G. L. Response of Paddy to Fe and Zn. *Indian J. Agron.* **1964**, *9*, 273–276.

Singh, A.; Shivay, Y. S. Residual Effect of Summer Green Manure Crops and Zn Fertilization and Quality and Zn Concentration of Durum Wheat (*Tritcum durum* Desf.) Under a Basmati Rice-durum Wheat Cropping System, Biological Agriculture and Horticulture. *Int. J. Sustain. Product. Syst.* **2013**, *29* (4), 271–287.

Singh, T. N.; Singh, H. P.; Singh, G. Zinc Requirement for Rice Wheat Sequence. *Int. Rice Res. Newslett.* **1987,** *12* (4), 64–65.

Singh, K.; Raj, L.; Ghose, R. L. Zinc Translocation to Wheat Roots and Its Implications for a Phosphorus/Zinc Interaction in Wheat Plants. *J. Plant Nutr.* **1996,** *13*, 1499–1512.

Singh, A. P.; Raha, P.; Yadav, P. K. Effect of Sulphur and Manganese Fertilization on Zn Content and Quality of Rice Bran Oil. *Indian J. Agric. Biochem.* **2005,** *18* (1), 19–23.

Singh, K.; Deo, C.; Bohra, J. S.; Singh, J. P.; Singh, R. N. Effect of Iron Carrier Sand Compost Application on Rice Yield and Their Residual Effect on Succeeding Wheat Crop in Entisol. *Ann. Agric. Res.* **1992,** *13* (2), 181–183.

Singh, U.; Praharaj, C. S.; Singh, S. S.; Singh, N. P. *Biofortification of Food Crops*; Singh, U., Praharaj, C. S., Singh, S. S., Singh, N. P., Eds.; Springer: India, 2016; pp.492.

Sinha, R. *The National Seminar on Importance of Zinc in Human Health*; Organized by International Life Sciences Institute (ILSI)-India and ILSI Human Nutrition Institute, Washington in Association with Indian Council of Medical Research and National Institute of Nutrition, Oct 25–26, 2004, New Delhi, India. http://indianpediatrics.net/dec2004/dec-1213-1217.htm (accessed April 11, 2016).

Slaton, N. A.; Gbur, J.; Wilson, E. E.; Jr, C. E.; Norman, R. J. Rice Response to Granular Zinc Sources Varying in Water-soluble Zinc. *Soil Sci. Am. J.* **2005,** *69*, 443–452.

Sommer, A. L.; Lipman, C. B. *Agronomy Terminology*, Indian Society of Agronomy, Division of Agronomy, IARI: New Delhi, India, 1926; pp 307–312.

Sridevi, G.; Rajkannan, B.; Surendran, U. Effect of Zn Enriched FYM and Cowdung on Yield of Rice (ADT-45). *Crop Res.* **2010,** *40* (1–3), 25–28.

Subbaiah, V. V.; Sreemannarayana, B.; Sairam, A.; Kumar, P. R. P.; Prasadini, P. P. Effect of Zinc Levels and Its Relative Proportion to Iron and Manganese Content in 3rd Leaf on Zn Deficiency and Grain Yield of Lowland Rice. *J. Res. APAU* **1994,** *22* (3–4), 135–136.

Suri, I. K.; Devkumar, C. Zincated Urea, A New Zinc Micronutrient Fertilizer (Indian). *J. Fertil.* **2006,** *2* (7), 19–22.

Suzuki, M.; Morikawa, K. C.; Nakanishi, H.; Takahashi, M.; Saigusa, M.; Mori, S.; Nishizawa, N. K. Transgenic Rice Lines That Include Barley Genes have Increased Tolerance to Low Iron Availability in a Calcareous Paddy Soil. *Soil Sci. Plant Nutr.* **2008,** *54*, 77–85.

Swarup, A. Iron, Zinc and Manganese Nutrition of Wetland Rice (*Oryza sativa* L.) on a Gypsum Amended Sodic Soil. *Plant Soil* **1993,** *155/156*, 477–480.

Takkar, P. N.; Singh, T. Zinc Nutrition of Rice as Influenced by Rates of Gypsum and Zn Fertilization of Alkali Soils. *Agron. J.* **1978,** *70*, 447–450.

Tripathi, A. K.; Tripathi, H. N. Studies on Zinc Requirement of Rice (*Oryza sativa*) in Relation to Different Modes of Zinc Application in Nursery and Rates of $ZnSO_4$ in Field. *Haryana J. Agron.* **2004,** *20* (1–2), 77–79.

Tripathi, S. D. Micronutrient Status of Rice Growing Belt of Varanasi Division with Special Reference to Zinc and Iron, PhD Desertation, Department of Soil Science and Agricultural Chemistry, Institute of Agricultural Sciences, Banaras Hindu University, Varanasi, India, 1993.

Ugurluoglu, H.; Kacar, B. Effect of Different Zn Sources on the Growth of Rice (*Oryza sativa*, L.). *Turkistan J. Agric. Forest.* **1996,** *20*, 473–478.

Varshney, P.; Singh, S. K.; Srivastava, P. C. Frequency and Rates of Zinc Application Under Hybrid Rice–Wheat Sequence in a Mollisol of Uttrakhand. *J. Indian Soc. Soil Sci.* **2008,** *56* (1), 92–96.

Verma, D. K. Physico-chemical and Cooking Characteristics of Azad Basmati (CSAR 839-3): A Newly Evolved Variety of Basmati Rice (*Oryza sativa* L.), M.Sc. (Agriculture) Dissertation, Department of Agricultural Biochemistry, College of Agriculture, CSA University of Agriculture and Technology, Kanpur 208002, Uttar Pradesh, India, 2011.

Verma, A. Food Fortification: A Complementary Strategy for Improving Micronutrient Malnutrition (MNM) Status. *Food Sci. Res. J.* **2015,** *6* (2), 381–389.

Verma, T. S.; Neue, H. U. Effect of Soil Salinity Level and Zinc Application on Growth, Yield and Nutrient Composition of Rice. *Plant Soil* **1984,** *82,* 3–14.

Verma, D. K.; Mohan, M.; Asthir, B. Physicochemical and Cooking Characteristics of Some Promising Basmati Genotypes. *Asian J. Food Agro-Ind.* **2013,** *6* (2), 94–99.

Verma, D. K.; Mohan, M.; Prabhakar, P. K.; Srivastav, P. P. Physico-chemical and Cooking Characteristics of Azad Basmati. *Int. Food Res. J.* **2015,** *22* (4), 1380–1389.

Verma, D. K.; Mohan, M.; Yadav, V. K.; Asthir, B.; Soni, S. K. Inquisition of Some Physico-Chemical Characteristics of Newly Evolved Basmati Rice. *Environ. Ecol.* **2012,** *30* (1), 114–117.

Welch, R. M.; Graham, R. D. A New Paradigm for World Agriculture: Meeting Human Needs: Productive, Sustainable, Nutritious. *Field Crops Res.* **1999,** *60,* 1–10.

Wissuwa, M.; Ismail, A. M.; Graham, R. D. Rice Grain Zinc Concentrations as Affected by Genotype, Native Soil-zinc Availability, and Zinc Fertilization. *Plant Soil* **2008,** *306,* 37–48.

Wissuwa, M.; Ismail, A. M.; Yanagihara, S. Effects of Zinc Deficiency on Rice Growth and Genetic Factors Contributing to Tolerance. *Plant Physiol.* **2006,** *142,* 731–41.

Yadav, G. S. Shivaya, Y. S.; Kumar, D.; Babu, S. Agronomic Evaluation of Mulching and Iron Nutrition on Productivity, Nutrient Uptake, Iron Use Efficiency and Economics of Aerobic Rice-wheat Cropping System. *J. Plant Nutr.* **2016,** *39* (1), 116–135.

Yao-Zheng, Z.; TaiHalchen, Y. Z.; Zhao, J. Y.; Tai, H. C. Effect of Two Fermented Organic Fertilizers on Some Soil, Trace Elements. *Acta Agric. Shanghai* **1998,** *9* (4), 53–58.

Yilmez, A. H.; Ekiz, B.; Torun, I.; Guttekin, S.; Karanlik, S. A.; Bagci and Cakmak, I. Effect of Different Zinc Application Methods on Grain Yield and Zinc Concentration in Wheat Cultivars Grown on Zinc Deficient Calcareous Soils. *J. Plant Nutr.* **1997,** *20,* 461–471.

Yoshida, S. *Climate and Rice*; International Rice Research Institute: Manila, Philippines, 1972; p 211.

Zhang, F. S.; Barrow, N. J. In *Mobilization of Iron and Manganese by Plant-borne and Synthetic Metal Chelators. Plant Nutrition from Genetic Engineering to Field Practice*, Proceeding, Held on Sept 21–26, 1993, Perth, Australia, 1993, pp 115–118.

CHAPTER 6

INTEGRATED NUTRIENT MANAGEMENT IN TRANSPLANTED RICE BY PELLETING TECHNIQUE

MANISH KUMAR SHARMA[1]* and PARMESHWAR KUMAR SAHU[2]

[1]*Department of Agronomy, Indira Gandhi Krishi Vishwavidyalaya, Raipur 492012, Chhattisgarh, India*

[2]*Department of Genetics and Plant Breeding, Indira Gandhi Krishi Vishwavidyalaya, Raipur 492012, Chhattisgarh, India*

Corresponding author. E-mail: mksharma003@gmail.com

ABSTRACT

Rice—the vital source of food—plays a key role in the food security system which in turn requires the NPK for optimum growth and yield. The practices of integrated nutrient management are highly important for sustained and better quality rice production because it involves the judicious use of organic, chemical, and microbial sources for improving the soil health and enhancing the yield through the economically and ecological friendly ways thus helpful to reduce the issues of nutrient mining. Therefore, the integrated nutrient management is the need of hour to maintain the soil fertility and productivity along with farmers' benefit by the integration of organic manures, inorganic fertilizers, and biofertilizers. Pelleting is the technique of nutrient management where nutrients are slowly released with long-term effects, reduced losses, and enhanced uptake of nutrients due to the condensed form of manures into a pellet. Such a pellet is also easy to handle, transport, store, and enriched with chemicals which contains all essential major nutrients and micronutrients and releases these nutrients slowly in root zones, making them available for long time and reducing their losses in different forms.

6.1 INTRODUCTION

Rice is the most important cereal crop in the world and considered as primary source of food. Rice crop is grown in about 154 million ha every year, that is, on approximately 11% area of total cultivated land. India is next to China in terms of production and cultivation over an area of 45.50 million ha (Gujja and Thiyagarajan, 2012). In India, rice covers about one-fourth area of total cropped area and contributed to about 40–43% of total food grain. Rice plays a key role in the food security system (Verma and Srivastav, 2017).

The major plant nutrients N, P, and K are among essential inputs of necessary to the crop. These are necessary for maximum growth and good harvest of rice. The requirement of major plant nutrients cannot be achieved by inorganic or organic sources of nutrients alone. Chemical fertilizers are well known for their effects on the yield increment (Prakash et al., 2002). An appropriate set of organic and inorganic sources of nutrients, that is, integration can help in achieving maximum yield of rice. Efforts are to be made to increase the yield of rice along with maintain their quality by integrated nutrient management practices.

The role of plant nutrients would be extremely important from sustainability point of view. With the increasing trend in price of fertilizers and reduction in the use of chemical fertilizers it has become important to manage integration of inorganic and organic sources of nutrients in judicious way. Therefore, information needs to be generated with respect to proper dose of organic manures along with inorganic fertilizers. So, there is need to generate a suitable combination of nutrient management practice that enables to bring better quality and productivity of rice along with increment in nutrient use efficiency (NUE).

It is well-known fact that alone organic manure or chemical fertilizer can bring stability in the crop yield in intensive farming system. Contrary to detrimental effects of inorganic fertilizers, organic manures are available indigenously which improve soil health resulting in enhanced crop yield. Secondly, alone organic manure cannot provide appropriate amount of plant nutrient because they contain less amount of plant nutrient and are very bulky. Hence, to provide all plant nutrients in available form and to keep soil health in good condition, it is essential to use organic sources in combination with chemical fertilizer to achieve higher yield (Rama Lakshmi et al., 2012).

Use of organic manure has been found to be promising in arresting the decline in productivity by overcoming on deficiency of necessary secondary and micronutrients and its beneficial influence on soil properties. Organic production systems maintained and improved the soil health through stimulating the activity of soil organisms and organic manures are helpful in decreasing deficiency of secondary and micronutrients and are able to sustain crop production. The manures improve the physical properties of soil and increase NUE of supplied nutrients (Pandey et al., 2007).

Keeping the increment in cost of chemical fertilizer in view, it is necessary to use organic manure in large quantity. In India, which is a tropical country, use of organic manures in soil is very necessary to maintain soil health and soil fertility to obtain good crop production (Umashankar et al., 2005). Currently, escalation in cost of fertilizer, environmental issues, and conservation of energy have created interest to use the organic manures as a nutrient source and also in integrated nutrient management system (Medhi et al., 1996).

Integrated nutrient management approach of plant nutrition can play important role in solving problem of nutrient mining. In simple words, it includes the judicious use of organic, chemical, and microbial sources to improve soil health and to achieve optimum yield which are economically and practically feasible and ecological friendly (Biswas et al., 1987).

Integration of plant nutrient involves improvement and maintenance of fertility of soil, crop production, and farmers' benefit through integration of organic manures, inorganic fertilizers, and biofertilizers. The NUE is greatly influenced by the use organic sources of plant nutrients. Using organic resources such as farmyard manure (FYM), poultry manure (PM), green manure, etc., deserves priority for sustained production and better resource utilization in integrated nutrient management. In this system, the use of chemicals is kept at its minimum, that is, to the level of bare necessity. Compared to chemical farming, this method was self-sufficient, self-dependent, and relies on more biological inputs (Singh et al., 2001).

Pelleting technique of integrated nutrient management; it is a type of slow-releasing source of nutrients with long-term effects including reduced losses and enhanced uptake of nutrients. The pellet preparation is capable of converting manures into a condensed form which is convenient in transportation, their handling and storage (Bhattacharya et al., 1989), and adjusting the nutrient content by adding chemical fertilizers. The small mass of pellet contains all essential major nutrients and micronutrients and releases these nutrients slowly in root zones, making them available for long time and reducing their losses in different forms. Thus, increases NUE.

Keeping these points in view, a field experiment was conducted at Agriculture College Research Farm, Indira Gandhi Krishi Vishwavidyalaya, Raipur, Chhattisgarh, India to study the "Alternate Strategies of Integrated

Nutrient Management by Innovative Pelleting Techniques to Increase Growth, Yield, and Nutrient Use Efficiency in Transplanted Rice" with the objectives to determine the effect of conventional and pelleted integrated nutrient management on morphophysiological traits and yield of rice; to determine the effect of integrated nutrient management on NUE and uptake of nutrients by transplanted rice and to work out the benefit:cost (B:C) ratio of integrated nutrient management treatments.

6.2   MATERIALS AND METHODS

6.2.1   WEATHER CONDITION DURING THE CROP GROWTH PERIOD

During crop growth period, the maximum temperature ranges between 25.8°C and 31.9°C. The minimum temperature ranged between 25.8°C in the last week of July and 17.3°C in the first week of November. Crop received sunshine for 0.0–8.6 hours per day. The maximum and minimum humidity during the crop period was 95% and 37%, respectively. A total of 1382.1-mm rainfall was received during the crop period. The wind velocity ranged between 0.76 and 11.1 km per hours. Thus, except sunshine hours, which are a general feature of rainy season, other weather parameters were favorable during crop growth period.

6.2.2   EXPERIMENTAL SITE

The experiment was carried out at Research Farm, College of Agriculture, Indira Gandhi Krishi Vishwavidyalaya, Raipur, India in bunded rice block. The rice followed by chickpea crop rotation was being used since last 5 years.

6.2.3   PHYSICOCHEMICAL PROPERTIES OF EXPERIMENTAL SOIL

Random soil samples were collected up to 20 cm soil depth from five places to determine the physical and chemical properties of the experimental soil. The method adopted for analysis of physical and chemical properties of soil and their values obtained are presented in Table 6.1.

TABLE 6.1   Physicochemical Properties of the Experimental Soil.

No.	Particulars	Values	Rating	Methods used
A. Physical properties				
1.	**Mechanical composition**			
	Sand (%)	53.5	Inceptisol	International pipette method (Black, 1965)
	Silt (%)	24.5		
	Clay (%)	22.0		
B. Chemical composition				
1.	Organic carbon (%)	0.47	Low	Walkley and Black's rapid titration method (Jackson, 1967)
2.	Available N (kg ha^{-1})	201	Low	Alkaline permanganate method (Subbaiah and Asija, 1956)
3.	Available P (kg ha^{-1})	19.6	Medium	Olsen's method (Olsen, 1954)
4.	Available K (kg ha^{-1})	282	High	Flame photometric method (Jackson, 1967)
5.	pH (1:2.5, soil:water)	7.44	Neutral	Glass electrode pH meter (Piper, 1967)
6.	Electrical conductivity (m mhos m^{-1} at 25°C)	0.39	Normal	Solubridge method, (Black, 1965)
C. Physical composition				
1.	Bulk density (g cc^{-1}) 0–30 cm of soil depth	1.20		Soil core method (Bodman, 1942)

6.2.4   TESTED VARIETY

Rice variety MTU-1010 was used as test crop in the experiment. It is a semi-dwarf variety which is a cross of Krishnaveni and IR-64. It is recommended for direct seeding and transplanting in upland and lowland ecosystem of different states (including Chhattisgarh), respectively. The crop matures in about 115–125 days. This variety is medium-grain type and has yield potential of 5–6 t ha^{-1}.

6.2.5   TREATMENTS

In all 13 treatments, inorganic, organic, and integrated nutrients management treatments were carried out in randomized block design in three replications. The treatments used in experiment are summarized in Table 6.2.

TABLE 6.2   List of Treatments Used in the Study (Name of These Treatments is Used as T_1, T_2, and T_3 like This at Further Part/Figures/Tables of This Paper).

Treatments	
T_1	Conventional 100:60:40 kg N:P$_2$O$_5$:K$_2$O ha^{-1}
T_2	Conventional 2.5 t FYM + 80:50:30 kg N:P$_2$O$_5$:K$_2$O ha^{-1}
T_3	Conventional 2.5 t FYM + 60:40:20 kg N:P$_2$O$_5$:K$_2$O ha^{-1}
T_4	Conventional 5 t FYM + 50:30:20 kg N:P$_2$O$_5$:K$_2$O ha^{-1}
T_5	Conventional 1 t PM + 50:30:20 kg N:P$_2$O$_5$:K$_2$O ha^{-1}
T_6	Conventional 2.5 t FYM + 0.5 t PM + 50:30:20 kg N:P$_2$O$_5$:K$_2$O ha^{-1}
T_7	Pelleted 100:60:40 kg N:P$_2$O$_5$:K$_2$O ha^{-1}
T_8	Pelleted 2.5 t FYM + 80:50:30 kg N:P$_2$O$_5$:K$_2$O ha^{-1}
T_9	Pelleted 2.5 t FYM + 60:40:20 kg N:P$_2$O$_5$:K$_2$O ha^{-1}
T_{10}	Pelleted 5 t FYM + 50:30:20 kg N:P$_2$O$_5$:K$_2$O ha^{-1}
T_{11}	Pelleted 1 t PM + 50:30:20 kg N:P$_2$O$_5$:K$_2$O ha^{-1}
T_{12}	Pelleted 2.5 t FYM + 0.5 t PM+50:30:20 kg N:P$_2$O$_5$:K$_2$O ha^{-1}
T_{13}	Control (no fertilizer)

FYM, farmyard manure; K, K$_2$O; N, nitrogen; P, P$_2$O$_5$; PM, poultry manure.

Note: Quantity in kilogram per hectare.

6.2.6   PELLET PREPARATION

The pellet of chemical fertilizer was first prepared by mixing chemical fertilizer with inert material. The FYM/PM or both (as per the treatment) were coated over the pellet of chemical fertilizer with the help of binding agent. The pellets were made by hand, though can be made by machine. The pellets of marble or pea size were applied in the root zone before transplanting on the line of marked with ropes for transplanting. A simple structure of pellet is given in Figure 6.1.

6.2.7   NUTRIENTS MANAGEMENT PRACTICES

Nutrients management practices (chemical fertilizers and organic manures) were applied according to combination of treatments. Nutrients were applied through urea, diammonium phosphate (DAP), single superphosphate (SSP), muriate of potash (MOP), FYM, and PM, respectively. The total amount of DAP, SSP, and MOP was applied as basal dressing, while urea was applied in three splits first as basal, second at tillering, and last at panicle stage as

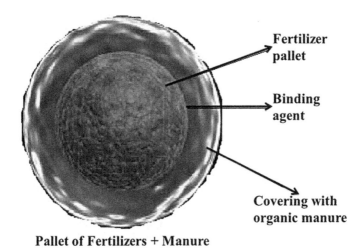

Pallet of Fertilizers + Manure

FIGURE 6.1 General structure of pellet.

per the treatment dose. The basal dose were given in the form of pellets by blending fertilizers with FYM and PM in pelleted treatments, while in conventional method chemical fertilizers were blended with FYM and PMs and broadcasted in the field before transplanting. The combinations of fertilizers and organic manures were applied after the transplanting of seedlings.

6.3 RESULTS AND DISCUSSION

6.3.1 WEATHER CONDITIONS

The vegetative growth and yield of crop is mostly governed by the genotype, agromanagement, and the environment to which the crop is exposed. Among the environmental factors, the prevailing weather condition such as rainfall, relative humidity, temperature, evaporation, solar radiation, etc., play crucial role in the performance of crop under a particular farming situation.

The weather elements, namely, temperature, solar radiation, and rainfall affect the rice yield directly by affecting the physiological processes involved in grain production and indirectly through diseases and insects. In general, the temperature regimes of tropics are favorable for rice growth throughout the year. Extreme temperatures are destructive to plant growth. Depending on stages of crop growth, injury to rice may occur when the daily mean temperature drops below 20°C or rises above 35°C (Yoshida, 1981).

However, during crops period, the weakly mean maximum and minimum temperature were 43.4 °C and 17.3°C, respectively. The rainfall was 1382.1 mm, which was supposed to be sufficient for rice growth but its intensity and distribution was uncommon and therefore, irrigation were given to the crop as per need to bring the saturated condition in the field. The intermittent bright sunshine favored the crop growth and development. The weekly average of sunshine hours from initial reproduction to harvest varied from 0 to 8.6 h day^{-1} during *Kharif*, 2012. All these parameters were congenial for good rice crop growth; as a result, the crop was normal and there was not any adverse effect on rice due to weather.

6.3.2 PREHARVEST OBSERVATIONS

6.3.2.1 PLANT HEIGHT (cm)

The data on mean plant height at various stages of rice growth are presented in Figure 6.2. The plant height under different conventional and pelleted integrated nutrient management treatments increased with the advancement of crop age. The Pelleted 2.5 t FYM + N_{80}:P_{50}:K_{30} (T_8) treatment registered significantly highest plant height at 30 days after treatment (DAT), 60 DAT, and at harvest; Pelleted 5 t FYM + N_{50}:P_{30}:K_{20} (T_{10}) recorded statistically similar plant height. Treatments conventional N_{100}:P_{60}:K_{40} (T_1) and conventional 2.5 t FYM + N_{80}:P_{50}:K_{30} (T_2) at 60 DAT and at harvest and pelleted 2.5 t FYM + N_{60}:P_{40}:K_{20} (T_9) at harvest also recorded comparable plant height. The minimum plant height at all the stages was recorded in control treatment (T_{13}). The increment in plant height might be due to slow release of nutrients by integration of sources of nutrients especially due to organic manures and chemical fertilizers. The variation in plant height might be due to variation in availability of major nutrients (Siavoshi, 2011). Sharma et al. (2015b) reported similar result in pelleted integrated nutrient management in transplanted rice and Muhammad et al. (2008) observed similar results with application of organic manure and compost in rice.

6.3.2.2 NUMBER OF TILLERS PER HILL

The data related to number of tillers per hill are presented in Figure 6.3. The pelleted 2.5 t FYM + N_{80}:P_{50}:K_{30} (T_8) treatment produced significantly highest number of tillers per hill at all stages of growth. At all the stages,

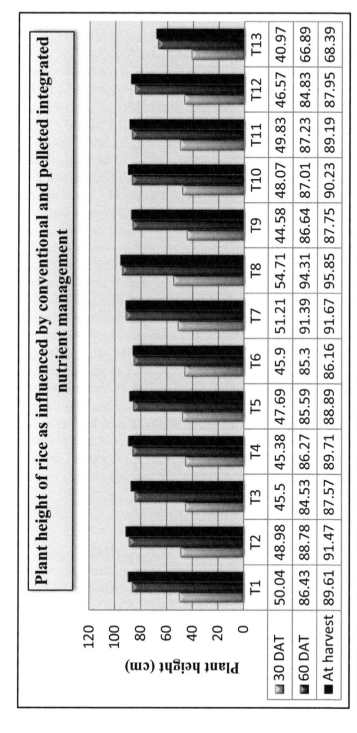

FIGURE 6.2 (See color insert.) Plant height of rice as influenced by conventional and pelleted integrated nutrient management.

treatments pelleted $N_{100}:P_{60}:K_{40}$ (T_7) and pelleted 5 t FYM + $N_{50}:P_{30}:K_{20}$ (T_{10}) recorded comparable number of tillers per hill to that of highest performing treatment. At 30 DAT, treatments conventional $N_{100}:P_{60}:K_{40}$ (T_1), conventional 2.5 t FYM + $N_{80}:P_{50}:K_{30}$ (T_2), conventional 5 t FYM + $N_{50}:P_{30}:K_{20}$ (T_4), pelleted 2.5 t FYM + $N_{60}:P_{40}:K_{20}$ (T_9), and pelleted 2.5 t FYM + 0.5 t PM+ $N_{50}:P_{30}:K_{20}$ (T_{12}); at 60 DAT, conventional $N_{100}:P_{60}:K_{40}$ (T_1), conventional 2.5 t FYM + $N_{80}:P_{50}:K_{30}$ (T_2), and pelleted 2.5 t FYM + $N_{60}:P_{40}:K_{20}$ (T_9); and at harvest, treatments conventional $N_{100}:P_{60}:K_{40}$ (T_1), conventional 2.5 t FYM + $N_{80}:P_{50}:K_{30}$ (T_2), conventional 5 t FYM + $N_{50}:P_{30}:K_{20}$ (T_4), conventional 2.5 t FYM + 0.5 t PM + $N_{50}:P_{30}:K_{20}$ (T_6), pelleted 2.5 t FYM + $N_{60}:P_{40}:K_{20}$ (T_9), pelleted 1 t PM + $N_{50}:P_{30}:K_{20}$ (T_{11}), and pelleted 2.5 t FYM + 0.5 t PM+ $N_{50}:P_{30}:K_{20}$ (T_{12}) also recorded at par number of tillers per hill as compared to highest performing treatment pelleted 2.5 t FYM + $N_{80}:P_{50}:K_{30}$ (T_8). The least number of tillers per hill was recorded in control treatment (T_{13}), which was significantly inferior to all other treatments at different stages of crop growth. It might be due to insufficient supply of nutrients (Jha et al., 2004). The higher number of tillers per hill observed in above treatments may be due to the fact that application of organic sources of nutrients and their combination with inorganic sources of nutrients leads to greater availability of nutrients to plants at all the stages of crop growth. Miller (2007) reported that organic sources offer more balanced nutrition to the plants, especially micronutrients which positively affect number of tillers in plants. Increased plant height helped in increasing the photosynthetic area for photosynthesis in plant, which in turn helped in formation of new tillers. Similar results were also obtained by Sarawgi and Sarawgi (2004) and Jha et al. (2006). Sharma et al. (2015b) reported similar result in pelleted integrated nutrient management in transplanted rice.

6.3.2.3 NUMBER OF LEAVES PER PLANT

The number of leaves was recorded at 30 DAT, 60 DAT, and at harvest (Fig. 6.4). The application of conventional and pelleted integrated nutrient management treatments significantly influenced production of leaves at all crop stages. The pelleted 2.5 t FYM + $N_{80}:P_{50}:K_{30}$ (T_8) produced significantly highest number of leaves per plant at 30 DAT, 60 DAT, and at harvest, which were comparable to pelleted $N_{100}:P_{60}:K_{40}$ (T_7) and pelleted 5 t FYM + $N_{50}:P_{30}:K_{20}$ (T_{10}) at all the stages of crop growth. The conventional 2.5 t FYM + $N_{80}:P_{50}:K_{30}$ (T_2) also recorded similar number of leaves comparable to highest number of leaves producing treatment (pelleted 2.5 t FYM

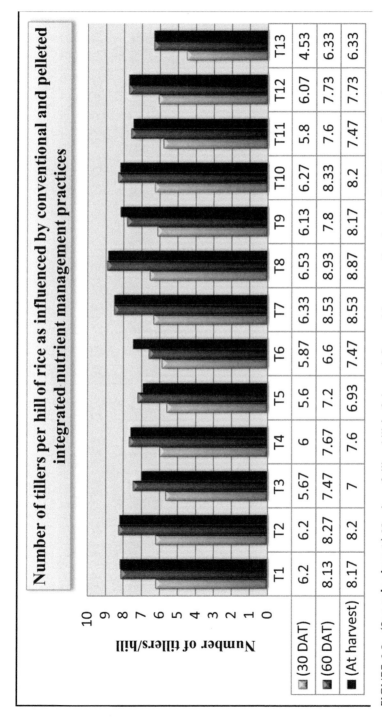

Number of tillers per hill of rice as influenced by conventional and pelleted integrated nutrient management practices

	T1	T2	T3	T4	T5	T6	T7	T8	T9	T10	T11	T12	T13
(30 DAT)	6.2	6.2	5.67	6	5.6	5.87	6.33	6.53	6.13	6.27	5.8	6.07	4.53
(60 DAT)	8.13	8.27	7.47	7.67	7.2	6.6	8.53	8.93	7.8	8.33	7.6	7.73	6.33
(At harvest)	8.17	8.2	7	7.6	6.93	7.47	8.53	8.87	8.17	8.2	7.47	7.73	6.33

FIGURE 6.3 **(See color insert.)** Number of tillers hill[-1] of rice as influenced by conventional and pelleted integrated nutrient management practices.

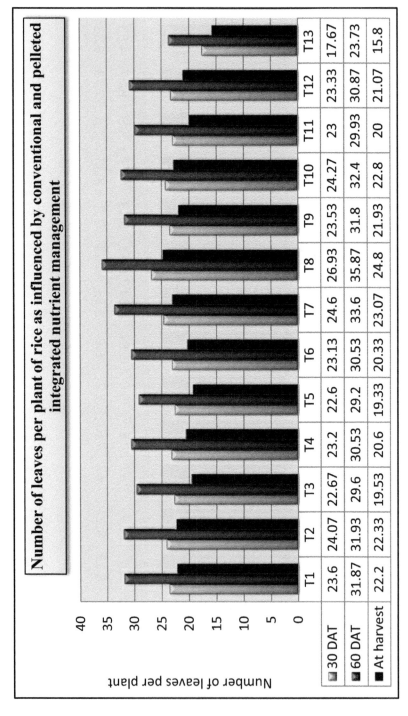

FIGURE 6.4   (See color insert.) Number of leaves plant[-1] of rice as influenced by conventional and pelleted integrated nutrient management.

+ $N_{80}:P_{50}:K_{30}$ (T_8)) at 30 DAT. The lowest numbers of leaves were observed under control treatment (T_{13}) at all stages of crop growth. At harvesting stage, number of leaves was, in general, decreased.

The production of leaves is a function of number of tillers produced in crop. The rice under pelleted 2.5 t FYM + $N_{80}:P_{50}:K_{30}$ (T_8) provided sufficient nutrients for plant and root growth and proliferation enhanced uptake and mobilization of nutrients and produced more number of tillers, which finally increased the number of leaves (Raju et al., 1989). Singh and Agarwal (2001) also explained that increase in number of leaf and size of leaf may be due to sufficient availability of nutrients that increased nutrient absorption capacity of plant resulting in better root growth and development and increased translocation of carbohydrates from source to growing points. The treatments which increased number of tillers also produced more number of leaves as also reported by Pandey et al. (1998). Sharma et al. (2015b) reported similar result in pelleted integrated nutrient management in transplanted rice.

6.3.2.4   LEAF AREA PER PLANT

The leaf area per plant was significantly affected due to conventional and pelleted integrated nutrient management treatments (Fig. 6.5). The pelleted 2.5 t FYM + $N_{80}:P_{50}:K_{30}$ (T_8) produced significantly highest leaf area per plant at 30 DAT, 60 DAT, and at harvest, which was at par to that pelleted $N_{100}:P_{60}:K_{40}$ (T_7) at 30 DAT and 60 DAT; and pelleted 5 t FYM + $N_{50}:P_{30}:K_{20}$ (T_{10}) at 30 DAT and 60 DAT; and conventional 2.5 t FYM + $N_{80}:P_{50}:K_{30}$ (T_2) also produced comparable leaf area to that highest performing pelleted 2.5 t FYM + $N_{80}:P_{50}:K_{30}$ (T_8) treatment. The control treatment (T_{13}) observed lowest leaf area at all growth stages of crop.

The increased number of leaves enhanced the leaf area per plant. The pelleted 2.5 t FYM + $N_{80}:P_{50}:K_{30}$ (T_8) helped in increasing the nutrient supply, photosynthetic efficiency, and formation of new leaves. Sharma et al. (2015b) reported similar result in pelleted integrated nutrient management in transplanted rice.

6.3.2.5   DRY MATTER ACCUMULATION PER HILL (g HILL^{-1})

The dry matter accumulation per hill was increased with the advancement of crop age (Fig. 6.6). The integration of organic and inorganic sources of nutrients significantly influenced the dry matter accumulation at all the

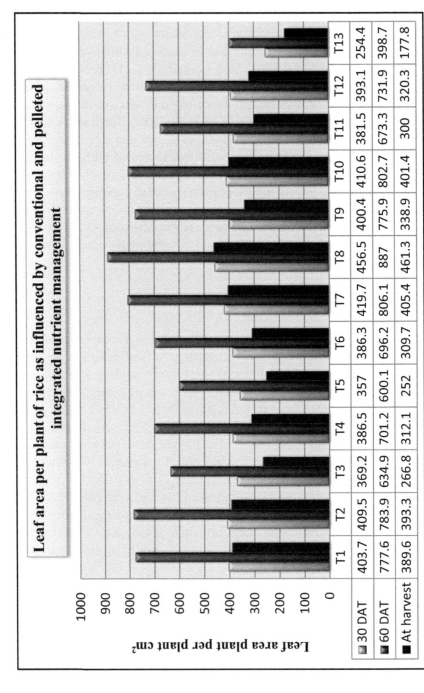

FIGURE 6.5   (See color insert.) Leaf area plant[-1] of rice as influenced by conventional and pelleted integrated nutrient management.

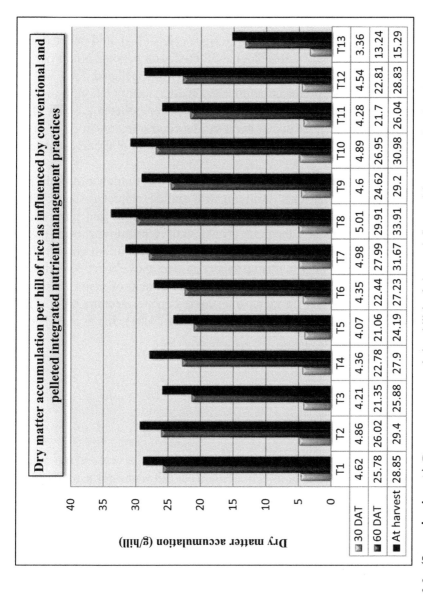

FIGURE 6.6 **(See color insert.)** Dry matter accumulation hill^{-1} of rice as influenced by conventional and pelleted integrated nutrient management practices.

stages of crop growth. At 30 DAT, pelleted 2.5 t FYM + N_{80}:P_{50}:K_{30} (T_8) registered significantly highest dry matter accumulation which was at par with conventional N_{100}:P_{60}:K_{40} (T_1), conventional 2.5 t FYM + N_{80}:P_{50}:K_{30} (T_2), conventional 5 t FYM + N_{50}:P_{30}:K_{20} (T_4), conventional 2.5 t FYM + 0.5 t PM + N_{50}:P_{30}:K_{20} (T_6), pelleted N_{100}:P_{60}:K_{40} (T_7), pelleted 2.5 t FYM + N_{60}:P_{40}:K_{20} (T_9), pelleted 5 t FYM + N_{50}:P_{30}:K_{20} (T_{10}), and pelleted 2.5 t FYM + 0.5 t PM+ N_{50}:P_{30}:K_{20} (T_{12}) at 30 DAT; and pelleted N_{100}:P_{60}:K_{40} (T_7) at 45 DAT. The pelleted N_{100}:P_{60}:K_{40} (T_7) and pelleted 5 t FYM + N_{50}:P_{30}:K_{20} (T_{10}) also produced similar dry matter accumulation to that of highest dry matter accumulation producing treatment pelleted 2.5 t FYM + N_{80}:P_{50}:K_{30} (T_8) at 60 DAT and at harvest. The control treatment (T_{13}) gave lowest dry matter accumulation at 30 DAT, 60 DAT, and at harvest stages of crop.

The higher dry matter accumulation might be due to higher availability and translocation of nutrients during growth and development stages. The dry matter accumulation depends upon the photosynthesis and respiration rate, which finally increase the plant growth with respect to increased plant height, tillers, etc. The treatment which attained the maximum growth also accumulated the higher dry matter up to harvest. The above better performing treatment as noticed released sufficiently higher amount of major nutrients which might have helped in increasing the dry matter accumulation. The results have been confirmed by Murali and Setty (2001), Jha et al. (2006), and Sharma et al. (2015b).

6.3.3 POSTHARVEST OBSERVATIONS

6.3.3.1 NUMBER OF EFFECTIVE TILLERS PER HILL

The number of effective tillers per hill were counted at harvest and presented in Figure 6.7. The number of effective tillers per hill was significantly influenced due to different integrated nutrient management treatments. The treatment, pelleted 2.5 t FYM + N_{80}:P_{50}:K_{30} (T_8) produced significantly higher number of effective tillers per hill as compared to others; however, it was comparable to that of conventional N_{100}:P_{60}:K_{40} (T_1), conventional 2.5 t FYM + N_{80}:P_{50}:K_{30} (T_2), pelleted N_{100}:P_{60}:K_{40} (T_7), pelleted 2.5 t FYM + N_{60}:P_{40}:K_{20} (T_9), pelleted 5 t FYM + N_{50}:P_{30}:K_{20} (T_{10}), and pelleted 2.5 t FYM + 0.5 t PM+ N_{50}:P_{30}:K_{20} (T_{12}). The lowest number of effective tillers per hill was obtained under control treatment (T_{13}).

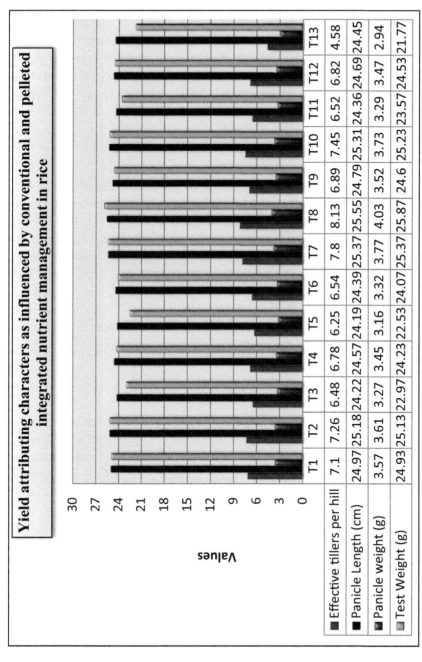

FIGURE 6.7 **(See color insert.)** Yield attributing characters as influenced by conventional and pelleted integrated nutrient management in rice.

Increase in plant height and number of tillers in successive growth stages helped in increasing the effective tillers. It may be due to higher photosynthesis efficiency and net assimilation, which helped in increased overall growth of the plant. All the treatments, except control, recorded higher number of effective tillers which might be due to higher and balanced availability of nutrients to the plants. Similar results have also been found by Murali and Setty (2001), Mandal et al. (2004), Mhaskar et al. (2005), and Sharma et al. (2015b).

6.3.3.2   PANICLE LENGTH (cm)

Data pertaining to length of panicle are presented in Figure 6.7. The findings revealed that length of panicle varied greatly due to different integrated nutrient management. The treatment, pelleted 2.5 t FYM + $N_{80}:P_{50}:K_{30}$ (T_8) produced significantly longest panicle, while treatments conventional $N_{100}:P_{60}:K_{40}$ (T_1), conventional 2.5 t FYM + $N_{80}:P_{50}:K_{30}$ (T_2), conventional 5 t FYM + $N_{50}:P_{30}:K_{20}$ (T_4), pelleted $N_{100}:P_{60}:K_{40}$ (T_7), pelleted 2.5 t FYM + $N_{60}:P_{40}:K_{20}$ (T_9), pelleted 5 t FYM + $N_{50}:P_{30}:K_{20}$ (T_{10}), and pelleted 2.5 t FYM + 0.5 t PM+ $N_{50}:P_{30}:K_{20}$ (T_{12}) were at par with that of treatment pelleted 2.5 t FYM + $N_{80}:P_{50}:K_{30}$ (T_8). The control treatment (T_{13}) recorded shortest panicle length.

6.3.3.3   WEIGHT OF PANICLE (g)

The data of panicle weight are presented in Figure 6.7. The results revealed that treatment pelleted 2.5 t FYM + $N_{80}:P_{50}:K_{30}$ (T_8) gained significantly highest panicle weight; however, it was statistically similar to that of pelleted $N_{100}:P_{60}:K_{40}$ (T_7) and pelleted 5 t FYM + $N_{50}:P_{30}:K_{20}$ (T_{10}). The control treatment (T_{13}) registered lowest panicle weight.

6.3.3.4   TEST WEIGHT (g)

The result of test weight is presented in Figure 6.7. The pelleted 2.5 t FYM + $N_{80}:P_{50}:K_{30}$ (T_8) recorded significantly higher test weight than others; however, it was comparable to that of conventional $N_{100}:P_{60}:K_{40}$ (T_1), conventional 2.5 t FYM + $N_{80}:P_{50}:K_{30}$ (T_2), conventional 5 t FYM + $N_{50}:P_{30}:K_{20}$ (T_4), conventional 2.5 t FYM + 0.5 t PM + $N_{50}:P_{30}:K_{20}$, pelleted $N_{100}:P_{60}:K_{40}$ (T_7), pelleted 2.5 t FYM + $N_{60}:P_{40}:K_{20}$ (T_9), pelleted 5 t FYM

+ N_{50}:P_{30}:K_{20} (T_{10}) and pelleted 2.5 t FYM + 0.5 t PM+ N_{50}:P_{30}:K_{20} (T_{12}). The control treatment (T_{13}) recorded lowest test weight among the entire integrated nutrient management treatments. Bagheri et al. (2011) reported that in corn production, the increase in test weight is due to slow release nitrogen from pelleted fertilizer and their prolonged availability for plant uptake in seed stage. The same results were reported by El-Kramany (2001) and Amany et al. (2006) in wheat and maize, respectively, and Sharma et al. (2015b) in pelleted integrated nutrient management in transplanted rice.

6.3.3.5 NUMBER OF SPIKELETS PER PANICLE (FILLED, UNFILLED, AND TOTAL)

The data related to total spikelets per panicle, filled spikelets per panicle, and unfilled spikelets per panicle are presented in Figure 6.8. Pelleted 2.5 t FYM + N_{80}:P_{50}:K_{30} (T_8) produced significantly higher number of total spikelets per panicle and filled spikelets per panicle as compared to others, while it was at par with the pelleted N_{100}:P_{60}:K_{40} (T_7), pelleted 2.5 t FYM + N_{60}:P_{40}:K_{20} (T_9), and conventional 2.5 t FYM + N_{80}:P_{50}:K_{30} (T_2). Further, pelleted 2.5 t FYM + 0.5 t PM+ N_{50}:P_{30}:K_{20} (T_{12}) was also at par to that of treatment pelleted 2.5 t FYM + N_{80}:P_{50}:K_{30} (T_8) with regards to number of total spikelets per panicle. The number of unfilled spikelets per panicle was observed minimum in pelleted 2.5 t FYM + N_{60}:P_{40}:K_{20} (T_9), while maximum number of unfilled spikelets per panicle was found in treatment pelleted 2.5 t FYM + 0.5 t PM+ N_{50}:P_{30}:K_{20} (T_{12}).

The application of organic sources of nutrients alone or in combination with inorganic sources of nutrients in conventional and pelleted form might have helped in improving the nutrient availability for a prolonged period during crop growth and development stages; ultimately it influenced the reproductive stage and resulted in more number of grains, filled grains per panicle, and test weight. The nutrient availability and their influence by organic sources and their combination with inorganic sources on crop growth has been confirmed by Sarawgi and Sarawgi (2004), Mandal et al. (2004), Chandrakar et al. (2004), and Paraye et al. (2006).

6.3.3.6 STERILITY PERCENTAGE

The data related to sterility percentage are presented in Figure 6.8. The treatment pelleted 2.5 t FYM + N_{80}:P_{50}:K_{30} (T_8) recorded least sterility percent

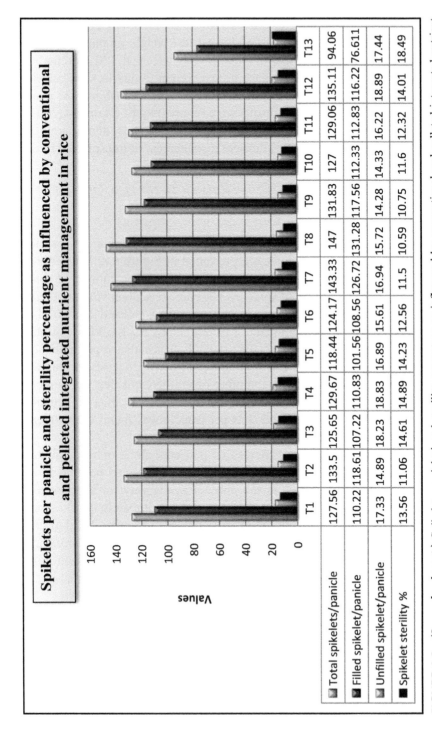

The figure presents the data in the following table:

	T1	T2	T3	T4	T5	T6	T7	T8	T9	T10	T11	T12	T13
Total spikelets/panicle	127.56	133.5	125.65	129.67	118.44	124.17	143.33	147	131.83	127	129.06	135.11	94.06
Filled spikelet/panicle	110.22	118.61	107.22	110.83	101.56	108.56	126.72	131.28	117.56	112.33	112.83	116.22	76.611
Unfilled spikelet/panicle	17.33	14.89	18.23	18.83	16.89	15.61	16.94	15.72	14.28	14.33	16.22	18.89	17.44
Spikelet sterility %	13.56	11.06	14.61	14.89	14.23	12.56	11.5	10.59	10.75	11.6	12.32	14.01	18.49

Spikelets per panicle and sterility percentage as influenced by conventional and pelleted integrated nutrient management in rice

FIGURE 6.8 (See color insert.) Spikelets panicle^{-1} and sterility percentage as influenced by conventional and pelleted integrated nutrient management in rice.

of spikelets per panicle. The highest sterility percentage was recorded under control treatment (T_{13}).

6.3.3.7 GRAIN YIELD (t ha^{-1})

The yield of grain was significantly influenced due to conventional and pelleted techniques of integrated nutrient management (Fig. 6.9). The pelleted 2.5 t FYM + N_{80}:P_{50}:K_{30} (T_8) produced the significantly highest grain yield, which was at par to that of pelleted N_{100}:P_{60}:K_{40} (T_7) and pelleted 5 t FYM + N_{50}:P_{30}:K_{20} (T_{10}). The increase in yield of grain was mainly associated with significant increase in effective tillers per panicle, filled spikelets per panicle, and test weight. The increase in above yield components was mainly due to proper supply and efficient utilization of nutrients by plants. The higher grain yield may be due to the application of organic sources of nutrients alone or in combination with inorganic sources of nutrients resulted to greater availability of essential nutrients to plants, improvement of soil environment which facilitated in better root proliferation leading to higher absorption of water, and nutrients and their translocation from source to sink (Ebaid et al., 2007), ultimately resulting in higher yield.

It is well known fact that blending of nitrogen with FYM helps in continuous supply of the nutrients, reduced nutrient loss, and enhanced the NUE and yield. The similar results have also been found by Pandey and Nandeha (2004), Sarawgi and Sarawgi (2004), Jha et al. (2006), Sharma et al. (2014), and Sharma et al. (2015b).

6.3.3.8 STRAW YIELD (t ha^{-1})

The results of straw yield as influenced by different treatments are presented in Figure 6.9. The treatment, pelleted 2.5 t FYM + N_{80}:P_{50}:K_{30} (T_8) produced significantly higher straw yield as compared to others; however, it was comparable to that obtained under pelleted N_{100}:P_{60}:K_{40} (T_7), pelleted 2.5 t FYM + N_{60}:P_{40}:K_{20} (T_9), pelleted 5 t FYM + N_{50}:P_{30}:K_{20} (T_{10}), pelleted 2.5 t FYM + 0.5 t PM + 50:30:20 kg N:P_2O_5:K_2O ha^{-1} (T_{12}), conventional 2.5 t FYM + N_{80}:P_{50}:K_{30} (T_2), and conventional N_{100}:P_{60}:K_{40} (T_1). The lowest straw yield was recorded under control treatment (T_{13}). Sharma et al. (2015a) reported similar result in pelleted integrated nutrient management in transplanted rice.

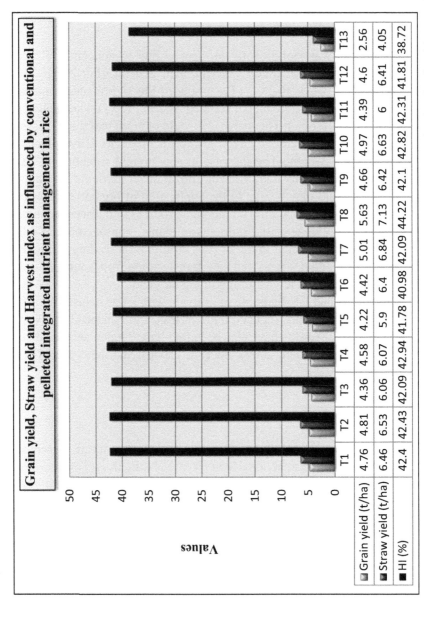

FIGURE 6.9 (**See color insert.**) Grain yield, straw yield, and harvest index as influenced by conventional and pelleted integrated nutrient management in rice.

6.3.3.9   HARVEST INDEX

Harvest index influenced significantly due to the application of organic sources of nutrients alone or in combination with inorganic sources of nutrients in conventional and pelleted techniques (Fig. 6.9). The treatment pelleted 2.5 t FYM + N_{80}:P_{50}:K_{30} (T_8) recorded significantly higher harvest index than control treatment (T_{13}), which recorded lowest harvest index, all the treatments were found to be similar to the highest performing harvest index treatment. Rizwan et al. (2003) found that harvest index is less sensitive to N fertilizer timing and splitting. Sharma et al. (2014) reported similar result in pelleted integrated nutrient management in transplanted rice.

6.3.4   NUTRIENT STATUS

6.3.4.1   NUTRIENT UPTAKE (kg ha^{-1})

The uptake of nutrients is the function of dry matter production and concentration of nutrients. The data presented in Figure 6.10 revealed that pelleted 2.5 t FYM + N_{80}:P_{50}:K_{30} (T_8) produced significantly highest uptake of total N (93.17 kg ha^{-1}) and total P (19.18 kg ha^{-1}), which was at par to pelleted N_{100}:P_{60}:K_{40} (T_7). The pelleted 2.5 t FYM + N_{80}:P_{50}:K_{30} (T_8) produced significantly highest total K uptake which was at par to that of pelleted N_{100}:P_{60}:K_{40} (T_7), pelleted 5 t FYM + N_{50}:P_{30}:K_{20} (T_{10}), and conventional 2.5 t FYM + N_{80}:P_{50}:K_{30} (T_2). The control treatment (T_{13}) recorded least total N, P, and K uptake. The increased nitrogen and phosphorus uptake by grain and straw was mainly due to increased grain and concentration of the nutrients, which has also been observed by Natarajan (1994) and Makarim et al. (2005). Sharma et al. (2014) reported similar result in pelleted integrated nutrient management in transplanted rice.

6.3.4.2   NUTRIENT USE EFFICIENCY (kg kg^{-1} NUTRIENT APPLIED)

The NUE for nitrogen, phosphorus, and potassium has been calculated and presented in Figure 6.11. The data showed that the highest NUE due to nitrogen was obtained in pelleted 2.5 t FYM + N_{80}:P_{50}:K_{30} (T_8) treatment followed by pelleted 5 t FYM + N_{50}:P_{30}:K_{20} (T_{10}) and pelleted 2.5 t FYM + N_{60}:P_{40}:K_{20} (T_9). The NUE due to phosphorus was obtained by pelleted 5 t FYM + N_{50}:P_{30}:K_{20} (T_{10}) followed by pelleted 2.5 t FYM + N_{80}:P_{50}:K_{30}

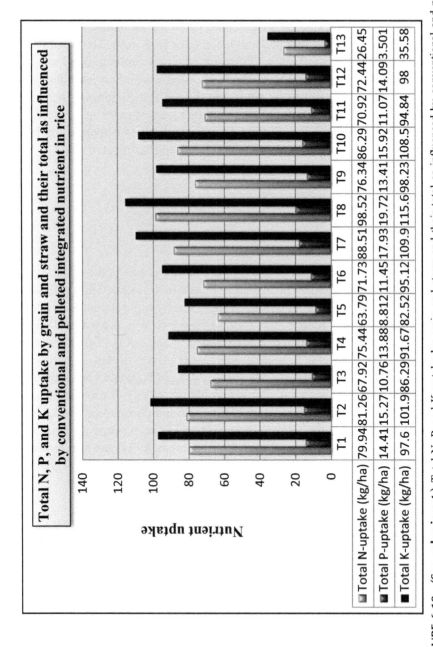

FIGURE 6.10   (See color insert.) Total N, P, and K uptake by grain and straw and their total as influenced by conventional and pelleted integrated nutrient in rice.

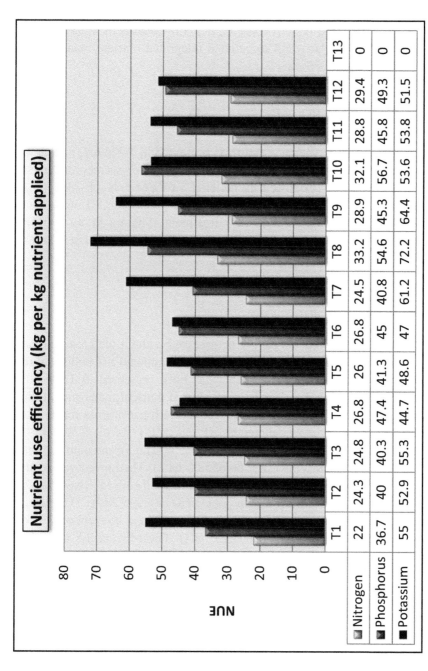

FIGURE 6.11 **(See color insert.)** Nutrient use efficiency of rice as influenced by conventional and pelleted integrated nutrient management.

(T_8) treatment. The highest NUE due to potassium was found under pelleted 2.5 t FYM + $N_{80}:P_{50}:K_{30}$ (T_8) treatment followed by pelleted 2.5 t FYM + $N_{60}:P_{40}:K_{20}$ (T_9) and pelleted $N_{100}:P_{60}:K_{40}$ (T_7) treatment. Sharma et al. (2015a) reported similar result in pelleted integrated nutrient management in transplanted rice.

6.3.5 GRAIN PRODUCTION EFFICIENCY

The grain production efficiency (GPE) is presented in Table 6.3, indicating great variation in these parameters. The maximum GPE was recorded under control treatment (T_{13}) followed by pelleted 5 t FYM + $N_{50}:P_{30}:K_{20}$ (T_{10}) and pelleted 2.5 t FYM + $N_{80}:P_{50}:K_{30}$ (T_8). The minimum GPE was found under conventional $N_{100}:P_{60}:K_{40}$ (T_1) treatment. Sharma et al. (2015a) reported similar result in pelleted integrated nutrient management in transplanted rice.

6.3.6 ECONOMICS

Effect of different treatments cannot be assessed without the gross and net profit from that treatment. The economics of conventional and pelleted integrated nutrient management treatments has been presented in Table 6.3. Among the conventional and pelleted integrated nutrient management, the pelleted 2.5 t FYM + $N_{80}:P_{50}:K_{30}$ (T_8) resulted in highest gross return (Rs. 77,505.00 ha^{-1}) followed by pelleted $N_{100}:P_{60}:K_{40}$ (T_7) (Rs. 69,465.00 ha^{-1}), pelleted 5 t FYM + $N_{50}:P_{30}:K_{20}$ (T_{10}) (Rs. 68,755.00 ha^{-1}), and conventional 2.5 t FYM + $N_{80}:P_{50}:K_{30}$ (T_2) (Rs. 66,655.00 ha^{-1}). The highest net return (Rs. 52,157.28 ha^{-1}) was recorded under pelleted 2.5 t FYM + $N_{80}:P_{50}:K_{30}$ (T_8) followed by pelleted $N_{100}:P_{60}:K_{40}$ (T_7) (Rs. 44,095.28 ha^{-1}), pelleted 5 t FYM + $N_{50}:P_{30}:K_{20}$ (T_{10}) (Rs. 43,836.88 ha^{-1}), and conventional 2.5 t FYM + $N_{80}:P_{50}:K_{30}$ (T_2) (Rs. 41,607.28 ha^{-1}). The pelleted 2.5 t FYM + $N_{80}:P_{50}:K_{30}$ (T_8) recorded higher B:C ratio (2.06) followed by pelleted 5 t FYM + $N_{50}:P_{30}:K_{20}$ (T_{10}) 1.76. The lowest value of total input cost (Rs. 22,267.00 ha^{-1}), gross return (Rs. 36,050.00 ha^{-1}), net return (Rs. 14,883.00 ha^{-1}), and B:C ratio (0.70) was recorded under control treatment (T_{13}). Saha and Mondal (2006) reported that maximum net returns and B:C ratio was found under treatment 75% recommended dose of fertilizer + pelleted form of organic manure. Sharma et al. (2015a) reported similar result in pelleted integrated nutrient management in transplanted rice.

TABLE 6.3 Economics and Grain Production Efficiency of Rice as Influenced by Conventional and Pelleted Integrated Nutrient Management.

Treatment		Gross return (Rs. ha^{-1})	Net return (Rs. ha^{-1})	B:C ratio	Grain production efficiency (Q.MJ × 10^{-3})
T$_1$	Conventional 100:60:40 kg N:P$_2$O$_5$:K$_2$O ha^{-1}	65,960.00	40,890.28	1.63	4.91
T$_2$	Conventional 2.5 t FYM + 80:50:30 kg N:P$_2$O$_5$:K$_2$O ha^{-1}	66,655.00	41,607.28	1.66	5.25
T$_3$	Conventional 2.5 t FYM + 60:40:20 kg N:P$_2$O$_5$:K$_2$O ha^{-1}	60,560.00	36,284.17	1.49	5.53
T$_4$	Conventional 5 t FYM + 50:30:20 kg N:P$_2$O$_5$:K$_2$O ha^{-1}	63,320.00	38,701.88	1.57	5.70
T$_5$	Conventional 1 t PM + 50:30:20 kg N:P$_2$O$_5$:K$_2$O ha^{-1}	58,650.00	35,081.88	1.49	5.53
T$_6$	Conventional 2.5 t FYM + 0.5 t PM + 50:30:20 kg N:P$_2$O$_5$:K$_2$O ha^{-1}	61,650.00	37,557.88	1.56	5.64
T$_7$	Pelleted 100:60:40 kg N:P$_2$O$_5$:K$_2$O ha^{-1}	69,465.00	44,095.28	1.74	5.15
T$_8$	Pelleted 2.5 t FYM + 80:50:30 kg N:P$_2$O$_5$:K$_2$O ha^{-1}	77,505.00	52,157.28	2.06	6.12
T$_9$	Pelleted 2.5 t FYM + 60:40:20 kg N:P$_2$O$_5$:K$_2$O ha^{-1}	67,170.00	42,594.17	1.73	5.88
T$_{10}$	Pelleted 5 t FYM + 50:30:20 kg N:P$_2$O$_5$:K$_2$O ha^{-1}	68,755.00	43,836.88	1.76	6.17
T$_{11}$	Pelleted 1 t PM + 50:30:20 kg N:P$_2$O$_5$:K$_2$O ha^{-1}	60,875.00	37,006.88	1.55	5.73
T$_{12}$	Pelleted 2.5 t FYM + 0.5 t PM+50:30:20 kg N:P$_2$O$_5$:K$_2$O ha^{-1}	63,910.00	39,516.88	1.62	5.84
T$_{13}$	Control	36,050.00	14,883.00	0.70	7.59

B:C ratio, benefit to cost ratio.

6.4 CONCLUSION

The nitrogen, phosphorus, and potassium are essential inputs required for better growth and yield of rice. Hence, there is need for application of these nutrients in balanced quantity with proper method. The nutrients are often subjected to different kinds of losses; therefore, proper form and method decides their efficiency. A lot of forms and modifications have been proposed for controlling the various kinds of losses of nutrients such as granular, prilled, pelleted, encapsulated form, etc. Pelleted form of nutrient management can be a good option for integrated nutrient management (INM). The pelleted form of INM contains all essential macronutrients and micronutrients that plant requires and pelleted form helps in slow release of those nutrients which assures continued supply of nutrients for longer periods. This method may also be useful for other cereals, pulses, oilseeds, vegetables, fruits, and forest crops. Pelleted 2.5 t FYM + $N_{80}:P_{50}:K_{30}$ produced highest grain yield (5.63 t ha^{-1}) which was comparable to that of pelleted $N_{100}:P_{60}:K_{40}$. The pelleted INM in transplanted rice found to be better than conventional INM methods for grain yield and nutrient uptake. The highest net return (Rs. 52,157.28 ha^{-1}) and B:C ratio (2.06) was observed under pelleted 2.5 t FYM + $N_{80}:P_{50}:K_{30}$.

6.5 SUMMARY

The nitrogen, phosphorus, and potassium are essential inputs required for better growth and yield of rice. Hence, there is need for application of these nutrients in balanced quantity with proper method. The nutrients are often subjected to different kinds of losses therefore proper form and method decides their efficiency. A lot of forms and modifications have been proposed for controlling the various kinds of losses of nutrients such as granular, prilled, pelleted, encapsulated form, etc. In the same continuation, an experiment was conducted in *kharif*-2012 at Instructional-cum-Research Farm at Indira Gandhi Krishi Vishwavidyalaya, Raipur, Chhattisgarh, India to investigate effect of INM by pelleting techniques to increase growth, yield, and NUE in transplanted rice. The treatment, pelleted 2.5 t FYM + 80:50:30 kg $N:P_2O_5:K_2O$ ha^{-1} produced highest plant height, number of tillers hill^{-1}, leaf area plant^{-1}, leaf area index, and dry matter accumulation at all the stages of crop growth except that of 15 DAT, where conventional 100:60:40 kg $N:P_2O_5:K_2O$ ha^{-1} performed better over other treatments. The results revealed that pelleted 2.5 t FYM + 80:50:30 kg $N:P_2O_5:K_2O$ ha^{-1} produced highest grain and straw yield; harvest index; nitrogen, phosphorus, and potassium content in grain and straw; uptake of N, P, and K; and nitrogen and potassium use efficiency, respectively. However, the highest potassium content in straw and potassium use efficiency was recorded under pelleted 5 t FYM + 50:30:20 kg $N:P_2O_5:K_2O$ ha^{-1}. The treatment received pelleted 2.5 t FYM + 80:50:30 kg $N:P_2O_5:K_2O$ ha^{-1} gave the highest B:C ratio (2.06) and highest net returns (Rs. 52,157.28 ha^{-1}). The pelleted form of INM contains all essential macronutrients and micronutrients that plant requires and pelleted form helps in slow release of those nutrients which assures continued supply of nutrients for longer periods. This method may also be useful for other cereals, pulses, oilseeds, vegetables, fruits, and forest crops.

KEYWORDS

- soil health
- solar radiation
- farmyard manure
- nutrient use efficiency
- productivity
- yield

REFERENCES

Amany, A.; Bahr, Z. M. S.; Hozayn, M. Yield and Quality of Maize (*Zea mays* L.) as Affected by Slow-release Nitrogen in Newly Reclaimed Sandy Soil. *Am.-Eurasian J. Agric. Environ. Sci.* **2006,** *1* (13), 239–242.

Bagheri, R.; Akbari, G. A.; Kianmehr, M. H.; Sarvastani, Z. A. T.; Hamzekhanlu, M. Y. The Effect of Pellet Fertilizer Application on Corn Yield and Its Components. *Afr. J. Agric. Res.* **2011,** *6,* 2364–2371.

Bhattacharya, S. C.; Sett, S.; Shrestha, R. M. State of the Art of Biomass Densification, Energy Sources, Division Energy Technology. *Energ. Sour.* **1989,** *11* (3), 161–186.

Biswas, B. S.; Yadav, D. S.; Maheshwari, S. Fertilizer Use in Cropping Systems. *Fertil. News* **1987,** *32,* 23–26.

Black, C. A. *Method of Soil Analysis*; Amar. Agcon. Inc.: Madison, Wisconsin, USA, 1965; pp 131–137.

Bodman, G. B. Nomograms for Rapid Calculation of Soil Density, Water Content and Total Porosity Relationship. *J. Am. Soc. Agron.* **1942,** *34,* 883–893.

Chandrakar, P. K.; Rastogi, N. K.; Sahu, L. In *Studies on Influence of Fertilizer Doses and Row Spacing on Seed Production and Seed Quality Parameters in Scented Rice cv. Indira Sugandhit Dhan-1*. International Symposium on Rainfed Rice Ecosystems: Perspective and Potential, Indira Gandhi Krishi Vishwavidyalaya, Raipur, India. Oct 11–13, 2004; p 116.

Ebaid, R. A.; El-Rafaee, I. S. In *Utilization of Rice Husk as an Organic Fertilizer to Improve Productivity and Water Use Efficiency in Rice Fields*, African Crop Science Conference Proceedings, Minia University, El-Minia, Egypt, 2007, Vol. 8, pp 1923–1928.

El-Kramany, M. F. Effect of Organic Manure and Slow Release N-fertilizers on the Productivity of Wheat (*Triticum aestivum* L.) in Sandy Soil. *Acta. Agronomica Hung.* **2001,** *49,* 379–85.

Gujja, B.; Thiyagarajan, T. M. *Knowledge and Practice SRI (System of Rice Intensification): Reducing Agriculture Foot Print and Ensuring Food Security*. National Consortium of SRI (NCS): Hyderabad, India, 2012; p 12.

Jha, S. K.; Tripathi, R. S.; Kumar, S.; Gupta, P. Effect of Integrated Nutrient Management on Production Potential, Economics and Energetics of Scented Rice (*Oryza sativa* L.). *Plant Arch.* **2004,** *4* (2), 503–505.

Jha, S. K.; Dewangan, Y. K.; Tripathi, R. S. In *Effect of Integrated Nutrient Management on Yield and Quality of Scented Rice in Chhattisgarh*, National Symposium on Conservation and Management of Agro-resources in Accelerating the Food Production for 21st Century, Indira Gandhi Krishi Vishwavidyalaya, Raipur, India, Dec 14–15, 2006, pp 248–250.

Rama Lakshmi, C. S.; Rao, P. C.; Sreelata, T.; Madahvi, M.; Padmaja, G.; Rao, P. V.; Sireesha, A. Nitrogen Use Efficiency and Production Efficiency of Rice Under Rice-pulse Cropping System with Integrated Nutrient Management. *J. Rice Res.* **2012,** *5* (1–2), 42–45.

Makarim, A. K. Shuartatik, E. Partial Efficiency Concept in New Rice Plant Type as Indicated by N Uptake. Paper Presented at International Rice Conference, Bali, Indonesia, Sept 12–14, 2005. Indonesian Agency for Agricultural Research and Development in Cooperation with the International Rice Research Institute.

Mandal, B. K.; Hajra, S. K.; Kundu, S.; Bose, P.; Ghose, S. In *Influence of Fertility Levels on the Yield of Scented Rice Cultivars*, International Symposium on Rainfed Rice Ecosystems: Perspective and Potential, Indira Gandhi Krishi Vishwavidyalaya, Raipur, India, Oct 11–13, 2004, p 99.

Medhi, B. D.; Datta, S. K.; Datyta, S. K. Nitrogen Use Efficiency and 15 N Balance Following Incorporation of Green Manure and Urea in Flooded Transplanted and Broadcast Seeded Rice. *J. Indian Soc. Soil Sci.* **1996,** *44* (3), 422–427.

Mhaskar, N. V.; Thorat, S. T.; Bhagat, S. B. Effect of Nitrogen Levels on Leaf Area, Leaf Area Index and Grain Yield of Scented Rice Varieties. *J. Soils Crops* **2005,** *15* (1), 218–220.

Miller, H. B. Poultry Litter Induces Tillering in Rice. *J. Sust. Agric.* **2007,** *31* (1), 151–160.

Muhammad, I. Response of Wheat Growth and Yield to Various Levels of Compost and Organic Manure. *Pak. J. Bot.* **2008,** *40* (5), 2135–2141.

Murali, M. K.; Setty, R. A. Grain Yield and Nutrient Uptake of Scented Rice Variety, Pusa Basmati-1, at Different Levels of NPK, Vermicompost and Tricontanol. *Oryza* **2001,** *38* (1–2), 84–85.

Natarajan, K.; Pushpavalli, R. Strategies for Higher Nitrogen Use Efficiency in Low Land Rice. *Ind. Fert. Sci. Ann.* **1994,** 92–93.

Olsen, S. R.; Cole, C. V.; Wantamable, F. S.; Dean, L. A. *Estimation of Available Phosphorus in Soils by Extraction with Sodium Bicarbonate*; USDA Circ. No. 939, United States Department of Agriculture: Washington, D.C., U.S.,1954; pp 1–19.

Pandey, T. D.; Nandeha, K. L. In *Response of Scented Rice (Oryza sativa) Varieties to FYM and Chemical Fertilizers in Bastar Plateau*, International Symposium on Rainfed Rice Ecosystems: Perspective and Potential, Indira Gandhi Krishi Vishwavidyalaya, Raipur, India, Oct 11–13, 2004, p 105.

Pandey, N.; Dhurandhar, R. L.; Tripathi, R. S. Grain Yield, Root Growth and N Uptake of Early Duration Rice Varieties as Influenced by Nitrogen Levels Under Rainfed Upland. *Oryza* **1998,** *37* (1), 60–62.

Pandey, N.; Verma; Anurag, A. K.; Tripathi, R. S. Integrated Nutrient Management in Transplanted Hybrid Rice. *Indian J. Agron.* **2007,** *52* (1), 40–42.

Paraye, M. P.; Bansasi, R.; Nair, S. K.; Pandey, D.; Soni, V. K. In *Response of Scented Rice (Oryza sativa) to Nutrient Management and Varieties*, National Symposium on Conservation and Management of Agro-resources in Accelerating the Food Production for 21st Century, Indira Gandhi Krishi Vishwavidyalaya, Raipur, India, Dec 14–15, 2006, pp 248–250.

Piper, C. S. *Soil and Plant Analysis*; The University of Adelaide, Australia, 1967; p 286.

Prakash, Y. S.; Bhadoria, P. B. S.; Amitava, R. Relative Efficiency of Organic Manure in Improving Milling and Cooking Quality of Rice. *IRRN* **2002**, *27* (1), 43–44.

Raju, R. A.; Reddy, G. C.; Nageswaran, R. M. Responses of Rice (*Oryza sativa* L.) to Algalization in Combination with Chemical Nutrients. *Tropic. Agric.* **1989**, *66* (4), 334–336.

Rizwan. M.; Maqsood, M.; Rafiq, M.; Saeed, M.; Ali, Z. Maize (*Zea mays* L.) Response to Split Application of Nitrogen. *Int. J. Agri. Bio.* **2003**, *5* (1), 19–21.

Saha, M.; Mondal, S.S. Influence of Integrated Plant Nutrient Supply on Growth, Productivity and Quality of Baby Corn (*Zea mays*) in Indo-Gangetic Plains. *Ind. J. Agron.* **2006**, *51*(3), 202–205.

Sarawgi, S. K.; Sarawgi, A. K. In *Effect of Blending of N with or Without FYM on Semi-dwarf, Medium to Long Slender Scented Rice Varieties in Lowland Alfisols of Chhattisgarh*, International Symposium on Rainfed Rice Ecosystems: Perspective and Potential, Indira Gandhi Krishi Vishwavidyalaya, Raipur, India. Oct 11–13, 2004,; pp 159–160.

Sharma, M. K.; Taunk, S. K.; Shrivastava, G. K.; Tomar, G. S. Studies on Nutrient Use Efficiency from Pelleting Techniques of Integrated Nutrient Management in Transplanted Rice. *Res. J. Agric. Sci.* **2014**, *5* (3), 539–542.

Sharma, M. K.; Nagre, S. K.; Dewangan, R.; Yadav, P.; Jangde, H. Economics and Energetics of Rice as Influenced by Pelleted Integrated Nutrient Management Under Transplanted Ecosystem. *Int. J. Tropic. Agric.* **2015a**, *33* (1), 89–94.

Sharma, M. K.; Taunk, S. K.; Thawait, D.; Banjare, A.; Tiwari, V. K. Effect of Slow Releasing Nutrients from Pelleted Integrated Nutrient Management on Growth and Yield of Transplanted Rice. *Ecol. Environ. Conserv.* **2015b**, *21* (Suppl.) 129–136.

Siavoshi, M. Effect of Organic Fertilizer on Growth and Yield Components in Rice (*Oryza sativa* L.). *J. Agric. Sci.* **2011**, *3* (3), 217–224.

Singh, R.; Agarwal, S. K. Analysis of Growth and Productivity of Wheat in Relation to Levels of FYM and Nitrogen. *Indian J. Plant Physiol.* **2001**, *6*, 279–283.

Singh, S. K.; Varma, S. C.; Singh, R. P. Effect of Integrated Nutrient Supply on Growth, Yield, Nutrient Uptake, Economics and Soil Fertility in Irrigated Rice. *Oryza* **2001**, *38* (1–2), 56–60.

Subbaiah, B. V.; Asija, G. L. A Rapid Procedure for the Determination of Available Nitrogen in Soils. *Curr. Sci.* **1956**, *31*, 196.

Umashankar, R.; Babu, C.; Kumar, P. S.; Prakash, R. Integrated Nutrient Management Practices on Growth and Yield of Direct Seeded Low Land Rice. *Asian J. Plant Sci.* **2005**, *4* (1), 23–26.

Verma, D. K.; Srivastav, P. P. Proximate Composition, Mineral Content and Fatty Acids Analyses of Aromatic and Non-aromatic Indian Rice. *Rice Sci.* **2017**, *24* (1), 21–31.

Yoshida, S. *Fundamental of Rice Crop Science*; International Rice Research Institute: Los Baños, Laguna, Philippines, 1981; p 269.

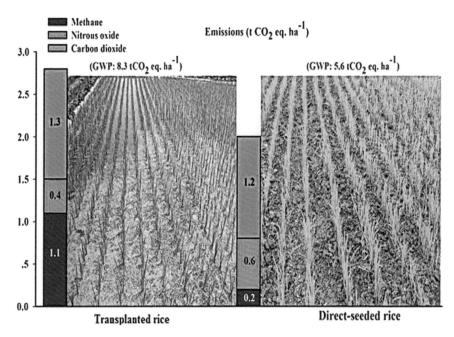

FIGURE 2.4 GHGs emission from transplanted and direct-seeded rice production systems.

Source: Reprinted from Hussain, S.; Peng, S.; Fahad, S.; Khaliq, A.; Huang, J.; Cui, K.; Nie, L. Rice Management Interventions to Mitigate Greenhouse Gas Emissions: A Review. *Environ. Sci. Pollut. Res.* **2015,** *22,* 3342–3360. Springer-Verlag: Berlin, Heidelberg, 2014. With permission from Springer.

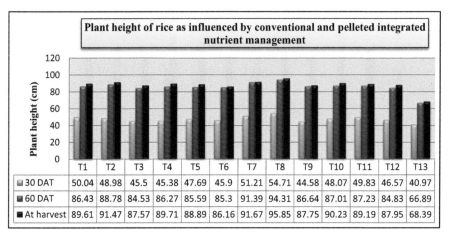

FIGURE 6.2 Plant height of rice as influenced by conventional and pelleted integrated nutrient management.

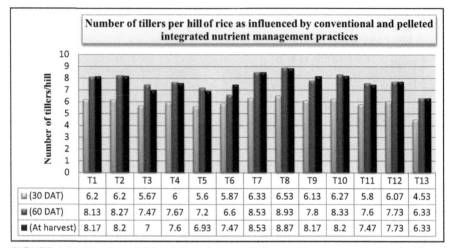

	T1	T2	T3	T4	T5	T6	T7	T8	T9	T10	T11	T12	T13
(30 DAT)	6.2	6.2	5.67	6	5.6	5.87	6.33	6.53	6.13	6.27	5.8	6.07	4.53
(60 DAT)	8.13	8.27	7.47	7.67	7.2	6.6	8.53	8.93	7.8	8.33	7.6	7.73	6.33
(At harvest)	8.17	8.2	7	7.6	6.93	7.47	8.53	8.87	8.17	8.2	7.47	7.73	6.33

FIGURE 6.3 Number of tillers hill^{-1} of rice as influenced by conventional and pelleted integrated nutrient management practices.

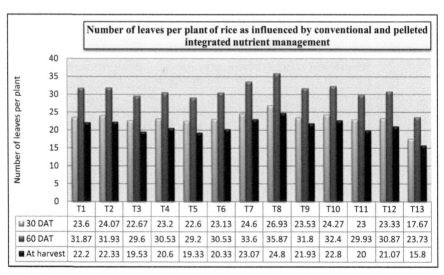

	T1	T2	T3	T4	T5	T6	T7	T8	T9	T10	T11	T12	T13
30 DAT	23.6	24.07	22.67	23.2	22.6	23.13	24.6	26.93	23.53	24.27	23	23.33	17.67
60 DAT	31.87	31.93	29.6	30.53	29.2	30.53	33.6	35.87	31.8	32.4	29.93	30.87	23.73
At harvest	22.2	22.33	19.53	20.6	19.33	20.33	23.07	24.8	21.93	22.8	20	21.07	15.8

FIGURE 6.4 Number of leaves plant^{-1} of rice as influenced by conventional and pelleted integrated nutrient management.

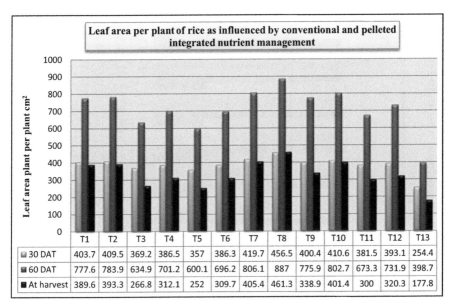

FIGURE 6.5   Leaf area plant^{-1} of rice as influenced by conventional and pelleted integrated nutrient management.

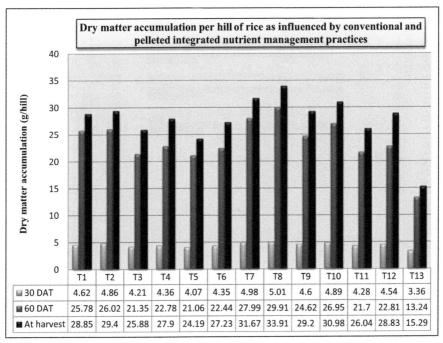

FIGURE 6.6   Dry matter accumulation hill^{-1} of rice as influenced by conventional and pelleted integrated nutrient management practices.

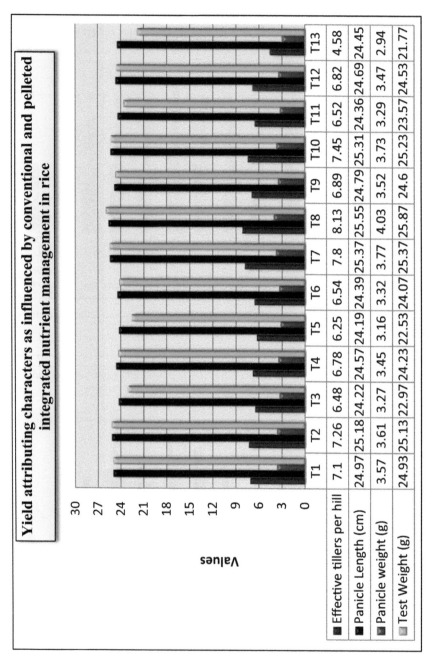

FIGURE 6.7 Yield attributing characters as influenced by conventional and pelleted integrated nutrient management in rice.

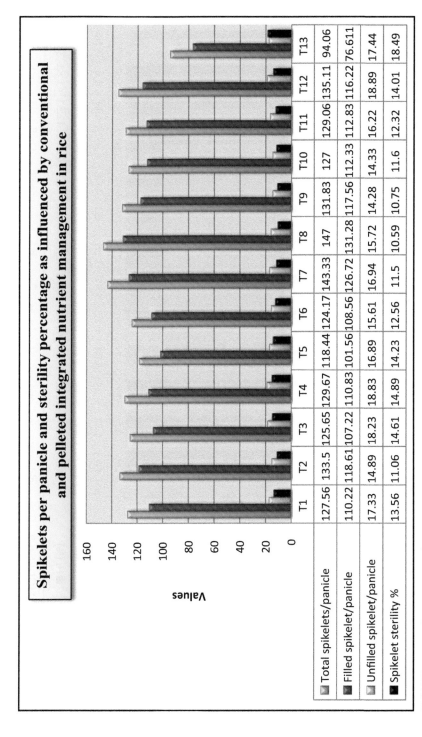

FIGURE 6.8 Spikelets panicle^{-1} and sterility percentage as influenced by conventional and pelleted integrated nutrient management in rice.

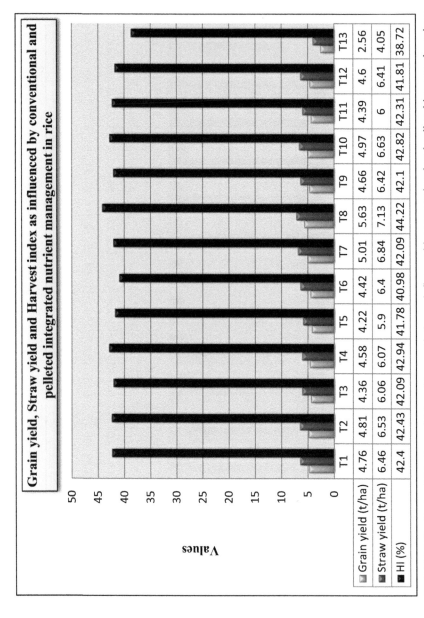

	T1	T2	T3	T4	T5	T6	T7	T8	T9	T10	T11	T12	T13
Grain yield (t/ha)	4.76	4.81	4.36	4.58	4.22	4.42	5.01	5.63	4.66	4.97	4.39	4.6	2.56
Straw yield (t/ha)	6.46	6.53	6.06	6.07	5.9	6.4	6.84	7.13	6.42	6.63	6	6.41	4.05
HI (%)	42.4	42.43	42.09	42.94	41.78	40.98	42.09	44.22	42.1	42.82	42.31	41.81	38.72

Grain yield, Straw yield and Harvest index as influenced by conventional and pelleted integrated nutrient management in rice

Values

FIGURE 6.9 Grain yield, straw yield, and harvest index as influenced by conventional and pelleted integrated nutrient management in rice.

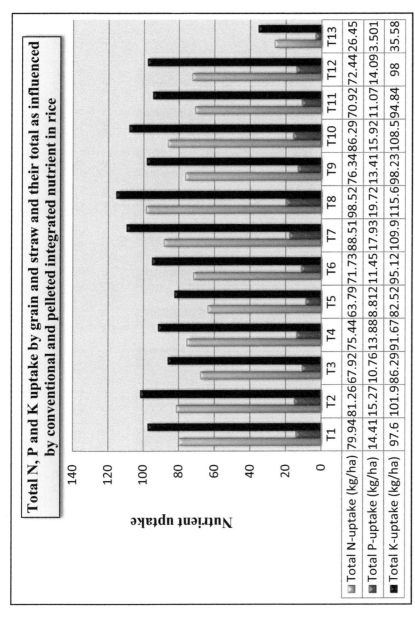

FIGURE 6.10 Total N, P, and K uptake by grain and straw and their total as influenced by conventional and pelleted integrated nutrient in rice.

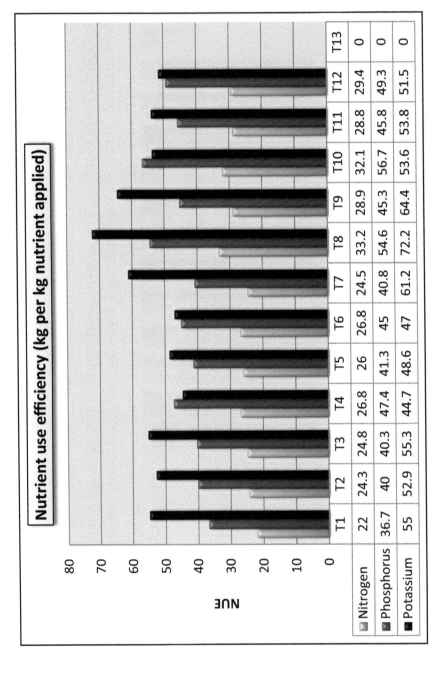

FIGURE 6.11 Nutrient use efficiency of rice as influenced by conventional and pelleted integrated nutrient management.

CHAPTER 7

INTEGRATED NUTRIENT MANAGEMENT: POTENTIAL PRACTICES FOR RICE PRODUCTIVITY AND YIELD

DEEPAK KUMAR VERMA[1*], ARNAB BANERJEE[2], MANOJ KUMAR JHARIYA[3], and DHIRAJ KUMAR YADAV[3]

[1]Department of Agricultural and Food Engineering, Indian Institute of Technology, Kharagpur 721302, West Bengal, India

[2]University Teaching Department, Department of Environmental Science, Sarguja Vishwavidyalaya, Ambikapur 497001, Chhattisgarh, India

[3]University Teaching Department, Department of Farm Forestry, Sarguja Vishwavidyalaya, Ambikapur 497001, Chhattisgarh, India

*Corresponding author.
E-mail: deepak.verma@agfe.iitkgp.ernet.in; rajadkv@rediffmail.com

ABSTRACT

Rice is the staple food globally which provides 60–70% calories of energy from carbohydrate. Recently, the production of rice has become a major challenge due to the climate variability, excess use of fertilizers, lack of land and soil resources, loss of agrobiodiversity, and presence of agricultural pollutions including pesticides and chemical fertilizers. In this scenario, the integrated nutrient management is the novel eco-friendly approach which involves the proper management of organic inputs as the nutrients source to improve the soil health, fertility, and productivity, thus providing a base for sustainable agriculture activities. In rice production system, the organic inputs such as farmyard manure, compost, biofertilizer, press mud, poultry litter, and blue-green algae are highly effective in the wetlands which boost up the rice productivity and yield maintaining soil health and quality. Therefore, this chapter focuses on the significance of integrated nutrient management in rice production, and effect of organic and inorganic fertilizers on the productivity, growth, and quality of rice.

7.1 INTRODUCTION

Paddy is a major food crop for more than half of the global population and have substantial role in global economy and maintaining social security throughout the world (Greenland, 1997; Evans, 1998; Dawe, 2000). Paddy has a major contribution in terms of human consumption rate in India (Hossain, 1997; Sharma et al., 2013). It stands as major food crop covering about 25% of the cropping area and feeding almost 50% of Indian population (Ballabh and Pandey, 1999). Areas receiving higher rainfall (1500 mm annually) is being predominated by paddy farming. In India, the eastern and southern part can be considered as the rice bowl having maximum area under paddy cultivation (Pingali et al., 1997; Muralidharan et al., 1998). India has a high resource base of paddy representing 4000 species in comparison to 10,000 varieties throughout the world (Vavilov, 1926; Chang, 1976; Choudhury et al., 2013). Asian people are very much dependent upon rice from nutritional perspectives that is procuring 60–70% calories of energy from carbohydrate. United Nations General Assembly (UNGA) has recommended 2004 as "International Year of Rice" (FAOUS, 2003; MSSRF, 2005; Verma and Srivastav, 2017).

India is the second largest producer of rice only after China (Sharma et al., 2013; Kaul, 2015; Verma et al., 2012, 2013, 2015). In India, around 45.0 million ha area is under rice cultivation with 106.29 million t production (Anonymous, 2014). Uttar Pradesh is the largest state in country known for rice farming only after West Bengal, which has about 5.29 million ha area to produce 14.41 million t rice (Anonymous, 2014) which contribute about 13.80% of total rice production in national agricultural economy of India. The unprecedented growth of human population has promoted a challenge toward production of rice under acute shortage of natural resources in terms of land, water, nutrients, labor, and other agricultural inputs (Chauhan and Mahajan, 2014). Modernization of agriculture has promoted human race in the clutches of environmental pollution, degradation of land, and other types of environmental perturbances.

This chapter features the importance and effect of integrated nutrient managements (INM), inorganic chemical fertilizers, and organic fertilizers (namely, farmyard manure, press mud, *Trichoderma* compost, and vermicompost) on growth, yield, and quality of rice and also states the economic values of such measures.

7.2   INM IN RICE: AT A GLANCE

INM is a judicious approach of eco-friendly agriculture which has its origin since prehistoric time (Bhandari et al., 1992; Bablad, 1999; Chandra et al., 2004; Mahajan et al., 2008; Mahajan and Gupta, 2009; Virdia and Mehta, 2010; Rahman et al., 2012; Dissanayake et al., 2014; Wolie and Admassu, 2016). Such an approach is mediated in such a way, so that it goes through in an environmental compatible manner keeping in view the economy gain as well as the soil health (Chandra et al., 2004; Saha et al., 2007; Mahajan et al., 2008; Mahajan and Gupta, 2009; Singh et al., 2009, 2014; Aulakh, 2010; Rahman et al., 2012; Tzudir and Ghosh, 2014; Roy et al., 2015; Sahu et al., 2015). The motto behind such policy is to provide sustainable agriculture system on long term basis (Chandra et al., 2004; Saha et al., 2007; Mahajan et al., 2008; Mahajan and Gupta, 2009; Rahman et al., 2012). The INM comprises of different forms of organic inputs as the nutrients source to improve the crop productivity in an eco-friendly way. This practice has got wider dimensions in terms of soil health, fertility, productivity, and social acceptability (Chandra et al., 2004; Mahajan et al., 2008; Mahajan and Gupta, 2009; Singh et al., 2009, 2014; Aulakh, 2010; Virdia and Mehta, 2010; Dissanayake et al., 2014; Tzudir and Ghosh, 2014; Roy et al., 2015; Sahu et al., 2015; Wolie and Admassu, 2016). As per Frinck (1998) process of fertilization should have a holistic approach which includes proper management of nutrients under integrated system and other associated production issues In this regard, INM is an approach which aims toward balanced fertilization for optimum crop yield with lesser environmental degradation (Kesavan and Swaminathan, 2008) along with Integrated Plant Nutrition System (IPNS) technique. Organic inputs such as farmyard manure (FYM), compost, blue-green algae (BGA) are very much effective in wetland rice agricultural practices (Begum et al., 2009).

7.3   INORGANIC AND ORGANIC FERTILIZERS IN INM FOR RICE: AN OVERVIEW

In the developing world and country like India, food security is a major challenge throughout the world (Pingali and Stringer, 2003). In the productivity sector, India is yet to achieve the target of producing 100 million t by 2015, although being a leading rice producer in the world (Mishra, 2004). To boost up the agricultural production area increment is a big problem. On the other hand boosting the productivity rate by technological innovation can only improve rice production scenario. INM reflects significant promise in terms of meeting up the growing demand for food as well as promote eco-friendly approaches for lesser environmental pollutions. Further INM approach leads toward sustainable crop production by managing resources and giving optimum yield and quality. Such approaches includes nutrient deficiency correction along with improvement in nutrient use and soil resource base (Bhandari et al., 1992; Bablad, 1999; Chandra et al., 2004; Singh et al., 2006a, 2012; Pal et al., 2007; Pandey et al., 2007; Pillai et al., 2007; Saha et al., 2007; Tripathi et al., 2007; Mahajan et al., 2008; Mahajan and Gupta, 2009; Sepehya, 2011; Senthivelu et al., 2009; Singh et al., 2009, 2014; Srivastava and Ngullie, 2009; Sudhakar, 2011; Aulakh, 2010; Virdia and Mehta, 2010; Rahman et al., 2012; Gautam et al., 2013; Dissanayake et al., 2014; Shah and Kumar, 2014; Suresh et al., 2013; Tzudir and Ghosh, 2014; Mondal et al., 2015; Roy et al., 2015; Sahu et al., 2015; Goutami and Rani, 2016; Wolie and Admassu, 2016).

Chemical fertilizers are becoming costly day by day resulting in no more useful measure in agriculture and due to which most of the farmers of India and other countries are not able to have enough money for these fertilizers (Hapse, 1993; Tripathi and Chaubey, 1996; Pender, 2008; Sebby, 2010; Singh and Manisha, 2015; GOI, 2017). For every ton harvest of rice grain, about 10–31 kg N, 1–5 kg P, 8–35 kg K, and 1–3 kg S ha^{-1} are removed from the soil (Dobermann et al., 1998). Environment friendly technology needs to be incorporated in the cultivation system of wetland rice aiming toward sustainable production. In this era of climate change the focal theme in agricultural sector is to improve the productivity in environmentally compatible way. Nonjudicious application of synthetic agrochemicals has declining effect on soil habitat as well as on rice productivity. As per research reports, soil acidity and compactness of upper layer of soil has lead toward reduction in cereal crop yield (Kang and Juo, 1986; Singh et al., 2006b). The hazardous effect of chemical fertilizer and agrochemicals can be minimized by optimum chemical use along with incorporation of organic amendments to meet the crop nutrient requirements in a sustainable way (Aktar et al., 2009; Diacono and Montemurro, 2010; SAFSSDSN, 2013; Usman et al., 2015). Nutrient release rate through organic amendments is usually little bit slower in comparison to synthetic fertilizers but from long-term sustainability basis, it could be matched with the ability of nutrient uptake by crop aiming toward lesser pollution (Manahan, 2006; Miller et al., 2015; Hazra, 2016). Organic inputs is not substantial for boosting up the productivity due to various reasons which includes lesser availability of various types of organic resources to meet nutrient demand of crops along with applicability of these sources and their effectiveness under diverse soil conditions (Ramesh et al., 2005; FAOUS, 2006; Badgley and Perfecto, 2007; Morris et al., 2007; Daniele et al., 2012; Pingali, 2012; Singh and Ryan, 2015; Lin and Hülsbergen, 2017). The decreasing impact of lesser production of rice may be due to higher nutrient leaching as well as lesser organic inputs in cultivation system. Such problem can be ameliorated through INM system (John et al., 2001; Virdia and Mehta, 2010).

In rice crop, about 30–35 kg N ha^{-1} contributed by organic manure which increase the economic yield of the rice (Lakpale and Shrivastava, 2012). Vermicompost liberated plant growth promoting substances such as amino acid, auxins (IAA), sugar, and vitamins (vitamin B$_{12}$, folic acid), which mobilizes unavailable P in the soil, improving soil physical health through better soil aggregation and increased water holding capacity as well as checking erosion (Edwards and Burrows, 1988; Kale and Bano, 1988; Werner and Cuevas, 1996; Alves and Passoni, 1997; Buckerfield et al., 1999; Atiyeh et al., 2000a, 2001; Kumari and Ushakumari, 2002; Arancon et al., 2003, 2005, 2006a, 2006b; Chaudhary et al., 2004; Edwards and Arancon, 2004a, 2004b; Edwards et al., 2004; Suthar, 2006; Sinha et al., 2009; Roy et al., 2010; Ievinsh, 2011; Bhadauria et al., 2014; Gupta et al. 2014a; Joshi et al., 2015). Vermicompost is an organic measure which can be a substitute of more than 50% N, P, K, and other micronutrients recommended for agricultural crops (Jadhav et al., 1997). Conjoint application of synthetic chemical fertilizer and biofertilizer provides similar yield as compare to 100% recommended dose of fertilizer. Such approaches improve the crops nutrient use efficiency (Singh et al., 2006a). Due consideration should be given regarding doses of application of chemical fertilizer in different cropping pattern. One needs to assess the nutrient level applied to a crop which may be carried over through the soil until the succeeding crop for optimum utilization of nutrient. Soil quality in the terms of nutrient level, biodiversity can be effectively managed by judicious application of INM approach. As per research reports, individual nutrient inputs from various sources appears to be nonsignificant in the terms of productivity. Also there is scarcity in the diversity of sources of organic fertilizers.

Dependency toward synthetic chemical fertilizer can be effectively minimized through judicious application of organic inputs such as biofertilizer (El-Hawary et al., 2002; Meshram et al., 2004; Gaur, 2006; Sahoo et al., 2013; Andrade et al., 2013; Lawal and Babalola, 2014; Vaghela et al., 2014; Naher et al., 2016; Trujillo-Tapia and Ramírez-Fuentes, 2016), compost (Schatz and Schatz, 1972; Luske, 2010), FYM (Yaduvanshi and Swarup, 2005; Reddy et al., 2007; Shekara and Sharnappa, 2010), green manure (Garrity and Flinn, 1987; Deshpande and Devasenapathy, 2011), and other organic inputs (Hapse, 1993; Sinha et al., 2009; Koushal et al., 2011). It not only acts as a source of nutrient but also provides micro-nutrient as well as modifies the soil physical behavior and increases the efficiency of applied nutrients (Pandey et al., 2007). Utilization of indig-enous organic sources, namely, FYM as well as unwanted plants species, mulches in the agricultural fields can act as suitable nutrient supplement to improve the productivity in paddy cultivation in India (Hobbs et al., 2008; Surekha et al., 2010; Sepehya, 2011; Lakpale and Shrivastava, 2012; Shetty et al., 2014). To improve the agroeconomy the conjoint application of synthetic fertilizers along with organic inputs are a suitable strategy for higher potential yield of crops leading to sustainable agroecosystem (Raju and Reddy, 2000; Mir et al., 2013). Integration of organic sources such as vermicompost and FYM may also help in the restoration of soil health (Hauck, 1978; Ghoshal and Singh, 1995; Pillai et al., 2007; Shalini et al., 2007; Chand et al., 2011; Adhikary, 2012; Ganiger et al., 2012; Chatterjee and Thirudasu, 2014; Mitran et al., 2017). Utilizing high-yielding vari-eties (HYV) which comprises hybrid genetic trait that leads to erosion of nutrient from the top soil and thus influencing the productivity in a nega-tive way. Day by day, the fertilizer economy is growing up to a signifi-cant level. This actually promotes the production cost of rice with lesser nutrient use by crops. The application of fertilizer to crop plants needs to be designed properly for optimum nutrient use; strategies such as using various organic inputs, chemical fertilizer, as well as applying fertilizer to the crop during its critical stage are to be followed. Organic inputs also supply micronutrients to crop plants, improves soil physical attributes as well as nutrient use efficiency of crops. Technology needs to be design to sustainable production of hybrid rice (Pandey et al., 2007). Rice yield can be optimized through combined applications of inorganic fertilizer substi-tuted by different grades of organic amendments for different crops and soil types (Singh and Sidhu, 2014; Baruah and Baruah, 2015; Usman et al., 2015).

7.4   INM IN DIFFERENT ATTRIBUTES OF RICE

7.4.1   GROWTH ATTRIBUTES

Early development of crop is a much sensitive issue in the terms of higher yield and productivity. It has been found during the early stage of development root extension, formation of new leaves is the pioneer processes in vegetative growth that maximizes the nutrient utilization by plants where lesser photosynthate promotes the yield attributes. Therefore, reproductive growth in early stage of development remain slow.

The growth parameters, namely, plant height, tillers hill^{-1}, and dry matter accumulation hill^{-1} were significantly affected by different bioorganic sources (Hasanuzzaman et al., 2010a, 2010b; Sharma et al., 2012a; Mahamud et al., 2013; Pramanik and Bera, 2013; Singh and Shivay, 2014; Rajesh et al., 2015). In an experimental trial combined application of vermicompost with *Trichoderma* compost have resulted into higher growth attributes in comparison to other treatments (Bablad, 1999; Gandhi and Sivakumar, 2010). Vermicompost has been found to have substances that promote plant growth. The growth attributes were positively influenced by application of worm casts which may be due to mutual biological effect of both vermicompost and metabolic product of microorganisms (Tomati et al., 1995).

Various essential nutrients are found in vermicompost necessary for proper growth and development of the plant (Orozco et al., 1996). Nitrogen (N) present in the vermicompost readily influences different plant growth attributes and higher biomass level. Research result revealed that FYM is less compatible in nutrient cycling in comparison to vermicompost due to higher presence of organic N_2. Further the lesser compatibility of FYM in nutrient release might be due to lesser efficiency of microbial population under FYM treatment (Jeyabal and Kuppuswamy, 2001). Vermicomposting induced synthesis of phenolic acids (Singh et al., 2003; Sarma et al., 2010; Barman et al., 2013). The induction of phenolic acids increased the degree of resistance in plants (Carrasco et al., 1978; Nicholson and Hammerschmidt, 1992; Velazhahan and Vidhyasekaran, 1994; Picinelli et al., 1995; Mayr et al., 1997; Bajaj, 1998; Maury, 2000; Mohammadi and Kazemi, 2002; Huiping et al., 2003; Prats et al., 2004; Usenik et al., 2004; Yamunarani et al., 2004; Treutter, 2005; Maurya et al., 2007; Daayf et al., 2012; Ewané et al., 2012; Barman et al., 2013). Thus, the growth of plants grown with vermicompost was much better than the growth of plants in the control (Atiyeh et al., 2000b; Singh et al., 2003; Chamani et al., 2008; Sinha et al., 2009; Junying et al., 2009; Adhikary, 2012; Farb, 2012; Pathma and Sakthivel, 2012; Gupta et al. 2014a; Kashem et al., 2015; Saikrithika et al., 2015; Jain, 2016).

In an experimental field trial mediated by Novoa and Loomis (1981) it was observed that vegetative growth in the terms of shoot elongation were higher under vermicompost application. Then such growth declined due to partitioning of photosynthate toward reproductive growth during crop maturity. Nitrogen proves to be an effective nutrient which promotes plant development and growth by accelerating various physiological processes. Higher canopy growth might be attributed toward higher biosynthesis of protein molecules present in the cell protoplasm (DaMatta et al., 2006). Availability of nutrient particularly N from organic sources might have favored greater assimilation of protein and carbohydrate (Lawlor, 2002). These two compounds, when present in meristematic region of the plant, induce rapid cell division and greater enlargement of the cells, which ultimately results in better growth performance. Consequently, the growth of crops is a function of availability of N in the soil (Uchida, 2000; Benincasa et al., 2011; Balamurugan and Sudhakar, 2012).

Steady increase in the number of tillers hill^{-1} up to 60 days after treatment (DAT) recorded which declined thereafter with the advancement in age (Panigrahi et al., 2015). The reduction in the number of tillers after 60 DAT resulted due to the ageing and senescence which was responsible for dying of the tertiary and secondary tillers. As per the genetic constitution, plants have a definite tillering period after which they enter into the shoot elongation and developmental stage and the young tillers do not get time to develop (Alberda, 1953; Kemp and Culvenor, 1994). Tillering is the product of the expansion of auxiliary buds (Xueyong et al., 2003; Zou et al. 2004, 2006; Kebrom and Mullet, 2015; Thomas and Hay, 2015) which is closely associated with the nutritional conditions of the mother culm because a tiller receives carbohydrate and nutrient from the mother culm during its early growth period which gets improved by the application of N (Yoshida, 1973; Zou et al., 2006; Dash et al., 2010; Sirvi et al., 2014). The increase in tillers hill^{-1} was probably because of the greater supply of nutrient particularly N through organic sources (Sowmya and Ramana, 2012; Yadav et al., 2013b; Crotty et al., 2014; Ray, 2014; Satapathya et al., 2015) with efficient utilization for cell multiplication and enlargement and also for the formation of nucleic acids and other vitally important organic compounds in the cell sap (Simons, 1982; Cai et al. 2009).

Application of vermicompost (5 t ha^{-1}) produced maximum LAI (Golchin et al. 2006; Berova and Karanastidis, 2008; Hasanuzzaman et al., 2010a,

2010b; Theunissen et al., 2010). Plants may become photosynthetically more active, which would contribute to improvement in yield attributes; these results are in conformity with finding of Gandhi and Sivakumar, 2010.

7.4.2 YIELD AND YIELD'S ATTRIBUTES

Yield attributes which determine yield, is the resultant of the vegetative development of the crop (Araus et al., 2008; Chen et al., 2008; Fraga et al., 2010; Alam and Kumar, 2015). All the attributes of yield, namely, number of panicles m^{-2}, length of panicle, panicle weight, and 1000-grains weight were significantly affected by bioorganic source. The better yield attributes were recorded among the different organic sources of nutrition. The application of vermicompost with *Trichoderma* compost recorded significantly higher yield parameter, namely, number of panicles m^{-2}, length of panicle, panicle weight, and 1000-grains weight. This might be due to the organic materials acting as slow release source of N and are expected to more closely match with N and supply of other nutrients with demand of rice crop which could reduce the N losses and also improved the nutrient use efficiency particularly of N.

N is an important constituent of chlorophyll and so, if supplied adequately stimulates photosynthesis in plants (Singh et al., 2002; Yildirim et al., 2009; Hokmalipour and Darbandi, 2011). Thus at higher N levels, there would have been more photosynthetic activity in plant (Evans, 1989; Murchie et al., 2002a, 2002b; Makino, 2003; Morales et al., 2012; Karki et al., 2013; Sun et al., 2016; Pan et al., 2016). In rice, the sink lies in panicle and grains (Ma et al., 2004; Gendua et al., 2009; Mohapatra et al., 2011; Mo et al., 2012; Zhan et al., 2015; Adriani et al., 2016; Bai et al., 2016; Fabre et al., 2016; Tian-Yao et al., 2016). Therefore, under adequate N supply throw vermicompost along with *Trichoderma* compost there would have been greater translocation of photosynthates from source to sink site. This resulted in production of longer and heavier panicle with more grains panicle/m^2 (Manzoor et al. 2006; Chaudhary et al. 2011). Higher rates of P markedly increased almost all the yield attributes (Nesgea et al., 2012; Usman, 2013; Kumar et al., 2016; Mitran et al., 2016; Rajput et al., 2016). Since P is a constituent of nucleic acid, phytin, and phospholipids (Hall and Hodges, 1966; Anonymous, 1998; Schachtman et al., 1998; Šarapatka, 2003) its increased uptake resulted into better growth of plant and yield attributing characters finally resulting into increased yield (Gebrekidan and Seyoum, 2006; Dash et al., 2011; Duarah et al., 2011; Stephen et al., 2015; Kumar et al 2016). P is essential in laying down the primordial for the reproductive parts of the plants (Anonymous, 1998; Schachtman et al., 1998; Šarapatka, 2003). Being constituent of majority of enzymes responsible for transformation of energy, carbohydrates, and fat metabolism, it also plays dominant role in respiration of plants. The maximum grain and straw yield was due to marked improvement in dry matter accumulation, yield attributes and greater nutrient content, and their uptake by rice crop (Gebrekidan and Seyoum, 2006; Senthivelu et al., 2009; Mondal et al., 2015; Panigrahi et al., 2015; Ram et al., 2016). The maximum grain and straw yield was due to marked improvement in dry matter accumulation, yield attributes and greater nutrient contents, and their removal by rice crop (Singh et al. 2000; Rajbhandari, 2007; Senthivelu et al., 2009; Ram et al., 2011; Vinod and Heuer, 2012; Mahajan et al., 2012; Gautam et al., 2013; Mondal et al., 2013; Pramanik and Bera, 2013; Zerin, 2013).

7.4.3   NPK CONTENT AND UPTAKE

Application of graded levels of N through organic sources increased N, P, and potassium (K) content in grain and straw of the rice at harvest (Sekhar, 2004; Saha et al., 2009; Dash et al., 2010; Hashim et al., 2015; Sultana et al., 2015; Parihar et al., 2015;). P content of the rice crop increases which might be due to organic anions and hydroxyl acids such as citric, malanic, malic, and tartaric acids, which liberated during the decomposition of organic manures (Fu, 1989; Mokolobate, 2000; Judge, 2001; Jagtap et al., 2006; Srivastava, 2013;) would have complexed or chelated iron, aluminum (Al), magnesium (Mg), and calcium (Ca) which prevented them from reacting with phosphate ions (PO_4^{3-}) to from insoluble phosphates enabling more P availability to rice plants (Siddaramappa et al., 1991; Saleque, and Kirk, 1995; Arcand and Schneider, 2006; Syers et al., 2008; Balemi and Negisho, 2012; Walpola and Yoon, 2012; Zhu et al., 2014). Increased K content in grain and straw of rice through FYM or vermicompost is quite likely as most of the K in plants remains in inorganic form (Banik and Bejbaruah, 1996; Sultana et al., 2015; Kandan and Subbulakshmi, 2015; Mahmud et al., 2016). The decomposing organic manures may have a solubilizing effect on native K in soil (Zhang et al., 2015; Wolie and Admassu, 2016).

Application of nutrient in rice through bioorganic sources found significant increase of N, P, and K uptake (Satheesh and Balasubramanian, 2003; Okamoto and Okada, 2004; Zhang and Wang, 2005; Arcand and Schneider, 2006; Senthivelu and Surya Prabha, 2007; Hanč et al., 2008; Rahaman and Sinha, 2013; Shah and Kumar, 2014; Goutami and Rani, 2016; Wolie and

Admassu, 2016). Report revealed that the combined application of vermi-compost along with *Trichoderma* compost was found significantly better with respect to NPK uptake (Kashem et al., 2015; Sultana et al., 2015). This might be due to increased efficiency and cumulative effect of combined application of bioorganic sources of nutrients resulting in increased uptake of nutrients (Goswami and Banerjee, 1978; Gill and Meelu, 1982; Pillai, 1990; Singh et al., 1999; Vennila et al., 2007; Diacono and Montemurro, 2010; Ram et al., 2011, 2014; Sudhakar, 2011; Balamurugan and Sudhakar, 2012; Anjaiah and Jeevan Rao, 2016). Increased N uptake in rice is described as a result of NPK application is understood in view of its N_2 fixing capacity (Zhang and Wang, 2005; Srivastava and Ngullie, 2009; Masclaux-Daubresse et al., 2010). Positive effect of NPK and vermicompost on N content and uptake by rice crop was reported (Khan et al., 1986; Sharma and Mitra, 1990; Bhandari et al., 1992; Ghosh et al., 1994; Prasad and Prasad, 1994; Kumar et al., 1995; Tripathi and Chaubey, 1996; Jadhav et al., 1997; Dixit and Gupta, 2000; Hemalatha et al., 2000; Subbiah and Kumaraswamy, 2000; Jeyabal and Kuppusamy, 2001; Benik and Bejbaruah, 2004; Mishra et al., 2005; Laxminarayana and Patiram, 2006; Sunitha et al., 2010; Urkurkar et al., 2010; Deshpande and Devasenapathy, 2011; Suresh et al., 2013).

7.4.4 EFFECT ON QUALITY

The milling and chemical parameters, namely, hulling recovery, milling recovery, protein content, and protein yield reported significant increase with vermicompost application along with *Trichoderma* compost in which, integration of bioorganic sources increased the N content of the rice plants which ultimately influenced to the protein content and protein yield (Gandhi and Sivakumar, 2010; Sangeetha et al. 2010). Application of higher dose of organic N either through vermicomposting or FYM brought out consider-able level of improvement on quality of paddy (Adhikary, 2012; Ramak-rishna Parama and Munawery, 2012; Srivastava, 2013). Elevated levels of protein have also been reported by paddy under organic treatment due to enhancement in the N level in grains.

The N is an essential constituent of protein and increasing levels of organic sources increased N removal leading thereby to enhanced protein content in grain and straw (Abedi et al., 2010; Yadav et al., 2013c; Chondie, 2015; Ladha et al., 2016). Many workers also obtained higher protein harvest with increasing levels of N applied through organic sources (Hemalatha et al., 1999; Yadav et al., 2013c; Rao et al., 2014; Sekhar et al., 2014; Anwar et al., 2015; Khatun et al., 2015; Sharma et al., 2015).

7.5   ROLE OF ORGANIC SOURCES IN INM PRACTICES FOR RICE: A CASE STUDY

Paddy has become a focal point of research in various parts of world along with different region in India. Rice is the most important crop in India and also the hub of food security of the global population to meet the challenges of food security for rapidly increasing population; intensive cropping pattern were adopted which resulted in declining nutrient status of soil and therefor use of chemical fertilizer are increasing day by day. Indiscriminate use of fertilizers adversely affect the physicochemical properties of the soil resulting in poor rice productivity. The decline response to fertilizer has been emerged as a major issue challenging the sustainability of rice-based cropping system. Among the external conditions such as different abiotic and biotic factors regulating crop growth are very much important for crop productivity and yield. Each of these factors has the potential to limit the crop productivity and yield. Since 1950s, the intensity of fertilizer use has been on increase as also the consumption of NPK fertilizers. While, from the point of increasing rice production in the country, these increases in consumptions of major nutrients are positive indicators, a deep cause of concern is declining of grain productivity per kg of added nutrients. The importance of organic manures and fertilizers application in the maintenance and improvement of soil fertility under rice cultivation in general and particularly under system of rice intensification has been well documented in the literature. Organic inputs obtained from farming industries are solely based on plant and animal system. These organic manures being slow in release of nutrients, assume greater significance in a cropping sequence than individual crops and their usefulness needs to be investigated on long term basis.

7.5.1   VERMICOMPOST IN INM SYSTEM

Vermicompost, being a rich source of macro- and micronutrients, vitamins, plant growth regulators, and beneficial microflora, appeared to be the best organic source in maintaining soil fertility on sustainable basis toward an eco-friendly environment (Buckerfield, 1999; Giraddi, 2000; Chaudhary et al., 2004; Edwards and Arancon, 2004b; Suthar, 2006; Sinha et al., 2009;

Mishra et al., 2005; Sarma et al., 2010; Theunissen et al., 2010; Adhikary, 2012; Pathma and Sakthivel, 2012; Gupta et al., 2014b; Joshi et al., 2015; Sultana et al., 2015; Jain, 2016; Junying et al., 2009). Vermicompost application to different crops such as balsam (Singh et al., 2003), brinjal (Najar et al., 2015), Chinese cabbage (Wang et al., 2010), Chrysanthemum (Sarojani et al., 2012), cucumber (Yardim et al., 2006), garlic (Argüello et al., 2006), ground nut (Ramesh, 2000), maize (Shirkhani and Nasrolahzadeh, 2016), paddy (Bhattacharjee et al., 2001), pea (Singh et al., 2003), sorghum (Reddy and Ohkura, 2004), spinach (Peyvast et al., 2008), strawberry (Singh et al., 2008), tomato (Yardim et al., 2006; Zaller, 2006, 2007), and wheat (Yousefi and Sadeghi, 2014) as well as other crops have been to reduce the requirement of chemical fertilizer without any reduction in crop yield. Hidlago et al. (2006) reported that the incorporation of earthworm increased plant growth, leaf growth, and root length. Vermicompost fertilizer is a suitable alternative to promote crop growth along with improvement in soil quality due to interaction with the soil colloid material.

Vermicompost application in soybean has reported to have stimulatory effects in terms of its vegetative growth with gradual increment of various yield attributes (Bablad, 1999). Application of vermicompost @ 10 t ha^{-1} proved to be positive in terms of yield attributes. Comparative applications of vermicompost and FYM @ 10 t ha^{-1} revealed 11.55% increment in the yield under vermicompost treatment (Devidayal and Agarwal, 1999). As per Cai et al. (2009) it was found that combined application of vermicompost and chemical N fertilizer positively influences yield attributes and progressive growth of rice root in comparison to N fertilizer. According to Gandhi and Sivakumar (2010) INM practices such as vermicompost can improve plant growth and reduced crop duration leading to early maturity of crop plants. Balamurugan and Sudhakar (2012) reported higher plant height under organic-based vermicompost treatment in comparison to control. In terms of plant height, vermicompost and press mud based application recorded highest value followed by sewage sludge-based vermicompost and FYM-based vermicompost. Minimum values were recorded for control that is without vermicompost application.

Sudha and Chandini (2002) worked on sandy loam soil and reported the yield potential of rice under 5 t vermicompost was found to be similar to that of 10 t FYM ha^{-1}. Mishra et al. (2006) reported the variable source of organic N improved yield of rice. Dahiphale et al. (2000) reported positive influence of organic nutrition on crop productivity and profitability under upland condition with variable organic nutrient sources in comparison to recommended fertilizer dose and control. As per Tripathi and Verma (2008) under silty clay loam soil condition it was found that vermicompost-treated plots had comparatively higher yield attributes in rice with subsequent lowest grain yield due to low number of ear-bearing shoots. Masciandaro et al. (2010) reported positive influence of organic inputs along with synthetic fertilizer as well as mixed treatment of organic and chemical fertilizer. Gandhi and Sivakumar (2010) reported that conjoint application of bioinoculants based on vermicompost significantly improve the yield attributes and productivity of rice in terms of higher grain yield and straw yield. Chaudhary et al. (2011) found the impact of sowing dates on paddy under application of organic fertilizer. Rice variety "Rajendra Suwasani" recorded higher vegetative growth along with yield and yield attributes under INM system in comparison to chemical fertilizer application. Under clayey textural soil it was found that vermicompost treatment promoted higher yield in comparison to controlled in successive seasons (Balamurugan and Sudhakar, 2012).

Nutrient use efficiency of rice plants under field conditions were significantly improved through press mud based vermicompost application (Balamurugan and Sudhakar, 2012). Hidayatullah (2016) conducted field experiments on INM practices through different combinations of organic and inorganic sources of N. The results revealed positive influence of such practices if there is conjoint application of organic and inorganic form of N. Poultry manure were proved to be the superior nutrient supplement with inorganic N fertilizer giving higher yield and productivity. Subsidy due to chemical fertilizer use can be minimized by supplementing through organic manures (Masarirambi et al. 2012). Sahrawat (2006) reported lower decomposition rate under water-logged condition of agricultural field. This therefore suggests that lowering of C/N ratio in wheat straw may boost up crop production under paddy cultivation. Higher test weights of grains were reported by crops under organic treatment due to higher nutrient availability and uptake by rice plants (Singh and Agarwal, 2001). Variability in nutrient source significantly varies rice growth and yield due to variable nutrient absorption rate depending upon the source (Ahmad et al. 2008). Hassanuzazzaman et al. (2010) reported effectiveness of poultry litter toward promoting enhanced crop growth and yields. As per the report of Garrity and Flinn (1987), the conjoint application of chemical fertilizer along with poultry manure can increase nutrient availability and thus maintain the soil health for optimum crop growth and yield. Gandhi and Sivakumar (2010) reported higher yield of rice under the treatment of vermicompost-based inoculant. Sangeetha et al. (2010) found organic amendment under various combinations of treatment along with synthetic fertilizer which significantly influenced rice quality parameters.

At Bangalore, application of vermicompost to summer paddy increased the N and P content of the soil. Application of vermicompost has a long-range influence on soil chemical and biological properties (Kale et al. 1991). Rajkhowa et al. (2003) reported the positive influence of organic fertilizer in improving the soil health and quality. Enhancement in the level of organic fertilizer in the soil physicochemical parameters is significantly improved in comparison to recommended dose of chemical fertilizer application. Higher conductivity values of soil were recorded under vermicompost treatments in comparison to untreated soil. Reductions in the soil pH value were also recorded in the similar treatment (Tharmaraj et al., 2011).

Sarangi and Lama (2013) reported under deep clayey soil condition under acidic pH in Meghalaya rice cultivation incurred higher expenditure with 7.5% lime substituents in comparison to 5% lime substituents, which were further followed by sole vermicompost application when compared to control condition. Vermicompost application with 5% lime substituents was found to be superior in comparison to other treatments. The benefit to cost (B:C) ratio were higher in vermicompost treatment in comparison to FYM and inorganic fertilizer treatments.

7.5.2 FARMYARD MANURE IN INM SYSTEM

FYM is the best option and has been considered by farmers and researchers as the nutrient supplement with the objective to maintain sustainability of agriculture and crop agricultural production since ages (Ghoshal and Singh, 1995; Stein-Bachinger, 1996; Dewes and Hunsche, 1998; Dixit and Gupta, 2000; Choudhary et al., 2004; Hansen et al., 2004; Yaduvanshi and Swarup, 2005; Reddy et al., 2007; Shekara and Sharnappa, 2010). Well-rotten FYM is a good source of various essential plant nutrients, that is, 0.5% N, 0.2% P_2O_5, and 0.5% K_2O (Das, 2013; Rana et al., 2014). In FYM, total amount of nutrient present is not available for soil and crop just after application. It is available to the first crop with about 30%, 60%, and 70% of NPK (Reddy and Reddy, 2003) which always employed to boost up the growth and increasing yield of agricultural crop in eco-friendly environment. In addition, FYM also influence the soil physical, chemical, and biological properties (Ganiger et al., 2012; Esmaeilzadeh and Ahangar, 2014; Nagar et al., 2016).

Bhattacharyya et al. (2003) reported higher plant height and biomass, and dry matter accumulation under FYM treatment @ 9.0 t ha^{-1} and higher dose

(20 t FYM ha^{-1}) at various growth stages of rice (Shekara and Sharnappa, 2010). Increment in the level of growth attributes of rice under organic manure application were reported by Kumar and Singh (2006). Mankotia et al. (2008) reported higher plant height of rice under 5 t ha^{-1} FYM application. Kharub and Chander (2008) reported nonsignificant level of difference between organic and inorganic fertilization in rice. As per Yadav et al. (2013a) significant level of influence were recorded in growth attributes and biomass at 50% lesser use of synthetic fertilizer dose along with organic inputs under variable conditions.

Verma et al. (2001) reported that FYM have reflected an increasing trend in productivity of paddy with increasing dose of FYM application. However, Sharma et al. (2012b) revealed that FYM @ 10 t ha^{-1} increased the grain yield of rice–wheat system by 1.2–1.3 t ha^{-1} and straw yield by 0.7–2.3 t ha^{-1}. According to the works of Mankotia et al. (2008) it was found that under clayey loam soil condition, application of FYM 5 t ha^{-1} reflected similar productivity when compared to FYM 5 t ha^{-1} + 1.25 t mushroom spent compost treatment, FYM 2.5 t ha^{-1} + gobhi–sarson straw 2.5 t ha^{-1} (4.72 t ha^{-1}) treatment. Similar reports of higher yield and productivity of rice and wheat under FYM application were done by Kharub and Chander (2008).

Research reports of Satheesh and Balasubharamanian (2003) reveal increase of N and P uptake through FYM application under paddy cultivation. One-fourth of recommended dose of nitrogenous chemical fertilizer along with application of FYM significantly improved N uptake by crops (Singh et al. 1998; Tripathi et al., 2007). Organic farming promoted higher N and P uptake under 10 t FYM ha^{-1} application (Quyen and Sharma, 2003). Maiti et al. (2006) reported higher nutrient status with recommended dose of fertilizer (RDF) and FYM application for both kharif and boro rice. As per Rasoon et al. (2007) the crop nutrient uptake were found to be higher under combined application of FYM with inorganic fertilizers in comparison to sole application of inorganic fertilizers. High protein content was reported in case of rice under FYM application (Tiwari et al., 2001; Prakash et al., 2002; Kharub and Chander, 2008). Combined application of FYM and RDF leads to better seed germination and elevated level of growth attributes under system of rice intensification techniques (Krishna et al., 2008).

Kharub and Chander (2008) reported the elevated level of NPK under higher dose of FYM application. K content reduced under inorganic fertilizer application due to higher crop uptake from soil. As per Singh et al. (1998), higher rate of FYM application leads to increment in the gross return (ha^{-1}) and variable expenses (ha^{-1}). Utility of vermicompost relies on soil physical properties and organic C level which promotes crop growth and development (Hidlago et al., 2006). Combined application of fly ash with FYM significantly improves B:C ratio (Reddy et al., 2007). Similar trend was also reported by Choudhary and Sinha (2007) under FYM treatment. Combined application of FYM and *Euputorium* improves mean rice-equivalent yield (REY) as well as FYM and *Alnus* application in highest net return and B:C ratio (Munda et al., 2008). Similar trend were also reported by Davari and Sharma (2010) under FYM treatment. Earlier research reports reveal (Shekara and Sharnappa, 2010) higher net B:C ratio and net economic gain under higher dose of FYM application. Combined application of FYM and BGA significantly improves the yield in comparison to sole application of FYM and BGA (Singh et al., 2013).

7.5.3 PRESS MUD IN INM SYSTEM

RDF and press mud application significantly improves the yield attributes and plant height of rice when compared to sole application of press mud and RDF (Singh and Sharma 2009). Khan et al. (2012) in Gomal University in Pakistan found that highest growth in terms of height of the plant were recorded for the treatment of N, P, and K along with 35.6 g kg^{-1} press mud. Enriched press mud compost significantly increases the yield attributes of rice at various stages of crop growth (Kalaivanan and Hattab, 2008). Khan et al. (2012) in Gomal University in Pakistan found that positive influence of press mud over yield attributes of rice.

Deshpande and Devasenapathy (2010) reported combined application of green manure along with poultry manure resulted into higher availability of soil nutrients to rice plant. Nutrient level in terms of available K increased in soil. Higher mobilizations in terms of crop uptake were recorded for N and P under green manure and poultry manure application. Combined application of green manure vermicompost and poultry manure reflected negative impact on soil electrical conductivity (EC) and pH but positively influenced soil organic carbon content. Combined application of inorganic and organic fertilizer positively influences the soil quality by improving the soil nutrient status. Macro- and micronutrient level in soil can be increased to a considerable level by such combination of organic and inorganic fertilizer. This therefore promotes sustainable agriculture practice under the aim of boosting of agricultural productivity (Ali et al. 2009). According to Singh et al. (2005) soil physical parameters varies significantly under organic treatment. Bulk density value shows a declining trend with enhanced level of porosity and water holding capacity value of soil under specific dose of FYM treatment. Gupta (1995) reported higher moisture percentage under combined treatment of FYM, pig manure, and poultry manure in comparison to tank clay and NPK fertilizer application. This therefore promotes higher productivity in alfisols due to higher moisture content in the root zone of paddy. Mathakiya and Meishri (2003) reported that application of press mud increase the EC of the soil in cabbage. Research work carried out at IARI, New Delhi reported highest economic gain under application of organic and inorganic fertilizer along with 25% recommended fertilizer dose in comparison to control (Singh and Sharma, 2009).

7.5.4   *Trichoderma* COMPOST IN INM SYSTEM

Doni et al. (2014) reported the positive influence of *Trichoderma* spp. on paddy cultivation under field condition. From the experimental field trial it appears that plant physiological process, growth attributes, biomass as well as yield and productivity of paddy significantly improved under *Trichoderma* treatment. These therefore suggest that *Trichoderma* could be effectively utilized as biofertilizer boosting up rice productivity and yield maintaining soil health and quality. Yield attributes such panicles/ m^2 revealed higher values under the treatment of compost inoculated with *Azotobacter* + *Bacillus polymyxa* as well as NPK fertilizer (Singh et al., 2006a). Composted paddy straw inoculated with *Trichoderma* spp. reveals higher organic carbon level into the soil (Son et al., 2008). Vermicompost with 5% lime supplement produced higher net return in case of paddy cultivation which was found to be superior over FYM, RDF, and without fertilizer application (Sarangi and Lama, 2013).

7.5.5   ORGANIC SOURCES AND CHEMICAL FERTILIZER COMBINATION IN INM SYSTEM

Soil health can be significantly improved through INM practices. Organic amendments provide a rich source of nutrients for the crop plants promoting higher yield and productivity. This approach can be utilized as aiming toward sustainability in agroecosystem. This is a serious concern as our soil resource is gradually depleting day by day due to nonjudicious application of chemical fertilizer. Organic inputs in terms of agricultural residues can be effectively utilized for improving soil health and maintain crop productivity. Such approaches promote microbial activity in the soil at higher rate which may increase pore space and enzyme activity in soil.

As per Babu et al. (2000) application of 5 t FYM + 50 kg N ha^{-1} increased yield and growth attributes. Pradhan and Sahu (2004) reported the combined application of fly ash, FYM, and 50% recommended fertilizer dose significantly influence the growth of rice plants. During various stages of crop growth, the elevated level of biomass accumulation were recorded under the application of FYM 12.5 t + 150, 100, and 50 kg NPK ha^{-1} (Senthivelu and Prabha, 2007).

Ramesh and Vayapuri (2008) reported maximum productivity under combined application of vermicompost and 100% recommended dose of nitrogen (RDN) in comparison to other sources of organic manure. Koushal et al. (2011) working on sandy loam in texture soil having pH 8.1 with rice variety "PC-19" at Jammu reported positive influence of vermicompost application along with 50% reduction in recommended dose of fertilizer on rice growth and yield attributes. Yadav et al. (2013a) working at Lembucherra, West Tripura on sandy clay loam soil found the maximum productivity of rice under different rice growing seasons with INM practices.

Tripathi et al. (2007) reported one-fourth increase of nutrient uptake by rice under supply of organic N in the form of FYM. Nutrient use efficiency of paddy plants in wet-seeded condition significantly improved under 75% inorganic N and 25% organic N source along with 100% RDFN (Vennila et al., 2007). According to Pal et al. (2007) application of N$_{120}$, P$_{60}$, and K$_{40}$ + FYM @ 10 t and 75% NPK (90–45–30 kg ha^{-1}) + 10 t FYM ha^{-1} improved crop nutrient uptake significantly. Similar trends of increased nutrient uptake were recorded by Singh et al. (2012) under 15 t of FYM + 120, 60, 30 kg NPK ha^{-1}.

Combined application of FYM, vermicompost, and BGA along with N and prilled urea promoted higher protein content in rice (Adhikari et al., 2005). Initially, higher protein content was reported under NPK fertilizer application. Further application of FYM and BGA also increased the level of protein content in rice grain (Dixit and Gupta, 2000). Influence of organic fertilizer over soil fertility was reported by Koushal et al. (2011). During the field trial, it was found that elevated level of organic carbon were reported under vermicompost and FYM application which supplied 100% RDN to the soil system. In the same experiment, it was found lesser N input from various other organic sources. Soil physical attributes such as bulk density value reduced under the organic fertilizer treatment in rice–wheat cultivation system.

Net return and B:C ratio were found to be highest under NPK treatment with FYM and zinc supplement in case of hybrid rice (Pandey et al., 2007). As per Ramesh and Vayapuri (2008) under clayey textural soil the economic gain in terms of rice yield was higher under vermicompost @ 5 t ha^{-1} along with 100% RDN treatment. Experiment conducted by Koushal et al. (2011) on sandy loam in texture soil (pH 8.1) with rice variety "PC-19" at Jammu, reported higher economic gain under 25% N fertilizer reduction dose through vermicompost application.

7.6   SUMMARY AND CONCLUSION

Agriculture is the base of most of the developing nations giving significant level of economic output and community well-being. In the present time, agriculture has become a major challenge for us due to climate variability, lack of land and soil resources, and loss of agrobiodiversity as well as different forms of agricultural pollutions. Under this context it needs to be mentioned that chemical fertilizer and pesticides pollution is the most hindering factor in agriculture sectors. As world population is increasing day by day improvement in the food production is an essential prerequisite. Keeping view in this to boost up the agricultural productivity nonjudicious use of chemical fertilizer and pesticides has led to different types of health hazards for human civilization.

Organic farming is such an approach which includes INM on one hand and pollution-free practices on the other hand. It also promote use of different forms of organic inputs such as green manure, FYM, vermicompost, biofertilizer, press mud, poultry litter, etc. Paddy is the major staple food crop of India. As India stands first in world population strength therefore, major focus is toward increasing rice productivity and yield. Under these circumstances the issues such as climatic variability, over use of fertilizers, use of HYV have reduced rice production throughout India to a significant level. This therefore also hinders Indian economy as agriculture is the backbone of Indian economy. Application of organic inputs in terms of FYM, poultry litter, vermicompost, *Trichoderma*, and biofertilizer have shown considerable promise toward boosting up rice productivity and yield maintaining soil health and quality. Combination of application of organic amendments along with NPK chemical fertilizer in the form of INM can be effectively utilized as a suitable strategy to combat the problems in agricultural sectors.

KEYWORDS

- chemical fertilizers
- conventional farming
- crop productivity
- environmental perturbances
- growth stages
- growth performance
- yield attributes

REFERENCES

Abedi, T.; Alemzadeh, A. Kazemeini, S. A. Effect of Organic and Inorganic Fertilizers on Grain Yield and Protein Banding Pattern of Wheat. *Aust. J. Crop Sci.* **2010,** *4* (6), 384–389.

Adhikari, N. P.; Mishra, B. N.; Mishra, P. K. Effect of Integrated Nitrogen Management on Quality of Aromatic Rice. *Ann. Agric. Res.* **2005,** *26* (2), 231–234.

Adhikary, S. Vermicompost, the Story of Organic Gold: A Review. *Agric. Sci.* **2012,** *3* (7), 905–917.

Adriani, D. E.; Dingkuhn, M.; Dardou, A.; Adam, H.; Luquet, D.; Lafarge, T. Rice Panicle Plasticity in Near Isogenic Lines Carrying a QTL for Larger Panicle is Genotype and Environment Dependent. *Rice* **2016,** *9* (28), 1–15.

Ahmad, R. M.; Naveed, M.; Aslam, Z. A.; Arshad, M. Economizing the Use of Nitrogen Fertilizer in Wheat Production Through Enriched Compost. *Rev. Agric. Food Sys.* **2008,** *23,* 243–249.

Aktar, M. W.; Sengupta, D.; Chowdhury, A. Impact of Pesticides Use in Agriculture: Their Benefits and Hazards. *Interdiscip. Toxicol.* **2009,** *2* (1), 1–12.

Alam, M. A.; Kumar, M. Effect of Zinc on Growth and Yield of Rice var. Pusa Basmati-1 in Saran District of Bihar. *Asian J. Plant Sci. Res.* **2015,** *5* (2), 82–85.

Alberda, T. Growth and Root Development of Lowland Rice and Its Relation to Oxygen Supply. *Plant Soil* **1953,** *5* (1), 1–28.

Ali, M. E.; Islam, M. R.; Jahiruddin, M, Effect of Integrated Use of Organic Manures with Chemical Fertilizers in the Rice–Rice Cropping System and Its Impact on Soil Health. *Bangladesh J. Agric. Res.* **2009,** *34* (1), 81–90.

Alves, W. L.; Passoni, A. A. Compost and Vermicompost of Urban Solid Waste in *Licania tomentosa* (Benth) Seedlings Production to Arborization. *Pesq. Agropec. Bras.* **1997,** *32,* 1053–1058.

Andrade, M. M. M.; Stamford, N. P.; Santos, C. E. R. S.; Freitas, A. D. S.; Sousa, C. A.; Lira-Junior, M. A. Effects of Biofertilizer with Diazotrophic Bacteria and Mycorrhizal Fungi in Soil Attribute, Cowpea Nodulation Yield and Nutrient Uptake in Field Conditions. *Sci. Hort.* **2013,** *162,* 374–379.

Anjaiah, T.; Jeevan Rao, K. Effect of Integrated Use of Municipal Compost, Sewage Sludge, Farm Yard Manure with Chemical Fertilizers on Soil Enzymatic Activity and Yield of Rice–Groundnut Cropping System. *Int. J. Trop. Agr.* **2016,** *34* (4), 997–1004.

Anonymous. Functions of Phosphorus in Plants. In *Better Crops with Plant Food*; Armstrong, D. L., Griffin, K. P., Eds.; Potash & Phosphate Institute: Norcross, Georgia, USA, 1998; Vol. 82 (3).

Anonymous. *Agriculture Statistics at a Glance*. Department of Agriculture and Co-operation, Ministry of Agriculture, Government of India, New Delhi, 2014.

Anwar, S. A.; Iqbal, M.; Anwar-ul-Hassan; Ullah, E. Growth and Yield of Rice–Wheat in Response to Integrated Use of Organic and Inorganic N Sources. *Soil Environ.* **2015,** *34* (2), 136–141.

Arancon, N. Q.; Lee, S.; Edwards, C. A.; Atiyeh, R. M. Effects of Humic Acids and Aqueous Extracts Derived from Cattle, Food and Paper-waste Vermicomposts on Growth of Greenhouse Plants. *Pedobiologia* **2003,** *47*, 741–744.

Arancon, N. Q.; Edwards, C. A.; Bierman, P.; Metzger, J. D.; Lucht, C. Effects of Vermicomposts Produced from Cattle Manure, Food Waste and Paper Waste on the Growth and Yields of Peppers in the Field. *Pedobiologia* **2005,** *49*, 297–306.

Arancon, N. Q.; Edwards, C. A.; Bierman, P. Influences of Vermicomposts on Field Strawberries: Part 2. Effects on Soil Microbiological and Chemical Properties. *Bioresour. Technol.* **2006a,** *97*, 831–840.

Arancon, N. Q.; Edwards, C. A.; Lee, S.; Byrne, R. Effects of Humic Acids from Vermicomposts on Plant Growth. *Eur. J. Soil Biol.* **2006b,** *45*, S65–S69.

Araus, J. L.; Slafer, G. A.; Royo, C.; Serret, M. D. Breeding for Yield Potential and Stress Adaptation in Cereals. *Crit. Rev. Plant Sci.* **2008,** *27*, 377–412.

Arcand, M. M, and Schneider, K. D. Plant- and Microbial-based Mechanisms to Improve the Agronomic Effectiveness of Phosphate Rock: A Review. *Ann. Braz. Acad. Sci.* **2006,** *78* (4), 791–807.

Argüello, J. A.; Ledesma, A.; Núñez, S. B.; Rodríguez, C. H.; Díaz Goldfarb, M. D. C. (2006). Vermicompost Effects on Bulbing Dynamics Nonstructural Carbohydrate Content, Yield, and Quality of 'Rosado Paraguayo' Garlic Bulbs. *Hortscience* **2006,** *41* (3), 589–592.

Atiyeh, R. M.; Dominguez, J.; Subler, S.; Edwards, C. A. Biochemical Changes in Cow Manure Processed by Earthworms (*Eisenia andreii*) and Their Effects on Plant Growth. *Pedobiologia* **2000a,** *44*, 709–724.

Atiyeh, R. M.; Subler, S.; Edwards, C. A.; Bachman, G.; Metzger, J. D.; Shuster, W. Effects of Vermicomposts and Composts on Plant Growth in Horticultural Container Media and Soil. *Pedobiologia* **2000b,** *44*, 579–590

Atiyeh, R. M.; Arancon, N. Q.; Edwards, C. A.; Metzger, J. D. The Influence of Earthworm-processed Pig Manure on the Growth and Productivity of Marigolds. *Bioresour. Technol.* **2001,** *81*, 103–108.

Aulakh, M. S. In *Integrated Nutrient Management for Sustainable Crop Production, Improving Crop Quality and Soil Health, and Minimizing Environmental Pollution*. 19th World Congress of Soil Science, Soil Solutions for a Changing World, Brisbane, Australia, Aug 1–6, 2010.

Bablad, H. B. Integrated Nutrient Management for Sustainable Production in Soybean Based Cropping Systems. Ph.D. Dissertation, University of Agricultural Sciences, Dharwad, India, 1999.

Babu, B. T. R.; Reddy, V. C. Effect of Nutrient Sources on Growth and Yield of Seeded Rice (*Oryza sativa*). *Crop Res.* (Hisar) **2000,** *19* (2), 189–193.

Badgley, C.; Perfecto, I. Can Organic Agriculture Feed the World? *Renew. Agric. Food Syst.* **2007,** *22* (2), 80–85.

Bai, P.; Bai, R.; Jin, Y. Characteristics and Coordination of Source-sink Relationships in Super Hybrid Rice. *Open Life Sci.* **2016,** *11*, 470–475.

Bajaj, K. L. Biochemical Basis for Disease Resistance-role of Plant Phenolics (C). In *Advances in Frontier Areas of Plant Biochemistry*; Singh, R.; Sawhney, S. K. Eds.; New Delhi: Prentica-Hall of India Private Limited, India, 1998; pp 487–510.

Balamurugan, R and Sudhakar, P. Influence of Planting Methods and Different Vermicompost on the Growth Attributes, Yield and Nutrient Uptake in Rice. *Int. J. Curr. Agric. Sci.* **2012,** *2* (6), 20–23.

Balemi, T.; Negisho, K. Management of Soil Phosphorus and Plant Adaptation Mechanisms to Phosphorus Stress for Sustainable Crop Production: A Review. *J. Soil Sci. Plant Nutr.* **2012,** *12* (3), 547–562.

Ballabh, V.; Pandey, S. Transitions in Rice Production Systems in Eastern India Evidence from Two Villages in Uttar Pradesh. *Econ. Political Weekly* **1999,** *34* (13), A11–A16.

Banik, P.; Bejbaruah, R. Effect of Vermicompost on Rice (*Oryza sativa*) Yield and Soil Fertility Status of Rainfed Humid Sub-tropics. *Indian J. Agric. Sci.* **1996,** *74* (9), 488–491.

Barman, L. K.; Kalita, R. B.; Jha, D. K. Inductions of Resistance in Brinjal (*Solanum melongenae* L.) by Aqueous Extract of Vermicompost Against Fusarium Wilt. *Int. J. Plant Anim. Environ. Sci.* **2013,** *3* (1), 141–148.

Baruah, A.; Baruah, K. K. Organic Manures and Crop Residues as Fertilizer Substitutes: Impact on Nitrous Oxide Emission, Plant Growth and Grain Yield in Pre-monsoon Rice Cropping System. *J. Environ. Prot.* **2015,** *6*, 755–770.

Begum, Z. N. T.; Mandal, R.; Islam, M. S. Effects of Organic Manure and Urea-N on Growth and Yield Performance of Traditional Variety of Rice. *J. Phytol. Res.* **2009,** *22* (2), 1660–1665.

Benik, P.; Bejbaruah, R. Effect of Vermicompost on Rice (*Oryza sativa*) Yield and Soil-fertility Status of Rainfed Humid Sub-tropics. *Indian J. Agric. Sci.* **2004,** *74*, 488–491.

Benincasa, P.; Guiducci, M.; Tei, F. The Nitrogen Use Efficiency: Meaning and Sources of Variation—Case Studies on Three Vegetable Crops in Central Italy. *HortTechnology* **2011,** *21* (3), 266–273.

Berova, M.; Karanatsidis, G. Physiological Response and Yield of Pepper Plants (*Capsicum annum* L.) to Organic Fertilization. *J. Central Eur. Agric.* **2008,** *9* (4), 715–722.

Bhadauria, T.; Kumar, P.; Maikhuri, R.; Saxena, K. G. Effect of Application of Vermicompost and Conventional Compost Derived from Different Residues on Pea Crop Production and Soil Faunal Diversity in Agricultural System in Garhwal Himalayas India. *Nat. Sci.* **2014,** *6*, 433–446.

Bhandari, A. L.; Sood, A.; Sharma, K. N.; Rana, D. S. Integrated Nutrient Management in Rice–Wheat system. *J. Indian Soc. Soil Sci.* **1992,** *40*, 742–747.

Bhattacharjee, G.; Chaudhuri, P. S.; Datta, M. Response of Paddy (Var. TRC-87- 251) Crop on Amendment of the Field with Different Levels of Vermicompost. *Asian J. Microbiol. Biotechnol. Environ. Sci.* **2001,** *3* (3), 191–196.

Bhattacharyya, P and Jain, R. K. Phosphorus Solubilizing Biofertilizers in the Whirlpool of Rock Phosphate-challenges and Opportunities. *Fertil. News* **2003**, *45*, 45–51.

Buckerfield, J. C.; Flavel, T.; Lee, K. E.; Webster, K. A. Vermicomposts in Solid and Liquid form as Plant-growth Promoter. *Pedobiologia* **1999**, *43*, 753–759.

Cai, S. M.; Qian, X. Q.; Bai, Y. C.; Lang, S. S.; Shan, Y. H. Effects of Vermicompost and Nitrogen Application on Growth and Root Vigor in Rice. *J. Yangzhou Univ. Agric. Life Sci.* **2009**, *30*, 4, 67–70.

Carrasco, A.; Boudet, A. M.; Marigo, G. Enhanced Resistance of Tomato Plants to Fusarium by Controlled Stimulation of Their Natural Phenolic Production. *Physiol. Plant Pathol.* **1978**, *12*, 225–232.

Chamani, E.; Joyce, D. C.; Reihanytabar, A. Vermicompost Effects on the Growth and Flowering of *Petunia hybrida* 'Dream Neon Rose'. *Am.-Eurasian J. Agric.; Environ. Sci.* **2008**, *3* (3), 506–512.

Chand, S.; Pandey, A.; Anwar, M.; Patra, D. D. Influence of Integrated Supply of Vermicompost, Biofertilizer and Inorganic Fertilizer on Productivity and Quality of Rose Scented Geranium (*Pelargonium species*). *Indian J. Natl. Prod. Resour.* **2011**, *2* (3), 375–382.

Chandra, D.; Nanda, P.; Singh, S. S.; Ghorai, A. K.; Singh, S. R. Integrated Nutrient Management For Sustainable Rice Production. *Arch. Agron. Soil Sci.* **2004**, *50* (2), 161–165.

Chang, T. Rice. *Evolution of Crop Plants*; Simmonds, N. W. Ed.; Longman: London, 1976; pp 98–104.

Chatterjee, R.; Thirumdasu, R. K. Nutrient Management in Organic Vegetable Production. *Int. J. Food Agric. Vet. Sci.* **2014**, *4* (3), 156–170.

Chaudhary, D. R.; Bhandari, S. C.; Shukla, L. M. Role of Vermicompost in Sustainable Agriculture—A Review. *Agric. Rev.* **2004**, *25*, 29–39.

Chaudhary, S. K.; Singh, J. P.; Jha, S. Effect of Integrated Nitrogen Management on Yield, Quality and Nutrient Uptake of Rice (*Oryza sativa*) Under Different Dates of Planting. *Indian J. Agron.* **2011**, *56* (3), 228–231.

Chauhan, B. S.; Mahajan, G. *Recent Advances in Weed Management*; Chauhan, B. S., Mahajan, G., Eds.; Springer Science + Business Media: New York, 2014; p 441.

Chen, S.; Zeng, F.-R.; Pao, Z.-Z.; Zhang, C.-P. Characterization of High-yield Performance as Affected by Genotype and Environment in rice. *J. Zhejiang Univ. Sci. B* **2008**, *9* (5), 363–370.

Chondie, Y. G. Effect of Integrated Nutrient Management on Wheat: A Review. *J. Biol. Agric. Healthc.* **2015**, *5* (13), 68–76.

Choudhary, S. K.; Sinha, N. K. Effect of Levels of Nitrogen and Zinc on Grain Yield and Their Uptake in Transplanted Rice. *Oryza Int. J. Rice* **2007**, *44* (1), 44–47.

Choudhary, O. P.; Josan, A. S.; Bajwa, M. S.; Kapur, M. L. Effect of Sustained Sodic and Saline-sodic Irrigations and Application of Gypsum and Farmyard Manure on Yield and Quality of Sugarcane Under Semi-arid Conditions. *Field Crops Res.* **2004**, *87*, 103–116.

Choudhury, B.; Khan, M. L.; Dayanandan, S. Genetic Structure and Diversity of Indigenous Rice (*Oryza sativa*) Varieties in the Eastern Himalayan Region of Northeast India. *Springerplus* **2013**, *2* (228), 1–10.

Crotty, F. V.; Fychan, R.; Theobald, V. J.; Sanderson, R.; Chadwick, D. R.; Marley, C. L. The Impact of Using Alternative Forages on the Nutrient Value within Slurry and Its Implications for Forage Productivity in Agricultural Systems. *PLoS One* **2014**, *9* (5), e97516. DOI: 10.1371/journal.pone.0097516.

Daayf, F.; El Hadrami, A.; El-Bebany, A. F.; Henriquez, M. A.; Yao, Z.; Derksen, H.; El-Hadrami, I.; Adam, L. R. Phenolic Compounds in Plant Defense and Pathogen Counter-defense Mechanisms. In *Recent Advances in Polyphenol Research*. Cheynier, V., Sarni-Manchado, P., Quideau, S., Eds.; Wiley-Blackwell: Oxford, UK, 2012; Vol. 3. DOI: 10.1002/9781118299753.ch8.

Dahiphale, V. V.; Gunjkar, M. U.; Jadhav, A. S.; Shinde, R. M. Quality Characters of Upland Irrigated Basmati Rice as Influenced by Nutrient Management. *J. Maharashtra Agric. Univ.* **2000**, *24* (3), 337–338.

DaMatta, F. M.; Ramalho, J. D. C. Impacts of Drought and Temperature Stress on Coffee Physiology and Production: A Review. *Braz. J. Plant Physiol.* **2006**, *18* (1), 55–81.

Daniele, G.; Scherr, S.; Nierenberg, D.; Hebebrand, C.; Shapiro, J.; Milder, J.; Wheeler, K. *Food and Agriculture: The Future of Sustainability*. A Strategic Input to the Sustainable Development in the 21st Century (SD21) Project. United Nations Department of Economic and Social Affairs, Division for Sustainable Development, New York, 2012; p 80.

Das, S. K. Integrated Nutrient Management Using Only Through Organic Sources of Nutrients. *Pop. Kheti* **2013**, *1* (4), 121–126.

Dash, D.; Patro, H.; Tiwari, R. C.; Shahid, M. Effect of Organic and Inorganic Sources of N on Yield Attributes, Grain Yield and Straw Yield of Rice (*Oryza sativa*). *Res. J. Agron.* **2010**, *4* (2), 18–23.

Dash, D.; Patro, H.; Tiwari, R. C.; Shahid, M. Effect of Organic and Inorganic Sources of N on Growth Attributes, Grain and Straw Yield of Rice (*Oryza sativa*). *Int. J. Pharm. Life Sci.* **2011**, *2* (4), 655–660.

Davari, M. R.; Sharma, S. N. Effect of Different Combinations of Organic Material and Biofertilizers on Productivity, Grain Quality and Economics in Organic Farming of Basmati Rice (*Oryza sativa* L.). *Indian J. Agron.* **2010**, *55*, 290–294.

Dawe, D. The Contribution of Rice Research to Poverty Alleviation. In *Redesigning Rice Photosynthesis to Increase Yield*; Sheehy, J. E. et al. Eds.; International Rice Research Institute: Elsevier Science, Makati City (Philippines), Amsterdam, 2000; pp 3–12.

Deshpande, H.; Devasenapathy, P. Influence of Green Manure and Different Organic Sources of Nutrients on Yield and Soil Chemical Properties of Rice (*Oryza sativa* L.) Grown Under Low Land Condition. *Int. J. Agric. Sci.* **2010**, 6 (2), 433–438.

Deshpande, H. H.; Devasenapathy, P. Effect of Different Organic Sources of Nutrients and Green Manure on Growth and Yield Parameters of Rice (*Oryza sativa* L.), Grown Under Low land Condition. *Crop Res.* **2011**, *41* (1–3), 1–5.

Devidayal; Agarwal, S. K. Response of Sunflower Genotype (*H. annus*) to Nutrient Management. *Ind. J. Agr. Sci.* **1999**, *69*, 10–13.

Dewes, T.; Hunsche, E. Composition and Microbial Degradability in the Soil of Farmyard Manure from Ecologically-managed Farms. *Biol. Agric. Hort.* **1998**, *16*, 251–258.

Diacono, M.; Montemurro, F. Long-term Effects of Organic Amendments on Soil Fertility. A Review. *Agron. Sustain. Dev.* **2010**, *30* (2), 401–422.

Dissanayake, D. M. D.; Premaratne, K. P.; Sangakkara, U. R. Integrated Nutrient Management for Lowland Rice (*Oryza sativa* L.) in the Anuradhapura District of Sri Lanka. *Trop. Agric. Res.* **2014,** *25* (2), 266–271.

Dixit, K. G.; Gupta, B. R. Effect of Farmyard Manure, Chemical and Bio-fertilizers on Yield and Quality of Rice (*Oryza sativa* L.) and Soil Properties. *J. Indian Soc. Soil Sci.* **2000,** *48* (4), 773–780.

Dobermann, A.; Cassman, K. G.; Mamaril, C. P.; Sheehy, J. E. Management of Phosphorus, Potassium and Sulphur in Intensive Irrigated Lowland Rice. *Field Crops Res.* **1998,** *56* (1–2), 113–138.

Doni, F.; Isahak, A.; Radziah, C.; Zain, M. C.; Yusoff, W. Physiological and Growth Response of Rice Plants (*Oryza sativa* L.) to *Trichoderma* spp. Inoculants. *AMB Express* **2014,** *4*, 1–7.

Duarah, I.; Deka, M.; Saikia, N.; Deka Boruah, H. P. Phosphate Solubilizers Enhance NPK Fertilizer Use Efficiency in Rice and Legume Cultivation. *3 Biotech.* **2011,** *1* (4), 227–238.

Edwards, C. A.; Burrows, I. The Potential of Earthworm Composts as Plant Growth Media. In *Earthworms in Environmental and Waste Management*; Edwards, C. A., Neuhauser, E. F., Eds.; SPB Academic Publ.: The Netherlands, 1988; pp 211–220.

Edwards, C. A.; Arancon, N. Q. Interactions Among Organic Matter, Earthworms and Microorganisms in Promoting Plant Growth. In *Functions and Management of Organic Matter in Agroecosystems*; Magdoff, F., Weil, R., Eds.; CRC Press: Boca Raton, FL, 2004a; pp 327–376.

Edwards, C. A.; Arancon, N. Q. The Use of Earthworms in the Breakdown and Management of Organic Wastes to Produce Vermicomposts and Feed Protein. In *Earthwom Ecology*, 2nd ed.; Edwards, C. A., Ed.; CRC Press: Boca Raton, FL, 2004b; pp 345–379.

Edwards, C. A.; Dominguez, J.; Arancon, N. Q. The Influence of Vermicomposts on Plant Growth and Pest Incidence. In *Soil Zoology for Sustainable Development in the 21st Century: A Festschrift in Honour of Prof. Samir I. Ghabbour on the Occasion of his 70th Birthday*; Hanna, S. H.; Mikhail, W. Z. A. Eds.; Geocities Publishers: Cairo, Egypt, 2004; pp 397–420.

El-Hawary, M. I.; Talman, I. E. I.; El-Ghamary, A. M.; Naggar, E. E. Effect of Application of Biofertilizer on the Yield and NPK Uptake of Some Wheat Genotypes as Affected by the Biological Properties of Soil. *Pak. J. Biol. Sci.* **2002,** *5*, 1181–1185.

Esmaeilzadeh, J.; Ahangar, A. G. Influence of Soil Organic Matter Content on Soil Physical, Chemical and Biological Properties. *Int. J. Plant Anim. Environ. Sci.* **2014,** *4* (4), 244–252.

Evans, J. R. Photosynthesis and Nitrogen Relationships in Leaves of C_3 Plants. *Oecologia* **1989,** *78*, 9–19.

Evans, L. T. *Feeding the Ten Billion: Plants and Population Growth*; Cambridge University Press: Cambridge, UK, 1998; p 247.

Ewané, C. A.; Lepoivre, P.; Bellaire, L. L.; Lassois, L. Involvement of Phenolic Compounds in the Susceptibility of Bananas to Crown Rot. A Review. *Biotechnol. Agron. Soc. Environ.* **2012,** *16* (3), 393–404.

Fabre, D.; Adriani, D. E.; Dingkuhn, M.; Ishimaru, T.; Punzalan, B.; Lafarge, T.; Clément-Vidal, A.; Luquet, D. The qTSN4 Effect on Flag Leaf Size, Photosynthesis and Panicle

Size, Benefits to Plant Grain Production in Rice, Depending on Light Availability. *Front Plant Sci.* **2016,** *7* (623), 1–12.

FAOUS (Food and Agriculture Organization of the United Nations). *Rice is Life: International Year of Rice-2004.* International Rice Commission Newsletter, Food and Agriculture Organization of the United Nations (FAOUS): Rome, Italy, 2003; Vol. 52. ftp://ftp.fao.org/docrep/fao/006/y5167E/y5167E00.pdf (accessed Feb 2, 2017).

FAOUS (Food and Agriculture Organization of the United Nations). In *Improving Plant Nutrient Management for Better Farmer Livelihoods, Food Security and Environmental Sustainability.* Proceedings of a Regional Workshop, Beijing, China, Dec 12–16, 2005; Regional Office for Asia and the Pacific, Food and Agriculture Organization of the United Nations (FAOUS): Bangkok, Thailand, 2006; p 247.

Farb, A. R. Effects of Vermicompost in Potting Soils and Extract Foliar Sprays on Vegetable Health and Productivity. Honors Dissertation, Department of Environmental Science, Dickinson College, Pennsylvania, USA, 2012.

Fraga, T. I.; Carmona, F. D. C.; Anghinoni, I.; Marcolin, E. Attributes of Irrigated Rice and Soil Solution as Affected by Salinity Levels of the Water Layer. *Rev. Bras. Ciênc. Solo,* **2010,** *34* (4), 1049–1057.

Frinck, A. Intergrtaed Nutrient Management: A Review. *Ann. Arid Zone* **1998,** *37*/B, 1–24.

Fu, M. Effect of pH and Organic Acids on Nitrogen Transformations and Metal Dissolution in Soils. Ph.D. Dissertation, Department of Agronomy, Iowa State University, USA, 1989.

Gandhi, A.; Sivakumar, K. Impact of Vermicompost Carrier Based Bio Inoculants on the Growth, Yield and Quality of Rice (*Oryza sativa* L.) cv. NLR 145. *Ecoscan* **2010,** *4* (1), 83–88,

Ganiger, V. M.; Mathad, J. C.; Madalageri, M. B.; Babalad, H. B.; Hebsur, N. S.; Yenagi, N. B. Effect of Organics on the Physico-chemical Properties of Soil After Bell Pepper Cropping Under Open Field Condition. *Karnataka J. Agric. Sci.* **2012,** *25* (4), 479–484.

Garrity, D. P.; Flinn, J. C. In *Farm-level Management Systems for Green Manure Crop in Asian Rice Environment.* Green Manures in Rice Farming: Proceedings of a Symposium. The Role of Green Manures in Rice Farming Systems, May 25–29, 1987; International Rice Research Institute: Manila, Philippines, 1987.

Gaur, A. C. *Blue-green Algae as Biofertilizer: Biofertilizers in Sustainable Agriculture.* Directorate of Information and Publications of Agriculture, Indian Council of Agricultural Research, Krishi Anusandhan Bhavan I, Pusa, New Delhi, India, 2006; pp 148–155.

Gautam, P.; Sharma, G. D.; Rana, R.; Lal, B. Effect of Integrated Nutrient Management and Spacing on Growth Parameters, Nutrient Content and Productivity of Rice Under System of Rice Intensification. *Int. J. Res. BioSci.* **2013,** *2* (3), 53–59.

Gebrekidan, H.; Seyoum, M. Effects of Mineral N and P Fertilizers on Yield and Yield Components of Flooded Lowland Rice on Vertisols of Fogera Plain, Ethiopia. *J. Agric. Rural Dev. Trop. Subtrop. Vol.* **2006,** *107* (2), 161–176.

Gendua, P. A.; Yamamoto, Y.; Miyazaki, A.; Yoshida, T.; Wang, Y. Responses of Yielding Ability, Sink Size and Percentage of Filled Grains to the Cultivation Practices in a Chinese Large-panicle-type Rice Cultivar, Yangdao 4. *Plant Prod. Sci.* **2009,** *12* (2), 243–256.

Ghosh, B. C.; Ghosh, R.; Mitra, B. N.; Mitra, A; Jana, M. R. Effect of Organic and Inorganic Fertilization on Growth and Nutrition of Rice and Fish in a Dual Culture System. *J. Agric. Sci.* **1994**, *122* (12), 41–45.

Ghoshal, N.; Singh. K. P. Effects of Farmyard Manure and Inorganic Fertilizer on the Dynamics of Soilmicrobial Biomass in a Tropical Dryland Agroecosystem. *Biol. Fert. Soils* **1995**, *19*, 231–238.

Gill, H. S.; Meelu, O. P. Studies on the Substitution of Inorganic Fertilizers with Organic Manure and Their Effect on Soil Fertility in Rice–Wheat Rotation. *Fert. Res.* **1982**, *3*, 304–314.

Giraddi, R. S. Influencing Vermicomposting Methods on the Bio-degradation of Organic Wastes. *Indian J. Agric. Sci.* **2000**, *70*, 663–666.

GOI (Government of India). 2017. *Report of the Working Group on Fertilizer Industry for the Twelfth Plan (2012–13 to 2016–17)*. Department of Fertilizers, Ministry of Chemicals & Fertilizers, Government of India, India. http://planningcommission.gov.in/aboutus/committee/wrkgrp12/wg_fert0203.pdf (accessed Feb 2, 2017).

Golchin, A.; Nadi, M.; Mozaffari, V. The Effects of Vermicomposts Produced from Various Organic Solid Wastes on Growth of Pistachio Seedlings. *Acta Hort.* **2006**, *726*, 301–306.

Goswami, N. N.; Banerjee, N. K. *Phosphorus, Potassium and Other Macro Elements in Soils and Rice*; IRRI: Los Banos, Philippines, 1978; pp 561–580.

Goutami, N.; Rani, P. P. Integrated Nutrient Management on Content of Nutrients by Rice-Fallow Sorghum. *Int. J. Agric. Innov. Res.* **2016**, *4* (5), 928–931.

Greenland, D. J. *The Sustainability of Rice Farming*; CAB International: Wallingford, UK, 1997; p 273.

Gupta, S. K. Effect of Organic Manures on Rice Yield and Moisture Retention in Alfisol Soil. *J. Hill Res.* **1995**, *8* (2), 169–173.

Gupta, R.; Yadav, A.; Garg, V. K. Influence of Vermicompost Application in Potting Media on Growth and Flowering of Marigold Crop. *Int. J. Recycl. Org. Waste Agric.* **2014a**, *3* (47), 1–7. DOI: 10.1007/s40093-014-0047-1.

Gupta, S.; Kushwah, T.; Yadav, S. Role of Earthworms in Promoting Sustainable Agriculture in India. *Int. J. Curr. Microbiol. Appl. Sci.* **2014b**, *3* (7), 449–460.

Hall, J. R.; Hodges, T. K. Phosphorus Metabolism of Germinating Oat Seeds. *Plant Physiol.* **1966**, *41*, 1459–1464.

Hanč, A.; Tlustoš, P.; Száková, J.; Balík, J. The Influence of Organic Fertilizers Application on Phosphorus and Potassium Bioavailability. *Plant Soil Environ.* **2008**, *54* (6), 247–254.

Hansen, E. M.; Thomsen, I. K.; Hansen, M. N. Optimising Farmyard Manure Utilization by Varying the Application Time and Tillage Strategy. *Soil Use Manag,* **2004**, *20*, 173–177.

Hapse, D. G. *Organic Farming in the Light of Reduction in Use of Chemical Fertilizers.* Proceedings of 43rd Annual Deccan Sugar Technology Association, Pune, Part 1, 1993, SA37–SA51.

Hasanuzzaman, M.; Ahamed, K. U.; Rahmatullah, N. M.; Akhter, N.; Nahar, K.; Rahman, M. L. Plant Growth Characters and Productivity of Wetland Rice (*Oryza sativa* L.) as Affected GY Application of Different Manures. *Emir. J. Food Agric.* **2010**, *22* (1), 46–58.

Hasanuzzaman, M.; Ahamed, K. U.; Nahar, K.; Akhter, N. Plant Growth Pattern, Tiller Dynamics and Dry Matter Accumulation of Wetland Rice (*Oryza sativa* L.) as Influenced by Application of Different Manures. *Nat. Sci.* **2010a**, *8* (4), 1–10.

Hasanuzzaman, M.; Ahamed, K. U.; Rahmatullah, N. M.; Akhter, N.; Nahar, K and Rahman, M. L. Plant Growth Characters Andproductivity of Wetland Rice (*Oryza sativa* L.) as Affected by Application of Different Manures. *Emir. J. Food Agric.* **2010b,** *22* (1), 46–58.

Hashim, M. M.; Yusop, M. K.; Othman, R.; Wahid, S. A. Characterization of Nitrogen Uptake Pattern in Malaysian Rice MR219 at Different Growth Stages Using [15]N Isotope. *Rice Sci.* **2015,** *22* (5), 250–254.

Hauck, F. W. *Organic Recycling to Improve Soil Productivity.* Paper Presented at the FAO/ SIDA Workshop on Organic Materials and Soil Productivity in the Near East; FAO Soils Bul. No. 45; United Nation Food and Agriculture Organization: Rome, Italy, 1978.

Hazra, G. Different Types of Eco-Friendly Fertilizers: An Overview. *Sust. Environ.* **2016,** *1* (1), 54–70.

Hemalatha, M.; Thirumurugan, V.; Balasubramanian, R. Influence of Organic, Biofertilizer and Inorganic Forms of Nitrogen on Rice Quality. *Int. Rice Res. Notes* **1999,** *24* (1), 33.

Hemalatha, M.; Thirumurugan, V.; Balasubramanian, R. Effect of Organic Sources of Nitrogen on Productivity, Quality of Rice (*Oryza sativa* L.) and Soil Fertility in Single Crop Wetlands. *Indian J. Agron.* **2000,** *45*, 564–567

Hidayatullah, A. Influence of Organic and Inorganic Nitrogen on Grain Yield and Yield Components of Hybrid Rice in Northwestern Pakistan. *Rice Res.* **2016,** *23* (6), 326–333.

Hidlago, P. R.; Matta, F. B.; Harkess, R. L. Physical and Chemical Properties of Substrates Containing Earthworm Castings and Effects on Marigold Growth. *Hortscience* **2006,** *41*, 1474–1476.

Hobbs, P. R.; Sayre, K.; Gupta, R. The Role of Conservation Agriculture in Sustainable Agriculture. *Philos. Trans. Royal Soc. Lon. Ser. B. Biol. Sci.* **2008,** *363* (1491), 543–555.

Hokmalipour, S.; Darbandi, M. H. Effects of Nitrogen Fertilizer on Chlorophyll Content and Other Leaf Indicate in Three Cultivars of Maize (*Zea mays* L.). *World Appl. Sci. J.* **2011,** *15*, 1780–1785.

Hossain, M. In Rice Supply and Demand in Asia: A Socioeconomic and Biophysical Analysis. *Applications of Systems Approaches at the Farm and Regional Levels*; Teng, P. S., et al. Eds.; IRRI Kluwer Academic Publ.: Dordrecht, 1997; Vol. 1, pp 263–279.

Huiping, L.; Zhigang, W.; Minsheng, Y.; Yanguang, Z.; Dazhuang, H.; Shihong, Z. The Relation Between Tannin and Phenol Constituents and Resistance to *Anoplophora glabripennis* of Various Poplar Tree Species. *J. Agric. Univ. Hebei* **2003,** *26* (1), 36–39.

Ievinsh, G. Vermicompost Treatment Differentially Affects Seed Germination, Seedling Growth and Physiological Status of Vegetable Crop Species. *Plant Growth Regul.* **2011,** *65*, 169–181.

Jadhav, A. D.; Talashilkar, S. C.; Powar, A. G. Influence of the Conjunctive Use of FYM, Vermicompost and Urea on Growth and Nutrient Uptake in Rice. *J. Maharastra Agric. Univ.* **1997,** *22* (2), 249–250.

Jagtap, P. B.; Patil, J. D.; Nimbalkar, C. A.; Kadlag, A. D. Effect of Micronutrient Fertilizers on Release of Nutrients in Saline-sorlic Soil. *J. Indian Soc. Soil Sci.* **2006,** *54* (4), 489–494.

Jain, N. Waste Management of Temple Floral offerings by Vermicomposting and its Effect on Soil and Plant Growth. *Int. J. Environ. Agric. Res. (IJOEAR)* **2016,** *2* (7), 89–94.

Jeyabal, A.; Kuppuswamy, G. Recycling of Organic Wastes for the Production of Vermicompost and Its Response in Rice–Legume Cropping System and Soil Fertility. *Eur. J. Agron.* **2001,** *15,* 153–170.

John, P. S.; George, M.; Zool, J. R. Nutrient Mining in Agroclimetic Zones of Kerala. *Fertil. News* **2001,** *46,* 45–52 and 55–57.

Joshi, R.; Singh, J.; Vig, A. P. Vermicompost as an Effective Organic Fertilizer and Biocontrol Agent: Effect on Growth, Yield and Quality of Plants. *Rev. Environ. Sci. Bio/Technol.* **2015,** *14* (1), 137–159.

Judge, A. The Effects of Surface-applied Poultry Manure on Top- and Subsoil Acidity and Selected Soil Fertility Characteristics. M.Sc. Ag. Dissertation, Faculty of Science and Agriculture, School of Applied Environmental Sciences, University of Natal, Pietermaritzburg, South Africa, 2001.

Junying, Z.; Yongli, X.; Fuping, L. In *Influence of Cow Manure Vermicompost on Plant Growth and Micrbes in Rhizosphere on Iron Tailing.* Proceedings of the 3rd International Conference on Bioinformatics and Biomedical Engineering (iCBBE 2009), Beijing, China, June 11–16, 2009. http://ieeexplore.ieee.org/stamp/stamp.jsp?arnumber=5162958 (accessed Feb 3, 2017).

Kalaivanan, D.; Hattab, K. Influence of Enriched Press Mud, Compost on Soil Chemical Properties and Yield of Rice. *Res. J. Microbiol.* **2008,** *3* (4), 254–261.

Kale, R. D.; Bano, K. Earthworm Cultivation and Culturing Techniques for Production of Vermicompost in Mysore. *J. Agric. Sci.* **1988,** *22,* 339–344.

Kale, R.; Bano, K.; Satyavati, G. P. *Influence of Vermicompost Application on Growth and Yield of Cereals, Vegetables and Ornamental Plants.* Final Report of KSCST Project No. 67–04/verm/34B (3478), Bangalore, India, 1991; p 87.

Kandan, T.; Subbulakshmi. Chemical Nutrient Analysis of Vermicompost and Their Effect on the Growth of SRI Rice Cultivation. *Int. J. Innov. Res. Sci. Eng. Technol.* **2015,** *4* (6), 4382–4388.

Kang, B. T.; Juo, A. S. R. Effect of Forest Clearing on Soil Chemical Properties and Crop Performance. In *Land Clearing and Development in the Tropics,* Lal, R.; Sanchez, P. A.; Cummings, Jr. R. W. (Eds.) Rotterdam A. A. Bakema: Netherlands, 1986; pp 383–394.

Karki, S.; Rizal, G.; Quick, W. P. Improvement of Photosynthesis in Rice (*Oryza sativa* L.) by Inserting the C_4 Pathway. *Rice* **2013,** *6* (28), 1–8.

Kashem, M. A.; Sarker, A.; Hossain, I.; Islam, M. S. Comparison of the Effect of Vermicompost and Inorganic Fertilizers on Vegetative Growth and Fruit Production of Tomato (*Solanum lycopersicum* L.). *Open J. Soil Sci.* **2015,** *5,* 53–58.

Kaul, V. India has Enough Land for Farming but There Are Other Bigger Issues to Worry About; 2015. http://www.firstpost.com/business/india-enough-land-farming-bigger-issues-worry-2032327.html (accessed Feb 2, 2017).

Kebrom, T. H.; Mullet, J. E. Photosynthetic Leaf Area Modulates Tiller Bud Outgrowth in Sorghum. *Plant Cell Environ.* **2015,** *38* (8), 1471–1478.

Kemp, D. R.; Culvenor, R. A. Improving the Grazing and Drought Tolerance of Temperate Perennial Grasses. *N. Z. J. Agric. Res.* **1994,** *37* (3), 365–378.

Kesavan, P. C.; Swaminathan, M. S. Strategies and Models for Agricultural Sustainability in Developing Asian Countries. *Philos. Trans. Royal Soc. Lon. Series B Biol. Sci.* **2008,** *363* (1492), 877–891.

Khan, S. K.; Mohanty, S. K.; Chalam, A. B. Integrated Management of Organic Manure and Fertilizer Nitrogen for Rice. *J. Indian Soc. Soil Sci.* **1986,** *24*, 505–509.

Khan, M. J.; Khan, M. Q.; Zia, M. S. Sugar Industry Press Mud as Alternate Organic Fertilizer Source. *Int. J. Environ. Waste Manag.* **2012,** *9* (1/2), 41–55.

Kharub, A. S.; Chander, S. Effect of Organic Farming on Yield, Quality and Soil-fertility Status Under Basmati Rice (*Oryza sativa*)–Wheat (*Triticum aestivum*) Cropping System. *Indian J. Agron.* **2008,** *53* (3), 172–177.

Khatun, A.; Sultana, H.; Jamiul Islam, A. B. M.; Bhuiya, M. S. U.; Saleque, M. A. Seed Yield and Quality of Lowland Rice (*Oryza sativa* L.) as Influenced by Nitrogen from Organic and Chemical Sources. *Agriculturists* **2015,** *13* (1), 109–118.

Koushal, S.; Sharma, A. K.; Singh, A. Yield Performance, Economics and Soil Fertility Through Direct and Residual Effects of Organic and Inorganic Sources of Nitrogen as Substitute to Chemical Fertilizer in Rice–Wheat Cropping System. *Res. J. Agric. Sci.***2011,** *43* (3), 189–193.

Krishna, A.; Biradarpatil, N. K.; Manjappa, K.; Channappagoudar, B. B. Evaluation of System of Rice Intensification Cultivation, Seedling Age and Spacing on Seed Yield and Quality in Samba Masuhri (BPT-5204) Rice. *Karnataka J. Agric. Sci.* **2008,** *21* (1), 20–25.

Kumar, A.; Yadav, D. S. Use of Organic Manure and Fertilizer in Rice (*Oryza sativa*)–Wheat (*Triticum aestivum*) Cropping System for Sustainability. *Indian J. Agric. Sci.* **1995,** *65* (10), 703–707.

Kumar, V.; Singh, O. P. Effect of Organic Manures, Nitrogen and Zinc Fertilization on Growth, Yield, Yield Attributes and Quality of Rice (*Oryza sativa L.*). *Int. J. Plant Sci.* **2006,** *1* (2), 311–314.

Kumar, R.; Singh, R. N.; Chaurasiya, P. C. Influence of Different Date of Planting of Rice on Yield and Crop Productivity in Vertisols in Chhattisgarh Plains. *Int. J. Multidiscip. Res. Adv.* **2016a,** *1* (1), 1–6.

Kumar, Y.; Dhyani, B. P.; Kumar, V.; Raj, R. Influence of Fertility Levels on Nutrient Uptake and Productivity of Rice Under Puddled and Unpuddled Conditions. *Ann. Agric. Res. New Ser.* **2016b,** *37* (2), 147–153.

Kumari, M. S. S.; Ushakumari, K. Effect of Vermicompost Enriched with Rock Phosphate on the Yield and Uptake of Nutrients in Cowpea (*Vigna unguiculata* L. Walp). *J. Trop. Agric.* **2002,** *40*, 27–30.

Ladha, J. K.; Tirol-Padre, A.; Reddy, C. K.; Cassman, K. G.; Verma, S.; Powlson, D. S.; van Kessel, C.; Richter, D. B. Chakraborty, D.; Pathak, H. Global Nitrogen Budgets in Cereals: A 50-year Assessment for Maize, Rice, and Wheat Production Systems. *Sci. Rep.* **2016,** *6*, 19355. DOI: 10.1038/srep19355.

Lakpale, R.; Shrivastava, G. K. Traditional System of Rice Cultivation in Chhattisgarh (*Biasi*): Limitation, Advantages and Practices for Improvement. In *Model Training Course on Rice Production Technology*; Feb 22–29, 2012; Directorate of Extension Services: IGKV, Raipur (CG), India, 2012; pp 1–180. http://igau.edu.in/pdf/pubdes4.pdf (accessed Feb 22–29, 2017).

Lawal, T. E.; Babalola, O. O. Relevance of Biofertilizers to Agriculture. *J. Hum. Ecol.* **2014,** *47* (1), 35–43.

Lawlor, D. W. Carbon and Nitrogen Assimilation in Relation to Yield: Mechanisms are the Key to Understanding Production Systems. *J. Exp. Bot.* **2002,** *53* (370), 773–787.

Laxminarayana, K.; Patiram. Effect of Integrated Use of Inorganic, Biological and Organic Manures on Rice Productivity and Soil Fertility in Ultisols of Mizoram. *J. Indian Soc. Soil Sci.* **2006,** *54* (2), 213–220.

Lin, H.-C.; Hülsbergen, K.-J. A New Method for Analyzing Agricultural Land-use Efficiency, and Its Application in Organic and Conventional Farming Systems in Southern Germany. *Eur. J. Agron.* **2017,** *83*, 15–27.

Luske, B. *Reduced GHG Emissions due to Compost Production and Compost Use in Egypt, Comparing Two Scenarios*; Louis Bolk Institute: the Netherlands, 2010; pp 30.

Ma, R. L.; Peng, S.; Akita, S.; Saka, S. Effect of Panicle Size on Grain Yield of IRRI-Released Indica Rice Cultivars in the Wet Season. *Plant Prod. Sci.* **2004,** *7* (3), 271–276.

Mahajan, A.; Gupta, R. D. *Integrated Nutrient Management (INM) in a Sustainable Rice–Wheat Cropping System*; Springer: the Netherlands, 2009.

Mahajan, A.; Bhagat, R. M.; Gupta, R. D. Integrated Nutrient Management in Sustainable Rice–Wheat Cropping System for Food Security in India. *SAARC Jn. of Agri.* **2008,** *6* (2), 149–163.

Mahajan, G.; Timsina, J.; Jhanji, S.; Sekhon, N. K.; Singh, K. Cultivar Response, Dry-Matter Partitioning, and Nitrogen-use Efficiency in Dry Direct-seeded Rice in Northwest India. *J. Crop Improv.* **2012,** *26* (6), 767–790.

Mahamud, J. A.; Haque, M. M.; Hasanuzzaman, M. Growth, Dry Matter Production and Yield Performance of Transplanted Aman Rice Varieties Influenced by Seedling Densities per Hill. *Int. J. Sust. Agric.* **2013,** *5* (1), 16–24.

Mahmud, A. J.; Shamsuddoha, A. T. M.; Issak, M.; Haque, M. N.; Achakzai, A. K. K. Effect of Vermicompost and Chemical Fertilizer on the Nutrient Content in Rice Grain, Straw and Post-harvest Soil. *Middle-East J. Sci. Res.* **2016,** *24* (2), 437–444.

Maiti, S.; Seha, M.; Banerjee, H.; Pal, S. Integrated Nutrient Management Under Hybrid Rice (*Oryza sativa*)—Hybrid Rice Cropping Sequence. *Ind. J. Agron.* **2006,** *51*, 157–159.

Makino, A. Rubisco and Nitrogen Relationships in Rice: Leaf Photosynthesis and Plant Growth. *Soil Sci. Plant Nutr.* **2003,** *49* (3), 319–327.

Manahan, S. E. The Geosphere, Soil, and Food Production: The Second Green Revolution. In *Green Chemistry*, 2nd ed.; Manahan, S. E., Ed.; ChemChar Research, Inc.: Columbia, USA, 2006; pp 251–272.

Mankotia, B. S.; Shekhar, J.; Thakur, R. C.; Negi, S. C. Effect of Organic and Inorganic Sources of Nutrients on Rice (*Oryza sativa*)—Wheat (*Triticum aestivum*) Cropping System. *Indian J. Agron.* **2008,** *53* (1), 32–36.

Manzoor, Z.; Awan, T. H.; Safdar, M. E.; Ali, R. I.; Ashraf, M. M.; Ahmed, M. *Effect of Nitrogen Levels on Yield and Yield Components of Basmati 2000*. Rice Research Institute: Kala Shah Kaku, Lahore, Pakistan, 2006; p 18.

Masarirambi, M. T.; Mandisodza, F. C.; Mashingaidze, A. B.; Bhebhe, E. Influence of Plant Population and Seed Tuber Size Ongrowth and Yield Components of Potato (*Solanum tuberosum*). *Int. J. Agric. Biol.* **2012,** *14*, 545–549.

Masciandaro, G.; Bianchi, V.; Macci, C.; Peruzzi, E.; Doni, S.; Ceccanti, B.; Iannelli, R. Ecological and Agronomical Perspectives of Vermicompost Utilization in Mediterranean Agro-ecosystems. (Special Issue: Vermitechnology II.) *Dyn. Soil Dyn. Plant* **2010,** *4* (Special Issue 1), 76–82.

Masclaux-Daubresse, C.; Daniel-Vedele, F.; Dechorgnat, J.; Chardon, F.; Gaufichon, L.; Suzuki, A. Nitrogen Uptake, Assimilation and Remobilization in Plants: Challenges for Sustainable and Productive Agriculture. *Ann. Bot.* **2010**, *105*, 1141–1157.

Mathakiya, H. V.; Meishri, M. B. Feasibility of Using some Solid Industrial Wastes on Cabbage and Its Effect on Yield, Nutrient Absorption and Soil Properties. *Indian J. Agric. Chem.* **2003**, *36*, 141–145.

Maury, S. Study of O-methyltransferases of the Phenylpropanoid Pathway in Tobacco and Modulation of Their Expression in Transgenic Tobaccos: Consequences on the Biosynthesis of Lignin and Other Phenolic Compounds and on Resistance to Pathogens. Ph.D. Dissertation, Université Louis Pasteur, Strasbourg, France, 2000.

Maurya, S.; Singh, R.; Singh, D. P.; Singh, H. B.; Srivastava, J. S.; Singh, U. P. Phenolic Compounds of *Sorghum vulgare* in Response to *Sclerotium rolfsii* Infection. *J. Plant Interact.* **2007**, *2* (1), 25–29,

Mayr, U.; Michalek, S.; Treutter, D.; Feucht, W. Phenolic Compounds of Apple and Their Relationship to Scab Resistance. *J. Phytopathol.* **1997**, *145*, 69–75.

Meshram, S. U.; Pande, S. S.; Shanware, A. S.; Kamdi, R. R. Efficacy of Biofertilizers Integrated with Chemical Fertilizers In Vivo in Soybean. *Fertil. Newslett.* **2004**, *12*, 7–10,

Miller, S. S.; Hott, C.; Tworkoski, T. Shade Effects on Growth, Flowering and Fruit of Apple. *J. App. Hort.* **2015**, *17* (2), 101–105.

Mir, M.; Hassan, G. I.; Mir, A.; Hassan, A.; Sulaimani, M. Effects of Bio-organics and Chemical Fertilizers on Nutrient Availability and Biological Properties of Pomegranate Orchard Soil. *Afr. J. Agric. Res.* **2013**, *8* (37), 4623–4627.

Mishra, M. S.; Rajani, K.; Sahu-Sanjat, K.; Padhy-Rabindra, N. Effect of Vermicomposted Municipal Solid Wastes on Growth, Yield and Heavy Metal Content of Rice. *Fresen. Environ. Bull.* **2005**, *14*, 584–590.

Mishra, A. K.; Singh, R.; Kaleem, M. In *Economic Feasibility of Supplementing Inorganic Fertilizers with Organic Manures in Basmati Rice (Oryza sativa L).* National Symposium on Conservation Agriculture and Environment, Banaras Hindu University, Varanasi, India Oct 26–28, 2006; pp 142–143.

Mitran, T.; Mani, P. K.; Basak, N.; Mazumder, D.; Roy, M. Long-term Manuring and Fertilization Influence Soil Inorganic Phosphorus Transformation Vis-a-vis Rice Yield in a Rice–Wheat Cropping System. *Arch. Agron. Soil Sci.* **2016**, *62* (1), 1–18.

Mitran, T.; Mani, P. K.; Basak, N.; Biswas, S.; Mandal, B. Organic Amendments Influence on Soil Biological Indices and Yield in Rice-based Cropping System in Coastal Sundarbans of India. *Commun. Soil Sci. Plant Anal.* **2017**, *48* (2), 170–185.

Mo, Y.-J.; Kim, K.-Y.; Park, H.-S.; Ko, J.-C.; Shin, W.-C.; Nam, J.-K.; Kim, B.-K.; Ko, J.-K. Changes in the Panicle-related Traits of Different Rice Varieties Under High Temperature Condition. *Aust. J. Crop Sci.* **2012**, *6* (3), 436–443.

Mohammadi, M.; Kazemi, H. Changes in Peroxidase and Polyphenol Activity in Susceptible and Resistant Wheat Heads Inoculated with *Fusarium graminearum* and Induced Resistance. *Plant Sci.* **2002**, *162*, 491–498.

Mohapatra, P. K.; Panigrahi, R.; Turner, N. C. Physiology of Spikelet Development on the Rice Panicle: Is Manipulation of Apical Dominance Crucial for Grain Yield Improvement?

In *Advances in Agronomy*; Sparks, D. L., Ed.; Academic Press: Burlington, 2011; Vol. 110, pp 333–359.

Mokolobate, M. S. An Evaluation of the Use of Organic Amendments to Ameliorate Aluminium Toxicity and Phosphorus Deficiency in an Acid Soil. M.Sc.Ag. (Soil Science) Dissertation, Faculty of Science and Agriculture, School of Applied Environmental Sciences, University of Natal, Pietermaritzburg, South Africa, 2000.

Mondal, S.; Bauri, A.; Pramanik, K.; Ghosh, M.; Malik, G. C.; Ghosh, D. C. Growth, Productivity and Economics of Hybrid Rice as Influenced by Fertility Level and Plant Density. *Int. J. Bio-Resour. Stress Manag.* **2013,** *4* (4), 547–554.

Mondal, S.; Mallikarjun, M.; Ghosh, M.; Ghosh, D. C.; Timsina, J. Effect of Integrated Nutrient Management on Growth and Productivity of Hybrid Rice. *J. Agric. Sci. Technol. B* **2015,** *5,* 297–308.

Morales, S. G.; Trejo-Téllez, L. I.; Merino, F. C. G.; Caldana, C.; Espinosa-Victoria, D.; Cabrera, B. E. H. Growth, Photosynthetic Activity, and Potassium and Sodium Concentration in Rice Plants Under Salt Stress. *Acta. Sci. Agron.* **2012,** *34* (3), 317–324.

Morris, M.; Kelly, V. A.; Kopicki, R. J.; Byerlee, D. *Fertilizer Use in African Agriculture: Lessons Learned and Good Practice Guidelines.* The World Bank: Washington, USA, 2007; p 114.

MSSRF (M. S. Swaminathan Research Foundation). *Gender, Rice and Food Security a Report on the IYR Programmes.* M. S. Swaminathan Research Foundation, Centre for Research on Sustainable Agricultural and Rural Development: Chennai, India, 2005.

Munda, G. C.; Islam, M.; Panda, B. B. Effects of Organics and Inorganics on Productivity and Uptake of Nutrients in Rice (*Oryza sativa*)–Toria (*Brassica campestris*) Cropping System. *Indian J. Agron.* **2008,** *53* (2), 107–111.

Muralidharan, K.; Rao, U. P.; Pasalu, I. C. P.; Reddy, A. P. K.; Singh, S. P.; Krishnaiah, K. *Technology for Rice Production.* Directorate of Rice Research; Hyderabad, India, 1998; p 31.

Murchie, E. H.; Hubbart, S.; Chen, Y.; Peng, S.; Horton, P. Acclimation of Rice Photosynthesis to Irradiance Under Field Conditions. *Plant Physiol.* **2002a,** *130,* 1999–2010.

Murchie, E. H.; Yang, J.; Hubbart, S.; Horton, P.; Peng, S. Are There Associations Between Grain filling Rate and Photosynthesis in the Flag Leaves of Field grown Rice? *J. Exp. Bot.* **2002b,** *53* (378), 2217–2224.

Nagar, R. K.; Goud, V. V.; Kumar, R.; Kumar, R. Effect of Organic Manures and Crop Residue Management on Physical, Chemical and Biological Properties of Soil Under Pigeonpea Based Intercropping System. *Int. J. Farm Sci.* **2016,** *6* (1), 101–113.

Naher, U. A. Panhwar, Q. A.; Othman, R.; Ismai, M. R.; Berahim, Z. Biofertilizer as a Supplement of Chemical Fertilizer for Yield Maximization of Rice. *J. Agric. Food Dev.* **2016,** *2,* 16–22.

Najar, I. A.; Khan, A. B.; Hai, A. Effect of Macrophyte Vermicompost on Growth and Productivity of Brinjal (*Solanum melongena*) Under Field Conditions. *Int. J. Recycl. Org. Waste Agric.* **2015,** *4,* 73–83.

Nesgea, S.; Gebrekidan, H.; Sharma, J. J.; Berhe, T. Effects of Nitrogen and Phosphorus Fertilizer Application on Yield Attributes, Grain Yield and Quality of Rain Fed Rice (NERICA-3) in Gambella, Southwestern Ethiopia. *East Afr. J. Sci.* **2012,** *6* (2), 91–104.

Nicholson, R. L.; Hammerschmidt, R. Phenolic Compounds and Their Role in Disease Resistance. *Ann. Rev. Phytopathol.* **1992,** 30, 369–389.

Novoa, R.; Loomis, R. S. Nitrogen and Plant Production. *Plant Soil* **1981**, *58* (1–3), 177–204.

Okamoto, M.; Okada, K. Differential Responses of Growth and Nitrogen Uptake to Organic Nitrogen in Four Gramineous Crops. *J. Exp. Bot.* **2004**, *55* (402), 1577–1585.

Orozco, S. H.; Cegarra, J.; Trujillo, L. M.; Roig, A. Vermicomposting of Coffee Pulp Using the Earthworm *Eisenia foetida*: Effects on C and N Contents and the Availability of Nutrients. *Biol. Fertil. Soils* **1996**, *22*, 162–166.

Pal, Y.; Singh, R. P.; Sharma, N. L.; Sachan, R. S. Effect of Integrated Nutrient Management Practices on Yield, N, P, K Uptake and Economics of Rice (*Oryza sativa L.*) in a Mollisol. *Pantnagar J. Res.* **2007**, *5* (2), 29–33.

Pan, S.; Liu, H.; Mo, Z.; Patterson, B.; Duan, M.; Tian, H.; Hu, S.; Tang, X. Effects of Nitrogen and Shading on Root Morphologies, Nutrient Accumulation, and Photosynthetic Parameters in Different Rice Genotypes. *Sci. Rep.* **2016**, *6* (32148), 1–13.

Pandey, N.; Verma, A. K.; Anurag; Tripathi, R. S. Effect of Integrated Nutrient Management in Transplanted Hybrid Rice (*Oryza sativa*). *Indian J. Agron.* **2007**, *52* (1), 40–42.

Panigrahi, T.; Garnayak, L. M.; Ghosh, M.; Ghosh, D. C. Growth Analysis of Basmati Rice Varieties and Its Impact on Grain Yield Under SRI. *Int. J. Plant Anim. Environ. Sci.* **2015**, *5* (3), 101–109.

Parihar, M.; Meena, R. K.; Jat, L. K.; Suryakant and Jatav, H. S. Effect of Inorganic Fertilizers with and Without FYM on Yield, Nutrient Uptake and Quality Parameters of Rice (*Oryza sativa L.*). *Environ. Ecol.* **2015**, 33 (4), 1480–1484.

Pathma, J.; Sakthivel, N. Microbial Diversity of Vermicompost Bacteria that Exhibit Useful Agricultural Traits and Waste Management Potential. *SpringerPlus* **2012**, *1* (26), 1–19. DOI: 10.1186/2193-1801-1-26.

Pender, J. *Agricultural Technology Choices for Poor Farmers in Less-favoured Areas of South and East Asia*. Asia and the Pacific Division, International Fund for Agricultural Development (IFAD): Rome, Italy, 2008.

Peyvast, G.; Olfati, J. A.; Madeni, S.; Forghani, A. Effect of Vermicompost on the Growth and Yield of Spinach (*Spinacia oleracea L.*). *J. Food Agric. Environ.* **2008**, *6*, 110–113.

Picinelli, A.; Dapena, E.; Mangas, J. J. Polyphenolic Pattern in Apple Tree Leaves in Relation to Scab Resistance. A Preliminary Study. *J. Agric. Food Chem.* **1995**, *4*, 2273–2278.

Pillai, K. G. An Integrated Approach to Nutrient Management in Rice. *Indian Farm.* **1990**, *40* (9), 15–18.

Pillai, S. P.; Geethakumari, V. L.; Rebecca, S. I. Balance-sheet of Soil Nitrogen in Rice (*Oryza sativa*)-based Cropping Systems Under Integrated Nutrient Management. *Indian J. Agron.* **2007**, *52* (1), 16–20.

Pingali, P. L. Green Revolution: Impacts, Limits, and the Path Ahead. *Proc. Natl. Acad. Sci.* **2012**, 109 (31), 12302–12308 (Clark, W. C., Ed.).

Pingali, P.; Stringer, R. Food Security and Agriculture in the Low Income Food Deficit Countries: 10 Years After the Uruguay Round. ESA Working Paper No. 3–18; Agricultural and Development Economics Division, The Food and Agriculture Organization of the United Nations (FAOUS): Rome, Italy, 2003.

Pingali, P. L.; Hossain, M.; Gerpacio, R. V. *Asian Rice Bowls: The Returning Crisis*? IRRI-CAB International: Philippines, 1997; p 341.

Pradhan, K. C.; Sahu, S. K. Direct and Residual Effects of Fly Ash Integrated with FYM and Chemical Fertilizers on Growth, Yield and Nutrient (Major) Uptake by Crop on

Rice–Groundnut Cropping System in Acid Soils of Orissa. *Environ. Ecol.* **2004,** *22* (2), 332–336.

Prakash, Y. S.; Bhadoria, P. B. S.; Rakshit, A. Relative Efficacy of Organic Manure in Improving Milling and Cooking Quality of Rice. *Int. Rice Res. Notes* **2002,** *27* (1), 43–44.

Pramanik. K.; Bera, A. K. Effect of Seedling Age and Nitrogen Fertilizer on Growth, Chlorophyll Content, Yield and Economics of Hybrid Rice (*Oryza sativa* L.). *Int. J. Agron. Plant Prod.* **2013,** *4* (S), 3489–3499.

Prasad, B.; Prasad, J. Integrated Nutrients Management for Specific Yield of Rice (Oryza sativa L) Based on Targeted Yield Concept and Soil Test Values in Old Alluvial Soils. *Oryza* **1994,** *31,* 140–143.

Prasanna, P. A. L.; Kumar, S.; Singh, A. Rice Production in India—Implications of Land Inequity and Market Imperfections. *Agr. Econ. Res. Rev.* **2009,** *22,* 431–442.

Prats, E.; Bazzalo, M. E.; Lean, A.; Jorrin, J. V. Accumulation of Soluble Phenolic Compounds in Sunflower Capitula Correlates with Resistance to *Sclerotinia sclerotiorum. Euphytica* **2004,** *132,* 321–329.

Quyen, N. V.; Sharma, S. N. Relative Effect of Organic and Conventional Farming on Growth Grain Quality of Scented Rice and Soil Fertility. *Arch. Agron. Soil Sci.* **2003,** *49,* 623–629.

Rahman, B.; Nath D.; Koyu, L. Response of Rice to an Integrated Nutrient Management Treatment in Soils Collected from the Long Term Fertility Experiment. *Int. J. Farm Sci.* **2012,** *2* (1), 105–110.

Rahaman, S.; Sinha, A. C. Effect of Water Regimes and Organic Sources of Nutrients for Higher Productivity and Nitrogen Use Efficiency of Summer Rice (*Oryza sativa*). *Afr. J. Agric. Res.* **2013,** *8* (48), 6189–6195.

Rajbhandari, R. System of Rice Intensification Under Different Plant Population and Levels of Nitrogen. M.Sc. Ag. (Agronomy) Dissertation, Institute of Agriculture and Animal Science Rampur, Chitwan, Nepal, 2007.

Rajesh, N.; Paulpandi, V. K.; Duraisingh, R. Enhancing the Growth and Yield of Pigeon Pea Through Growth Promoters and Organic Mulching: A Review. *Afr. J. Agric. Res.* **2015,** *10* (12), 1359–1366.

Rajkhowa, D. J.; Baroova, S. R.; Sharma, N. N. Sources and Levels of Phosphorus on Productivity, Economics and Phosphorus Balance in Rice–Rice Sequence in Acidic Soils of Assam. *J. Agric. Sci. Soc. North East India* **2003,** *12* (1), 13–17.

Rajput, A.; Sharma, S.; Rajput, S. S.; Jha, G. Impact of Different Crop Geometries and Depths of Planting on Growth and Yield of Rice in System of Rice Intensification. *Biosci. Biotech. Res. Comm.* **2016,** *9* (3), 512–516.

Raju, R. A.; Reddy, N. N. Integrated Management of Green Leaf Compost, Crop Residues and Inorganic Fertilizers in Rice (*Oryza sativa*)–Rice System. *Indian J. Agron.* **2000,** *45* (4), 629–635.

Ram, M.; Davari, M.; Sharma, S. N. Organic Farming of Rice (*Oryza sativa* L.)–Wheat (*Triticum aestivum* L.) Cropping System: A Review. *Int. J. Agron. Plant Prod.* **2011,** *2* (3), 114–134.

Ram, M.; Davari, M. R.; Sharma, S. N. Direct, Residual and Cumulative Effects of Organic Manures and Biofertilizers on Yields, NPK Uptake, Grain Quality and Economics of

Wheat (*Triticum aestivum* L.) Under Organic Farming of Rice–Wheat Cropping System. *J. Org. Syst.* **2014,** *9* (1), 16–30.

Ram, U. S.; Srivastava, V. K.; Hemantaranjan, A.; Sen, A.; Singh, R. K.; Bohra, J. S.; Shukla, U. Effect of Zn, Fe and FYM Application on Growth, Yield and Nutrient Content of Rice. *Oryza* **2016,** *50* (4), 351–357.

Ramakrishna Parama, V. R.; Munawery, A. Sustainable Soil Nutrient Management. *J. Indian Inst. Sci.* **2012,** *92* (1), 1–16.

Ramesh, P. Effects of Vermicomposts and Vermicomposting on Damage by Sucking Pests to Ground Nut (*Arachis hypogea*). *Indian J. Agric. Sci.* **2000,** *70* (5), 334.

Ramesh, S.; Vayapuri, V. Yield Potential and Economic Efficiency of Rice (*Oryza sativa*) as Influenced by Organic Nutrition Under Cauvery Delta Region of Tamil Nadu. *Plant Arch.* **2008,** *8* (2), 621–622.

Ramesh, P.; Singh, M.; Subba Rao, A. Organic Farming: Its Relevance to the Indian context. *Curr. Sci.* **2005,** *88* (4), 561–568.

Rana, R.; Badiyala, D.; Ramesh and Kaushal, S. Effect of Organic Manures on Sustainable Agriculture and Soil Quality. *Pop. Kheti* **2014,** *2* (2), 65–70.

Rao, K. T.; Rao, A. U.; Sekhar, D.; Ramu, P. S.; Rao, N. V. Effect of Different Doses of Nitrogen on Performance of Promising Varieties of Rice in High Altitude Areas of Andhra Pradesh. *Int. J. Farm Sci.* **2014,** *4* (1), 6–15.

Rasoon R.; Kukal, S. S.; Hira, G. S. Soil Physical Fertility and Crop Performance as Affected by Long Term Application of FYM and Inorganic Fertilizers in Rice–Wheat System. *Soil Tillage Res.* **2007,** *96,* 64–72.

Ray, A. Effect of Organic Techniques of Seed Crop Administration on Seed yield and Quality in Rice cv. ADT 43. *Adv. J. Seed Sci. Technol.* **2014,** *1* (1), 001–008.

Reddy, M. S.; Reddy, D. S. Nutrient Uptake and Quality of Rice as Influenced by Different Nitrogen Management Practices. *Res. Crops* **2003,** *4* (3), 291–294.

Reddy, M. V.; Ohkura, K. Vermicomposting of Rice-straw and Its Effects on Sorghum Growth. *Trop. Ecol.* **2004,** *45* (2), 327–331.

Reddy, T. P.; Umadevi, M.; Rao, P. C. Integrated Effect of Fly Ash and Farmyard Manure on Dry Matter Production, Yield and Economics of Rice in Andhra Pradesh. *J. Res. ANGRAU,* **2007,** *35* (4), 32–38.

Roy, S.; Arunachalam, K.; Dutta, B. K.; Arunachalam, A. Effect of Organic Amendments of Soil on Growth and Productivity of Three Common Crops viz. *Zea mays, Phaseolus vulgaris* and *Abelmoschus esculentus. Appl. Soil Ecol.* **2010,** *45,* 78–84.

Roy, B.; Sarkar, M. A. R.; Paul, S. K. Effect of Integrated Nutrient Management in Boro Rice Cultivation. *SAARC J. Agric.* **2015,** *13* (2), 131–140.

SAFSSDSN (Sustainable Agriculture and Food Systems of the Sustainable Development Solutions Network). *Solutions for Sustainable Agriculture and Food Systems: Technical Report for the Post-2015 Development Agenda*; Sustainable Development Solutions Network (SDSN): New York, USA, 2013; pp 1–98 (Sept 18, 2013). http://unsdsn.org/wp-content/uploads/2014/02/130919-TG07-Agriculture-Report-WEB.pdf (accessed Feb 2, 2017).

Saha, P. K.; Ishaque, M.; Saleque, M. A.; Miah, M. A. M.; Panaullah, G. M.; Bhuiyan, N. I. Long-term Integrated Nutrient Management for Rice-based Cropping Pattern: Effect on

Growth, Yield, Nutrient Uptake, Nutrient Balance Sheet, and Soil Fertility. *Commun. Soil Sci. Plant Anal.* **2007,** *38* (5–6), 579–610.

Saha, P. K.; Miah, M. A. M.; Hossain, A. T. M. S. Contribution of Rice Straw to Potassium Supply in Ricefallow–Rice Cropping Pattern. *Bangladesh J. Agril. Res.* **2009,** *34* (4), 633–643.

Sahoo, R. K.; Bhardwaj, D.; Tuteja, N. In *Biofertilizers: A Sustainable Eco-friendly Agricultural Approach to Crop Improvement. Plant Acclimation to environmental Stress*; Tuteja, N., Gill, S. S., Eds.; Springer: New York, USA, 2013; pp 403–432.

Sahrawat, K. L. Organic Matter and Mineralizable Nitrogen Relationships in Wetland Rice Soils. *Comm. Soil Sci. Plant Anal.* **2006,** *37*, 787–796.

Sahu, Y. K.; Chaubey, A. K.; Mishra, V. N.; Rajput, A. S.; Bajpai, R. K. Effect of Integrated Nutrient Management on Growth and Yield of Rice (*Oriza sativa* l.) in Inceptisol. *Plant Arch.* **2015,** *15* (2), 983–986.

Saikrithika, S.; Santhiya, K. R.; Veena, G. K. Effects of Different Substrates on Vermicomposting Using Eudrilus Eugenia on the Growth of *Vinca rosea. Int. J. Sci. Res. Pub.* **2015,** *5* (9), 1–11.

Saleque, M. A.; Kirk, G. J. D. Root-induced Solubilization of Phosphate in the Rhizosphere of Lowland Rice. *New Phytol.* **1995,** *129*, 325–336.

Sangeetha, S. P.; Balakrishnan, A.; Bhuvaneswari, J. Influence of Organic Nutrient Sources on Quality of Rice. *Madras Agric. J.* **2010,** *97* (7–9), 230–233.

Sarangi, S. K.; Lama, T. D. Composting Rice Straw Using Earthworm (*Eudrilus eugeniae*) or Fungal Inoculant (*Trichoderma viridae*) and Its Utilization in Rice (*Oryza sativa*)–Groundnut (*Arachis hypogaea*) Cropping System. *Indian J. Agron.* **2013,** *58* (2), 146–151.

Šarapatka, B. Phosphatase Activities (ACP, ALP) in Agroecosystem Soils, Ph.D. Dissertation, Department of Ecology and Crop Production Science, Swedish University of Agricultural Sciences, Uppsala, Sweden, 2003.

Sarma, B. K.; Singh, P.; Pandey, S. K.; Singh, H. B. Vermicompost as Modulator of Plant Growth and Disease Suppression. *Dyn. Soil Dyn. Plant* **2010,** 4 (Special Issue 1), 58–66.

Sarojani, J. K.; Hilli, J. S.; Devendrappa, S. Impact of Vermicompost on Quality and Yield of Chrysanthemum. *Int. J. Farm Sci.* **2012,** *2* (2), 48–53.

Satapathya, S. S.; Swaina, D. K.; Pasupalakb, S.; Bhadoriaa, P. B. S. Effect of Elevated (CO_2) and Nutrient Management on Wet and Dry Season Rice Production in Subtropical India. *Crop J.* **2015,** *3* (6), 468–480.

Satheesh, N.; Balasubramanian, N. Effect of Organic Manure on Yield and Nutrient Uptake Under Rice–Rice Cropping System. *Madras Agric. J.* **2003,** *90* (1–3), 41–46.

Schachtman, D. P.; Reid, R. J.; Ayling, S. M. Phosphorus Uptake by Plants: From Soil to Cell. *Plant Physiol.* **1998,** *116* (2), 447–453.

Schatz, A.; Schatz, V. Use of Compost Instead of Chemical Fertilizer to Avoid Fluorine Contamination of Soil, Water, and Food. *J. Nihon. Univ. Sch. Dent.* **1972,** *14* (3), 99–105.

Sebby, K. The Green Revolution of the 1960's and Its Impact on Small Farmers in India. B.A. and B.Sc. Dissertation, Faculty of Environmental Studies Program, University of Nebraska-Lincoln, Nebraska, USA, 2010.

Sekhar, D. Productivity and Quality of Rice (*Oryza sativa* l.) As Influenced by Nitrogen Source and Planting Method. M.Sc. Ag. Dissertation, Department of Agronomy,

Agricultural College, Acharya N. G. Ranga Agricultural University, Andhra Pradesh, India, 2004.

Sekhar, D.; Rao, K. T.; Rao, N. V. Effect of Integrated Nitrogen Management on Yield and Quality of Rice. *Int. J. Farm Sci.* **2014,** *4* (4), 1–7.

Senthivelu, M.; Prabha, A. C. S. Influence of Integrated Nitrogen Management Practices on Yield Attributes, Yield and Harvest Index of Wet Seeded Rice. *Int. J. Agric. Sci.* **2007,** *3* (2), 70–74.

Senthivelu, M.; Surya Prabha, A. C. Influence of Integrated Nutrient Management Practices on Dry Matter Production Yield and NPK Uptake of Wet Seeded Rice. *Asian J. Soil Sci.* **2007,** *2* (1), 45–50.

Senthivelu, M.; Pandian, B. J.; Surya, P. A. C. Dry Matter Production and Nutrient Removal in Wet Seeded Rice–Cotton Cropping Sequence Under Integrated Nutrient Management Practices. *ORYZA Int. J. Rice* **2009,** *46* (4), 279–289.

Sepehya, S. 2011. Long-term Effect of Integrated Nutrient Management on Dynamics of Nitrogen, Phosphorus and Potassium in Rice–Wheat System. Ph.D. Thesis, Department of Soil Science, CSK Himachal Pradesh Krishi Vishvavidyalaya, Palampur, India, 2011; p 179.

Shah, R. A.; Kumar, S. Direct and Residual Effect of Integrated Nutrient Management and Economics in Hybrid Rice Wheat Cropping System. *Am.-Eurasian J. Agric. Environ. Sci.* **2014,** *14* (5), 455–458.

Shalini, P. P.; Geethakumari, V. L.; Rebecca, I. S. Balance-sheet of Soil Nitrogen in Rice (*Oryza sativa*)-based Cropping Systems Under Integrated Nutrient Management. *Indian J. Agron.* **2007,** *52* (1), 16–20.

Sharma, A. R.; Mitra, B. N. Complementary Effect of Organic Bio and Mineral Fertilizers in Rice Based Cropping System. *Fertil. News* **1990,** *35* (2), 43–51.

Sharma, D.; Sagwal, P. K.; Singh, I.; Sangwan, A. Influence of Different Nitrogen and Phosphorus Levels on Profitability, Plant Nutrient Content, Yield and Quality in Basmati Cultivars. *Int. J. IT Eng. Appl. Sci. Res.* **2012a,** *1* (1), 1–4.

Sharma, R. K.; Rampal, S.; Sharma, K.; Gujar, G. T. Vermitechnology for Organic Farming: An Eco-friendly Approach for Sustainable Agriculture. *Indian Farm.* **2012b,** *62* (4), 16–19.

Sharma, V. Giri, S.; Rai, S. S. Supply Chain Management of Rice in India: A Rice Processing Company's Perspective. *Int. J. Manag. Value Suppl. Chains* **2013,** *4* (1), 25–36.

Sharma, G. D.; Thakur, R.; Chouhan, N.; Keram, K. S. Effect of Integrated Nutrient Management on Yield, Nutrient Uptake, Protein Content, Soil Fertility and Economic Performance of Rice (*Oryza sativa* L.) in a Vertisol. *J. Indian Soc. Soil Sci.* **2015,** *63* (3), 320–326.

Shekara, B. G.; Sharnappa, K. N. Effect of Irrigation Schedules on Growth and Yield of Aerobic Rice (*Oryza sativa*) Under Varied Levels of Farmyard Manure in Cauvery Command Area. *Indian J. Agron.* **2010,** *55* (1), 35–39.

Shetty, P. K.; Alvares, C.; Yadav, A. K. *Organic Farming and Sustainability*; Shetty, P. K., Alvares, C., Yadav, A. K., Eds.; National Institute of Advanced Studies, Indian Institute of Science Campus: Bangalore, India, 2014; p 288.

Shirkhani, A.; Nasrolahzadeh, S. Vermicompost and Azotobacter as an Ecological Pathway to Decrease Chemical Fertilizers in the Maize, *Zea mays. Biosci. Biotechnol. Res. Commun.* **2016,** *9* (3), 382–390.

Siddaramappa, R.; Jagadish, N. R.; Srinivasamurthy, C. A. In Efficiency of Rock Phosphate as Phosphatic Fertilizer to Rice in Acid Soil of Karnataka, India. *Plant–Soil Interactions at low pH*; Wright, R. J., Baligar, V. C., Murrmann, R. P., Eds.; Kluwer Academic Publishers: the Netherlands, 1991; pp 307–312.

Simons, R. G. Tiller and Ear Production of Winter Wheat. *Field Crops Abstract* **1982**, *35* (11), 857–870.

Singh, R.; Agarwal, S. K. Analysis of Growth and Productivityof Wheat in Relation to Levels of FYM and Nitrogen. *Indian J. Plant Physiol.* **2001**, *6*, 279–283.

Singh, K. K.; Sharma, S. K. Effect of Distillery Effluent Based Press Mud Compost on Seed Yield and Seed Quality Attribute of Paddy (*Oryza sativa*) Cultivar Pusa Basmati No.1. *J. Soil Water Conserv.* **2009**, *8* (4), 67–71.

Singh, A.; Shivay, Y. S. Enhancement of Growth Parameters and Productivity of Basmati Rice Through Summer Green Manuring and Zinc Fertilization. *Int. J. Bio-Resour. Stress Manag.* **2014**, *5* (4), 486–494.

Singh, Y.; Sidhu, H. S. Management of Cereal Crop Residues for Sustainable Rice–Wheat Production System in the Indo-Gangetic Plains of India. *Proc. Indian Natl. Sci. Acad.* **2014**, *80* (1), 95–114.

Singh, B.; Ryan, J. *Managing Fertilizers to Enhance Soil Health*; International Fertilizer Industry Association (IFIA): Paris, France, 2015; p 23.

Singh, H.; Manisha. Crisis in Agricultural Heartland: Farm Suicides in Malwa Region of Punjab, India. *Int. J. Adv. Res. Technol.* **2015**, *4* (2), 107–118.

Singh, V. B.; Rathi, K. S.; Shivay, Y. S.; Singh, R. Effect of FYM and NPK Fertilizers on Yield Attributes, Yield and Economics of Rice in the Field After Nursery. *Ann. Agric. Res.* **1998**, *19* (1), 22–25.

Singh, N. P.; Sachan, R. S.; Pandey, P. C.; Bisht, P. S. Effect of a Decade Long Fertilizer and Manure Application on Soil Fertility and Productivity of Rice–Wheat System in Mollisol. *J. Indian Soc. Soil Sci.* **1999**, *47*, 72–80.

Singh, M. K.; Thakur, R.; Verma, U. N.; Upasani, R. R.; Pal, S. K. Effect of Planting Time and Nitrogen on Production Potential of Basmati Rice (*Oryza sativa*) Cultivars in Bihar Plateau. *Indian J. Agron.* **2000**, *45* (2), 300–303.

Singh, B.; Singh, Y.; Ladha, J. K.; Bronson, K. F.; Balasubramanian, V.; Singh, J.; Khind, C. S. Chlorophyll-meter and Leaf Color Chart-based Nitrogen Management for Rice and Wheat in Northwestern India. *Agron. J.* **2002**, *94*, 821–829.

Singh, U. P.; Maurya, S.; Singh, D. P. Antifungal Activity and Induced Resistance in Pea by Aqueous Extract of Vermicompost and for Control of Powdery Mildew of Pea and Balsam. *J. Plant Dis. Prot.* **2003**, *110* (6), 544–553.

Singh, D.; Chhonkar, P. K.; Dwivedi, B. S. *Manual on Soil, Plant and Water Analysis*; Westville Publishing House: New Delhi, India, 2005; p 200.

Singh, R.; Singh, O. P.; Singh, R. G.; Mehta, R. K.; Kumar, V.; Singh, R. P. Effect of Integrated Nutrient Management on Yield and Nutrient Uptake of Rice (*Oryza sativa L.*): Wheat (*Triticum aestivum L.*) Cropping System in Lowland of Eastern Uttar Pradesh. *Indian J. Agron.* **2006a**, *51* (2), 85–88.

Singh, Y,; Singh, C. S.; Singh, T. K.; Singh, J. P. Effect of Fortified and Unfortified Rice-straw Compost with NPK Fertilizers on Productivity, Nutrient Uptake and Economics of Rice (*Oryza sativa*) Method for Measuring Microbial Biomass Carbon. *Soil Biol. Biochem.* **2006b**, *19*, 703–707.

Singh, R.; Sharma, R. R.; Kumar, S.; Gupta, R. K.; Patil, R. T. Vermicompost Substitution Influences Growth, Physiological Disorders, Fruit Yield and Quality of Strawberry (*Fragaria × ananassa* Duch.). *Bioresour. Technol.* **2008,** *99,* 8507–8511.

Singh, N. A. K.; Basumatary, A.; Barua, N. G. Influence of Integrated Nutrient Management on Yield, Nutrient Uptake and Economics in Rice-niger Cropping Sequence. *ORYZA Int. J.* **2009,** *46* (2), 160–162.

Singh, L.; Singh, P.; Kotru, R.; Hasan, B.; Chand, L.; Lone, B. A. Cow Urine Spray and Integrated Nutrient Management of Rice on Productivity and Energy Use Under Temperate Valley Conditions. *Indian J. Agric. Sci.* **2012,** *82* (7), 582–588.

Singh, K.; Singh, S. R.; Singh, J. K.; Rathore, R. S.; Singh, S. P.; Roy, R. Effect of Age of Seedling and Spacing on Yield, Economics, Soil Health and Digestibility of Rice (*Oryza sativa*) Genotypes Under System of Rice Intensification. *Indian J. Agric. Sci.* **2013,** *83* (5), 479–483.

Singh, N. A. K.; Basumatary, A.; Barua, N. G. Assessment of Soil Fertility Under Integrated Nutrient Management in Rice–Niger Sequence. *J. Krishi Vigyan* **2014,** *3* (1), 5–9.

Sinha, R. K.; Herat, S.; Valani, D. B.; Chauhan, K. A. Earthworms Vermicompost: A Powerful Crop Nutrient over the Conventional Compost and Protective Soil Conditioner Against the Destructive Chemical Fertilizers for Food Safety and Security. *Am.-Eurasian J. Agric. Environ. Sci.* **2009,** *5* (Suppl.): 14–22.

Sirvi, A. R.; Kumar, D.; Singh, N. Effect of Sulphur on Growth, Productivity and Economics of Aerobic Rice (*Oryza sativa*). *Indian J. Agron.* **2014,** *59* (3), 404–409.

Son, T. T. N.; Man, L. H.; Diep, C. N.; Thu, T. T. A.; Nam, N. N. Bioconversion of Paddy Straw and Biofertilizer for Sustainable Rice Based Cropping System. *Omonrice* **2008,** *16,* 57–70.

Sowmya, C.; Ramana, M. V. Performance of Hybrids/High Yielding Varieties and Nutrient Management in System of Rice Intensification: A Review. *Agri. Rev.* **2012,** *33* (1), 1–15.

Srivastava, A. K.; Ngullie, E. Integrated Nutrient Management: Theory and Practice. *Dyn. Soi Dyn. Plant* **2009,** *3* (1), 1–30.

Srivastava, V. K. Nutrient Management in Rice (oryza sativa l.) Under SRI Through Integration of NPK Levels with FYM and Vermicompost. Ph.D. Dissertation, Department of Agronomy, Institute of Agricultural Sciences, Banaras Hindu University, Varanasi 221005 India, 2013.

Stein-Bachinger, K. In *Strategies to Improve Yield and Crop Quality by Different Distribution of Limited Amounts of Farmyard and Liquid Manure Applied to Subsequent Crops After Grass-clover,* Proceedings of the 3rd Meeting of Fertilization Systems in Organic Farming, Copenhagen, Denmark, 1996; pp 64–74.

Stephen, J.; Shabanamol, S.; Rishad, K. S.; Jisha, M. S. Growth Enhancement of Rice (*Oryza sativa*) by Phosphate Solubilizing *Gluconacetobacter* sp. (MTCC 8368) and *Burkholderia* sp. (MTCC 8369) Under Greenhouse Conditions. *3 Biotech.* **2015,** *5* (5), 831–837.

Subbiah, S.; Kumaraswamy, K. Effect of Manure-fertilizers on the Yield and Quality of Rice and on Soil Fertility. *Fertil. News* **2000,** *45* (10), 61–68.

Sudha, B.; Chandini, S. Nutrient Management in Rice (*Oryza sativa L.*). *J. Trop. Agric.* **2002,** *40,* 63–64.

Sudhakar, C. Integrated Nutrient Management in Rice–Maize Cropping System. M.Sc.Ag. (Agronomy) Dissertation, Department of Agronomy, Agricultural College, Acharya N. G. Ranga Agricultural University, Andhra Pradesh, India, 2011.

Sultana, M. S.; Rahman, M. H.; Rahman, M. S.; Sultana, S.; Paul, A. K. Effect of Integrated Use of Vermicompost, Pressmud and Urea on the Nutrient Content of Grain and Straw of Rice (Hybrid Dhan Hira 2). *Int. J. Sci. Res. Pub.* **2015**, *5* (11), 765–770.

Sun, J.; Ye, M.; Peng, S and Li, Y. Nitrogen can Improve the Rapid Response of Photosynthesis to Changing Irradiance in Rice (*Oryza sativa* L.) Plants. *Sci. Rep.* **2016**, *6* (31305), 1–10.

Sunitha, B. P.; Prakasha, H. C.; Gurumurthy, K. T. Influence of Organics, Inorganics and Their Combinations on Availability, Content and Uptake of Secondary Nutrients by Rice Crop (*Oryza sativa* L.) in Bhadra Commend, Karnataka. *Mysore J. Agric. Sci.* **2010**, *44* (3), 509–516.

Surekha, K.; Jhansilakshmi, V.; Somasekhar, N.; Latha, P. C.; Kumar, R. M. Status of Organic Farming and Research Experiences in Rice. *J. Rice Res.* **2010**, *3* (1), 23–29

Suresh, K.; Reddy, G. R.; Hemalatha, S.; Reddy, S. N.; Raju, A. S.; Madhulety, T. Y. Integrated Nutrient Management in Rice: A Critical Review. *Int. J. Appl. Biol. Pharma. Technol.* **2013**, *4* (2), 47–53.

Suthar, S. Effect of Vermicompost and Inorganic Fertilizer on Wheat (*Triticum aesticum*) Production. *J. Nat. Environ. Pollut. Technol.* **2006**, *5*, 197–201.

Syers, J. K.; Johnston, A. E.; Curtin, D. *Efficiency of Soil and Fertilizer Phosphorus Use: Reconciling Changing Concepts of Soil Phosphorus Behaviour with Agronomic Information.* FAO Fertilizer and Plant Nutrition Bulletin 18, Food and Agriculture Organization of the United Nations (FAOUS): Rome, Italy, 2008; pp 1–63.

Tharmaraj, K., Ganesh, P.; Kumar, S. R.; Anandan, A.; Kolanjinathan, K. Vermicompost-A Soil Conditioner Cum Nutrient Supplier. *Int. J. Pharm. Biol. Sci. Arch.* **2011**, *2*, 1615–1620.

Theunissen, J.; Ndakidemi, P. A.; Laubscher, C. P. Potential of Vermicompost Produced from Plant Waste on the Growth and Nutrient Status in Vegetable Production. *Int. J. Phys. Sci.* **2010**, *5* (13), 1964–1973.

Thomas, R. G.; Hay, M. J. M. Regulation of Correlative Inhibition of Axillary Bud Outgrowth by Basal Branches Varies with Growth Stage in *Trifolium repens*. *J. Exp. Bot.* **2015**, *66* (13), 3803–3813.

Tian-Yao, M.; Huan-He, W.; Chao, L.; Qi-Gen, D.; Ke, X.; Zhong-Yang, H.; Hai-Yan, W.; Bao-Wei, G.; Hong-Cheng, Z. Morphological and Physiological Traits of Large-panicle Rice Varieties with High Filled-grain Percentage. *J. Integr. Agric.* **2016**, *15* (8), 1751–1762.

Tiwari, V. N.; Singh, H.; Upadhyay, R. M.; Effect of Biocides, Organic Manure and Blue Green Algae on Yield and Yield Attributing Characteristics of Rice and Soil Productivity Under Sodic Soil Condition. *J. Indian Soc. Soil Sci.* **2001**, *49* (2), 332–336.

Tomati, U.; Grappelli, A.; Galli, E. The Hormone like Effect of Earthworm Casts on Plant Growth. *Biol. Fert. Soils* **1995**, *5*, 288–294.

Treutter, D. Significance of Flavanoids in Plant Resistance and Enhancement of Their Biosynthesis. *Plant Biol.* **2005**, *7*, 581–591

Tripathi, B. N.; Chaubey, C. N. Effect of Organic Sources of Plant Nutrients in Conjunction with Chemical Fertilizers on Bulk Density, Yield and Uptake of Nutrients by Rice. *Oryza* **1996**, *33*, 200–207.

Tripathi, N.; Verma, R. S. Assessment of Grain Quality Attributes of Basmati Rice Produced by Organic System. *Pantnagar J. Res.* **2008,** *6* (2), 192–195.

Tripathi, H. P.; Mauriya, A. K.; Kumar, A. Effect of Integrated Nutrient Management on Rice–Wheat Cropping System in Eastern Plain Zone of Uttar Pradesh. *J. Farm. Syst. Res. Dev.* **2007,** *13* (2), 198–203.

Trujillo-Tapia, M. N.; Ramírez-Fuentes, E. Bio-fertilizer: An Alternative to Reduce Chemical Fertilizer in Agriculture. *J. Glob. Agric. Ecol.* **2016,** *4* (2), 99–103.

Tzudir, L.; Ghosh, R. K. Impact of Integrated Nutrient Management on Performance of Rice Under System of Rice Intensification (SRI). *J. Crop Weed* **2014,** *10* (2), 331–333.

Uchida, R. Essential Nutrients for Plant Growth: Nutrient Functions and Deficiency Symptoms. In *Plant Nutrient Management in Hawaii's Soils, Approaches for Tropical and Subtropical Agriculture;* Silva, J. A., Uchida, R. Eds.; College of Tropical Agriculture and Human Resources: Manoa, Hawaii, USA, 2000; pp 31–55.

Urkurkar, J. S.; Chitale, S.; Tiwari, A. Effect of Organic v/s Chemical Nutrient Packages on Productivity, Economics and Physical Status of Soil in Rice (*Oryza sativa*)-Potato (*Solanum tuberosum*) Cropping System in Chhattisgarh. *Indian J. Agron.* **2010,** *55* (1), 6–10.

Usenik, V.; Mikulic Petkovsek, M.; Solar, A.; Stampar, F. Flavanols of Leaves in Relation to Apple Scab Resistance. *J. Plant Dis. Plant Prot.* **2004,** *111,* 137–144.

Usman, K. Effect of Phosphorus and Irrigation Levels on Yield, Water Productivity, Phosphorus Use Efficiency and Income of Lowland Rice in Northwest Pakistan. *Rice Sci.* **2013,** *20* (1), 61–72.

Usman, M.; Madu, V. U.; Alkali, G. The Combined Use of Organic and Inorganic Fertilizers for Improving Maize Crop Productivity in Nigeria. *Int. J. Sci. Res. Pub.* **2015,** *5* (10), 1–7.

Vaghela, P. O.; Sutariya, D. A.; Prajapati, D. V.; Parmar, S. K. Biofertilizers: A Enemy of Chemical Fertilizer. *Rashtriya Krishi* **2014,** *9* (1), 15–16.

Vavilov, N. I. Studies on the Origin of Cultivated Plants. *Bull. Appl. Biol.* **1926,** *16,* 139–248.

Velazhahan, R.; Vidhyasekaran, P. Role of Phenolic Compounds, Peroxidase and Polyphenol-oxidase in Resistance of Groundnut to Rust. *Acta. Phytopathol. Entomol. Hung.* **1994,** *29,* 23–29.

Vennila, C.; Jayanthi, C.; Nalini, K. Nitrogen Management in Wet Seeded Rice: A Review. *Agric. Rev.* **2007,** *28* (4), 270–276.

Verma, D. K.; Srivastav, P. P. Proximate Composition, Mineral Content and Fatty Acids Analyses of Aromatic and Non-aromatic Indian Rice. *Rice Sci.* **2017,** *24* (1), 21–31.

Verma, C. P.; Tripathi, H. N.; Prasad, K. Effect of FYM and Zinc Sulphate on Yield and Yield Attribute of Rice Grown After Paddy Nursery. *Crop Res.* **2001,** *21* (3), 382–383.

Verma, D. K.; Mohan, M.; Yadav, V. K.; Asthir, B.; Soni, S. K. Inquisition of Some Physico-chemical Characteristics of Newly Evolved Basmati Rice. *Environ. Eco.* **2012,** *30* (1), 114–117.

Verma, D. K.; Mohan, M.; Asthir, B. Physicochemical and Cooking Characteristics of some Promising Basmati Genotypes. *Asian J. Food Agro-Indus.* **2013,** *6* (2), 94–99.

Verma, D. K.; Mohan, M.; Prabhakar, P. K.; Srivastav, P. P. Physico-chemical and Cooking Characteristics of Azad Basmati. *Int. Food Res. J.* **2015,** *22* (4), 1380–1389.

Vinod, K. K.; Heuer, S. Approaches Towards Nitrogen- and Phosphorus-efficient Rice. *AoB Plants* **2012,** pls028. DOI:10.1093/aobpla/pls028.

Virdia, H. M.; Mehta, H. D. Integrated Nutrient Management in Transplanted Rice (*Oryza sativa* L.). *Int. J. Agric. Sci.* **2010,** *6* (1), 295–299.

Walpola, B. C.; Yoon, M.-H. Prospectus of Phosphate Solubilizing Microorganisms and Phosphorus Availability in Agricultural Soils: A Review. *Afri. J. Microbiol. Res.* **2012,** *6* (37), 6600–6605.

Wang, D.; Shi, Q.; Wang, X.; Wei, M.; Hu, J.; Liu, J.; Yang, F. Influence of Cow Manure Vermicompost on the Growth, Metabolite Contents, and Antioxidant Activities of Chinese Cabbage (*Brassica campestris* ssp. chinensis). *Biol. Fert. Soils* **2010,** *46,* 689–696.

Werner, M.; Cuevas, R. *Vermiculture in Cuba Biocycle*; JG Press: Emmaus, USA, 1996; Vol. 37 (6), pp 61–62.

Wolie, A. W.; Admassu, M. A. Effects of Integrated Nutrient Management on Rice (Oryza sativa L) Yield and Yield Attributes, Nutrient Uptake and Some Physico-chemical Properties of Soil: A Review. *J. Biol. Agric. Healthc.* **2016,** *6* (5), 20–26.

Xueyong, L.; Qian, Q.; Fu, Z.; Wang, Y.; Xiong, G.; Zeng, D.; Wang, X.; Liu, X.; Teng, S.; Hiroshi, F.; Yuan, M.; Luo, D.; Han, B.; Li, J. Control of Tillering in Rice. *Nature* **2003,** *422,* 618–621.

Yadav, G. S.; Datta, M.; Babu, S.; Debnath, C and Sarkar, P. K. Growth and Productivity of Lowland Rice (*Oryza sativa*) as Influenced by Substitution of Nitrogen Fertilizer by Organic Sources. *Indian J. Agric. Sci.* **2013a,** *83* (10), 1038–1042.

Yadav, S. K.; Singh, Y.; Kumar, R. P.; Yadav, M. K.; Singh, K. Effect of Organic Nitrogen Sources on Yield Quality and Nutrient Uptake of Rice (*Oryza sativa*) Under Different Cropping System. *Vegetos: An Int. J. Plant Res.* **2013b,** *26* (1), 58–66.

Yadav, S. K.; Babu, S.; Yadav, M. K.; Singh, K.; Yadav, G. S.; Pal, S. A Review of Organic Farming for Sustainable Agriculture in Northern India. *Int. J. Agron.* **2013c,** *2013* (2013), 1–8. DOI: http://dx.doi.org/10.1155/2013/718145.

Yaduvanshi, N. P. S.; Swarup, A. Effect of Continuous Use of Sodic Irrigation Water with and Without Gypsum, Farmyard Manure, Press Mud and Fertilizer on Soil Properties and Yields of Rice and Wheat in a Long Term Experiment. *Nutr. Cycl. Agroecosyst.* **2005,** *73,* 111–118.

Yamunarani, K.; Jaganathan, R.; Bhaskaran, R.; Govindaraju, P.; Velazhahan, R. Induction of Early Blight Resistance in Tomato by *Quercus infectoria* Gall Extract in Association with Accumulation of Phenolics and Defense-related Enzymes. *Acta. Physiol. Plant* **2004,** *26,* 281–290.

Yardim, E. N.; Arancon, N. A.; Edwards, C. A.; Oliver, T. J.; Byrne, R. J. Suppression of Tomato Hornworm (*Manduca quinquemaculata*) and Cucumber Beetles (*Acalymma vittatum* and *Diabotrica undecimpunctata*) Populations and Damage by Vermicomposts. *Pedobiologia* **2006,** *50,* 23–29.

Yildirim, M.; Akinci, C.; Koc, M.; Barutcular, C. Applicability of Canopy Temperature Depression and Chlorophyl Content in Durum Wheat Breeding. *Anadolu Journal of Agric. Sci.* **2009,** *24* (3), 158–166.

Yoshida, S. Effects of Temperature on Growth of the Rice Plant (*Oryza sativa* L.) in a Controlled Environment. *Soil Sc. Plant Nutr.* **1973,** *19* (4), 299–310.

Yousefi, A. A.; Sadeghi, M. Effect of Vermicompost and Urea Chemical Fertilizers on Yield and Yield Components of Wheat (*Triticum aestivum*) in the Field Condition. *Int. J. Agric. Crop Sci.* **2014,** *7* (12), 1227–1230.

Zaller, J. G. Foliar Spraying of Vermicompost Extracts: Effects on Fruit Quality and Indications of Late-blight Suppression of Field-grown Tomatoes. *Biol. Agric. Hort.* **2006,** *24*, 165–180.

Zaller, J. G. Vermicompost as a Substitute for Peat in Potting Media: Effects on Germination, Biomass Allocation, Yields and Fruit Quality of Three Tomato Varieties. *Sci. Hort.* **2007,** *112*, 191–199.

Zerin, F. Effects of Rice Straw and Banana Plant Residue as Sources of Potassium on BRRI Dhan49 Production, M.Sc. (Soil Science) Dissertation, Department of Soil Science, Bangladesh Agricultural University, Mymensingh, Bangladesh, 2013.

Zhan, X.; Sun, B.; Lin, Z.; Gao, Z.; Yu, P.; Liu, Q.; Shen, X.; Zhang, Y.; Chen, D.; Cheng, S.; Cao, L. Genetic Mapping of a QTL Controlling Source–Sink Size and Heading Date in Rice. *Gene* **2015,** *571* (2), 263–270.

Zhang, Q.-C.; Wang, G.-H. Studies on Nutrient Uptake of Rice and Characteristics of Soil Microorganisms in a Long-term Fertilization Experiments for Irrigated Rice. *J. Zhejiang Univ. Sci. B.* **2005,** *6* (2), 147–154.

Zhang, Q.; Zhou, W.; Liang, G.; Wang, X.; Sun, J.; He, P.; Li, L. Effects of Different Organic Manures on the Biochemical and Microbial Characteristics of Albic Paddy Soil in a Short-term Experiment. *PLoS One* **2015,** *10* (4), 1–19. DOI: 10.1371/journal.pone.0124096.

Zhu, X. F.; Wang, Z. W.; Wan, J. X.; Sun, Y.; Wu, Y. R.; Li. G. X.; Shen, R. F.; Zheng, S. J. Pectin Enhances Rice (*Oryza sativa*) Root Phosphorus Remobilization. *J. Exp. Bot.* **2014,** *66* (3), 1017–1024.

Zou, L.; Stout, M. J.; Dunand, R. T. The Effects of Feeding by the Rice Water Weevil, *Lissorhoptrus oryzophilus* Kuschel, on the Growth and Yield Components of Rice, *Oryza sativa*. *Agric. For. Entomol.* **2004,** *6* (1), 47–54.

Zou, J.; Zhang, S.; Zhang, Z.; Li, G.; Chen, Z.; Zhai, W.; Zhao, X.; Pan, X.; Xie, Q.; Zhu, L. The Rice *HIGH-TILLERING DWARF*1 Encoding an Ortholog of *Arabidopsis* MAX3 Is Required for Negative Regulation of the Outgrowth of Axillary Buds. *Plant J.* **2006,** *48* (5), 687–698.

PART 3

Weed Management for Improved Rice Production

CHAPTER 8

INTEGRATED WEED MANAGEMENT PRACTICES IN ZERO-TILL DIRECT-SEEDED RICE

GAURAV S. K. VERMA[1*], V. K. VERMA[1], DEEPAK KUMAR VERMA[2], and RAM KUMAR SINGH[1]

[1]*Department of Agronomy, Institute of Agricultural Sciences, Banaras Hindu University, Varanasi 221005, Uttar Pradesh, India*

[2]*Department of Agricultural and Food Engineering, Indian Institute of Technology, Kharagpur 721302, West Bengal, India*

Corresponding author. E-mail: gauraviasbhu@gmail.com

ABSTRACT

Rice—the major crop of Asia is typically grown by the transplantation of seedlings in prepared land which demands increased manual power and large water quantity. Nowadays, many farmers are therefore moving toward the direct-seeded rice systems that are of two types—wet- and dry-seeded rice. The direct-seeded rice system offers numerous benefits to farmers but major limitation is the growth of weeds due to the absence of standing water and rice seedlings, thus causing a significant loss in rice yield. However, weeds can be controlled by application of herbicides or manually. Manual weeding is today less prevalent due to the increased labor cost and unavailability of labor. On the other side, increased use of herbicides may lead to shifts in weed populations, development of resistance in weeds, and pollution. Therefore a strong need is nowadays felt to integrate the weed management practices for controlling the weeds in direct-seeded rice systems. This chapter deals with the introduction of commonly found weeds, and traditional and recent weed management practices in direct-seeded rice systems.

8.1 INTRODUCTION

Rice (*Oryza sativa* L.) is the most main cereal crop of the world as it forms the staple diet of 70% of the world's population. Globally, it is grown on approximately 163 m ha out of which 145 m ha is in Asia. Ninety percent of the world's rice is produced as well as consumed in Asia (Solunke et al., 2006). India has the largest area under rice cultivation in the world and is the second largest producer of rice after China, contributing to about 20% of the world rice production. India, Pakistan, and Bangladesh have about 14.2 million ha direct-seeded rice (DSR) of the total world's rice area of 55.3 million ha. India has the largest area under rice cultivation, that is, 42.56 million ha and occupies second position in production (95.33 million t) next to China among the rice-growing countries of the world with average productivity of 2.24 t ha^{-1} (Singh et al., 2012). DSR occupies 26% of the total rice area in South Asia (Gupta et al., 2006).

Rice is stuff of life, it is necessary to increase its production and productivity in order to meet the growing demands of rice by our increasing population. In addition to milled rice need, about 114 million t is estimated by the rice scientists to be produced by 2035 to fulfill the global demand of rice. This target in next 25 years will be equal to 26% overall increase while the possibility of rice production to meet the global rice demand in the near future under the available arable land have limit. Therefore, this world's growing demand of rice is under the consideration of the term "productivity gain" and rice scientist and researchers are facing many challenges day by day to achieve the goal of this target. Some of them are very major such as agrochemicals, farm labors, and less irrigated water availability thereby ensures long-term sustainability.

In Asian countries, transplanting of rice seedlings into puddled soil (wet tillage) is very common cultivation practice (Fig. 8.1). This cultivation practice is less profitable for farmers from economic point of view because of complete dependency upon the labor, water, and energy which are increasingly scarce. This practice is also responsible for deteriorating the soil physical properties, adversely affecting the performance of succeeding upland crops, and causing emissions of methane gas. All these factors have posed a threat to the productivity and sustainability of rice-based systems and demand a major shift from the current system of transplanted rice production. In recent years, due to severe water and labor scarcity, farmers are changing their rice establishment method from transplanting to direct seeding. The DSR practice deals with many advantages as it is known for faster and easier planting, reduced cost of labor and their efficiency and less drudgery. This

FIGURE 8.1   Transplanting of rice seedlings done by men and women in Asia.

system provides the crop maturity within 7–10 days, that is, very earlier, more efficient in water use and higher tolerance for water-deficit condition along with less or no emission of methane gas, and higher profit comparatively in areas with an assured water supply. The additional benefits of DSR would be water conservation, soil temperature moderation, and built up of soil organic carbon status due to residue retention at the surface.

Herbicide use in DSR becomes more important in rice production when weeds and rice crop emerge simultaneously. Among them, some weeds such as *Echinochloa colona* and *Echinochloa crus-galli* are morphologically very similar to rice, which are difficult to be differentiated at early stages of growth. The use of appropriate herbicides in rice field in order to ensure the effective weed control, especially during labor shortage periods and when weeding coincides with other farm work. Development of resistance in weeds against commonly used herbicides has been reported from different parts of the world. So, overreliance on chemical herbicides for weed control should be avoided. Application of pendimethalin @ 1.0 kg ha^{-1} at preemergence has been quite effective and economical for DSR in reducing weed count and biomass whether applied as a sole treatment or followed in sequence with a postemergence herbicide (Jayadeva and Bhairappanavar, 2002). Postemergence application (15–25 DAS) of bispyribac @ 25 g ha^{-1} is effective in controlling all three types of weed flora (grasses, sedges, and broadleaf weeds). However, this herbicide is poor in controlling *Leptochloa* spp., *Eragrostis* spp., *Dactyloctenium aegyptium*, and *Cyperus rotundus* (Gopal et al., 2010). Also, it fails to control grasses emerging late in the season such as *Dactyloctenium aegyptium, Eleusine indica*, and *Sagitaria arvensis* (Khaliq et al., 2011). Many herbicides and efficient technologies have been reported for better control of weeds in DSR. But their random selection, combination, and adoption is not a rational approach for tackling the weed menace in DSR as it will depend on many factors such as climatic conditions, type of rice culture, availability of technology, socioeconomic conditions of farmers, and the cost of weed control in comparison with the estimated value of the resulting yield increase. Integration of appropriate weed management practices should reduce the actual production cost and minimize environmental pollution besides lowering of weed problems in DSR.

Keeping the above facts into consideration, the present entitled chapter is focused on DSR, how the weed flora and their growth under integrated weed management (IWM) practices in DSR; what are the effects of IWM on growth and yield of DSR; and the nutrient content and their uptake by rice and depletion by weeds in different treatments.

8.2   WEED AT A GLANCE

Weeds are no strangers to man. Weeds are considered as unwanted and undesirable plants which interfere with man's interest in utilizing land, nutrient, and water resources. Weeds compete with crop plants mainly for plant nutrient, soil moisture, and sunlight. In this chapter, an attempt has been made to review the work done on IWM practices in zero-till DSR in India as well as abroad to bring about critical evaluation in these aspects.

Weeds are undesirable, prolific, competitive, and often harmful or even poisonous to the total environment and occur in every rice field in the world. The type of dominating weed flora and degree of their infection is mostly governed by the factors such as method of raising crop and depth of standing water in the field. In a crop raised by direct seeding in dry condition, weeds are abundant. On the other hand, if the crop is raised by transplanting or by direct seeding sprouted seeds in puddled condition the incidence of grassy weed are less and the weed flora mostly consist of sedges. Some important weeds of rice field such as grasses sedges as well as broadleaf weeds are shown in Figure 8.2.

8.2.1   WEED FLORA IN DSR

For successful weed control program, weed flora, is an important aspect. A galaxy of workers reported that DSR weed flora consists of grasses, sedges, and broadleaf weeds (Table 8.1). Adopting DSR may also result in weed flora shifts toward more difficult to control competitive grasses and sedges. For example, associated with direct seeding is an inevitable shift in the weed flora toward competitive grasses, including *Echinochloa* spp., *Leptochloa chinensis*, and *Ischaemum rugosum* in wet-seeded rice (WSR) and in dry-seeded rice (D-SR) the perennial sedges *Cyperus rotundus*. These shifts have been reported in irrigated rice in Southeast Asia, where DSR has largely replaced transplanted rice. Management of such weeds requires farmers to have the ability to anticipate changes in weed populations and to reduce losses, exploit integrated strategies comprising tillage, water, and crop management to complement herbicide application. Khaliq et al. (2013) reported that weed flora *Alternanthera philoxeroides, Cynodon dactylon, Cyperus rotundus, Dactyloctenium aegyptium, Echinochloa colona, Echinochloa crus-galli, Eleusine indica, Trianthema portulacastrum,* and *Trianthema portulacastrum* were associated with DSR.

In India, Maity and Mukherjee (2011) observed during experimentation on DSR at Cooch Behar, West Bengal, India that the weed flora consisted of grasses such as *Cynodon dactylon* and *Echinochloa colona*; sedges such as *Cyperus iria, Cyperus rotundus,* and *Fimbristylis miliacea*; and broadleaf weeds such as *Ageratum conyzoides, Eclipta alba, Enhydra fluctuans, Ludwigia parviflora,* and *Spilanthes paniculata.* Kumar et al. (2010) conducted a field experiment in Pantnagar, Uttarakhand, India which revealed that *Echinochloa colona* among grasses; *Caesulia axillaris* and *Commelina benghalensis* among nongrasses; and *Cyperus rotundus* among sedges were

A) *Dactyloctenium aegyptium*,

B) *Echinochloa colona*,

C) *Echinochloa crus-galli*,

D) *Leptochloa chinensis*,

E) *Caesulia axillaris*,

F) *Commelina bengalensis*,

G) *Eclipta prostrate*,

H) *Euphorbia hirta*,

I) *Lindernia crustaceae*,

J) *Portulaca oleracea*,

K) *Trianthema portulacastrum*

FIGURE 8.2 Some important grassy weeds and broadleaf weeds of rice field. (A) *Dactyloctenium aegyptium*, (B) *Echinochloa colona*, (C) *Echinochloa crus-galli*, (D) *Leptochloa chinensis*, (E) *Caesulia axillaris*, (F) *Commelina bengalensis*, (G) *Eclipta prostrate*, (H) *Euphorbia hirta*, (I) *Lindernia crustacean*, (J) *Portulaca oleracea*, and (K) *Trianthema portulacastrum*.

TABLE 8.1   Some Important Weeds of Rice Field.

Scientific name	Family	Common name
Grassy weeds		
Dactyloctenium aegyptium	Poaceae	Crowfoot Grass or Egyptian crowfoot grass
Echinochloa colona	Poaceae	Awnless barnyard grass, corn panic grass, deccan grass, jungle rice, jungle ricegrass, and shama millet
Echinochloa crus-galli	Poaceae	Barnyard grass, barnyard millet, cockspur (or cockspur grass), common barnyard grass, Japanese millet, and water grass
Leptochloa chinensis	Poaceae	Asian sprangletop, Chinese sprangletop, and red sprangletop
Broadleaf weeds		
Caesulia axillaris	Asteraceae	Pink node flower
Commelina benghalensis	Commelinaceae	Benghal dayflower, tropical spiderwort, and wandering Jew
Eclipta prostrata	Asteraceae	Bhringraj, false daisy, and yerba de tago
Euphorbia hirta	Euphorbiaceae	Asthma plant
Lindernia crustacea	Linderniaceae	Brittle false pimpernel and Malaysian false pimpernel
Portulaca oleracea	Portulacaceae	Common purslane, little hogweed, pigweed, parsley, red root, and verdolaga
Trianthema portulacastrum	Aizoaceae	Black pigweed, desert horsepurslane, and giant pigweed

the predominant weed species in the experimental plot. Yadav et al. (2008) observed that *Ammannia baccifera, Commelina diffusa, Cynodon dactylon, Cyperus rotundus,* and *Echinochloa colona* were the dominant weed species in Kumarganj, Faizabad, Uttar Pradesh, India. Singh et al. (2007) identified the weeds associated with dry-DSR at Modipuram, Meerut, Uttar Pradesh, India and these included grassy and broadleaf weeds (Table 8.2). The density of both broadleaf and grassy weeds increased up to 45 days after sowing (DAS) followed by a decline at 75 DAS. However, their dry weight increased with the age of the crop. According to Sinha et al. (2005) in Bihar prominence, *Caesulia axillaris, Cynodon dactylon, Dactyloctenium aegyptium, Eleusine indica,* and *Phyllanthus niruri* were the predominant weed flora in DSR. In Pakistan, Mann et al. (2007) noted that *Cyperus difformis, Cyperus iria, Echinochloa crus-galli, Eclipta prostrate, Paspalum distichum,* and

TABLE 8.2 Major Weed Flora Associated with DSR Reported in Western Uttar Pradesh of India.

Researchers	Location	Weed flora associated with DSR
Singh et al. (2007)	Modipuram, Meerut, Uttar Pradesh, India	Dry-DSR Grassy Weeds: *Dactyloctenium aegyptium, Echinochloa colona, Echinochloa crus-galli,* and *Leptochloa chinensis*
		Broadleaf Weeds: *Caesulia axillaris, Commelina benghalensis* L., *Eclipta prostrate* Linn., *Euphorbia hirta* Linn., *Lindernia* spp., *Portulaca oleracea* Linn., and *Trianthema portulacastrum* Linn.
Naresh et al. (2011)	Western part of Uttar Pradesh, India	DSR *Ageratum conyzoides, Alternanthera sessilis, Amaranthus viridis, Ammannia auriculata, Ammannia baccifera, Brachiaria* spp., *Caesulia axillaris, Celosia argentea, Commelina benghalensis, Commelina diffusa, Convolvulus arvensis, Corchorus* spp., *Cynodon dactylon, Dactyloctenium aegyptium, Digera muricata, Digitaria ciliaris, Digitaria sanguinalis, Echinochloa colona, Echinochloa crus-galli, Eclipta prostrata, Eleusine indica, Eragrostis japonica, Eragrostis tenela, Euphorbia hirta, Euphorbia microphylla, Fimbristylis miliacea, Ischaemum rugosum, Leptochloa chinensis, Lindernia crustaceae, Ludwigia* spp., *Marsilea quadrifolia, Monochoria vaginalis, Panicum repens, Parthenium hysterophorus, Paspalum distichum, Phyllanthus fraternus, Physalis minima, Portulaca oleracea, Sorghum halepense, Sphenoclea zeylanica,* and *Trianthema portulacastrum*

DSR, direct-seeded rice.

Trianthema portulacastrum were the major weeds associated with dry-DSR at Kala Shah Kaku of Punjab state.

8.2.2 CRITICAL PERIOD OF CROP–WEED COMPETITION

The outcome of the process of interspecific competition among plant species is the differential acquisition of resources for growth and yield. In plant species, the most important competition is usually preemptive competition that occurs among seedling, since it is almost impossible for a seedling of one species to out compete an established adult of another species. The inherent size differences between rice seedling and emerging weed species that confers a competitive advantage to transplanted rice is removed when direct seeding is practiced.

Productivity of DSR is mainly dependent on effective and timely management of weed. This, in turns requires knowledge about critical period of crop–weed competition. Therefore, it is essential to establish critical duration of crop–weed competition and a limit for an acceptable presence of weeds so as to formulate an effective and economical weed management system for DSR. Information on the critical period of crop–weed competition in DSR could help to improve timing of postemergence herbicides application. Reduction in the number of herbicide application as a result of better timing and efficiency may reduce potential environment contamination and the development of herbicide resistant weeds (Khaliq and Matloob, 2011). Singh (2008) suggested that the longer the presence of crop–weed competition in DSR during the initial period the lower the yield, while at later stages yield might not change since the maximum damage has already occurred. The effective period of weed–crop competition in DSR occurs in two phases, that is, between 15 and 30 DAS and 45 and 60 DAS. The competition beyond 15 DAS may cause significant reduction in the grain yield. However, competition for only first 15 days may not have much adverse effect on crop. Juraimi et al. (2010) opined that DSR should be kept weed-free for 2–71 DAS in saturated condition and 15–73 DAS in flooded condition. In Philippines, Chauhan and Johnson (2011) estimated critical period for weed control of rice as between 18 and 52 DAS to obtain 95% of weed-free yield. In the field experiment conducted during 2006–2007 and 2007–2008 at Krishi Vigyan Kendra, Madurai, Tamil Nadu, India under the Zonal Project Directorate, Bangalore, India, Singh et al. (2012) found that first 40 DAS was the most critical period for crop–weed competition. The crop grown under weed-free condition for 40 DAS gave the maximum grain yield.

8.2.3 LOSSES CAUSED BY WEEDS

Weeds offer a serious competition with the rice plant for all critical growth factors, namely, space, sunlight, water, and nutrients, thus adversely affecting yield, quality, and cost of production. Weeds compete for moisture, nutrients, light, and space and as a consequence, weeds infestation in DSR results in yield losses. Out of the losses due to various biotic stresses, weeds are known to account for nearly one-third of losses. Weeds are major yield-limiting factor in rice production (Bastiaans et al., 1997). Globally, actual rice yield losses due to pest have been estimated at 40%, of which, weeds have the highest loss potential (32%). The worldwide estimated loss in rice yield is around 10% of the total production (Oerke and Dehne, 2004). In India, yield of DSR on raised beds was reduced by 65% when weeds were not controlled (Singh et al., 2006). Research has shown that in the absence of effective weed control options, yield losses are greater in DSR than in transplanted rice (Rao et al., 2007). Moorthy and Saha (2003) reported that loss in grain yield due to unchecked weed competition was estimated to be 33.5% and 51.9% during 1997 and 1998, respectively. Weeds in DSR cause 73% loss in yield and the farmers may end up using most of the labor saved by wet-seeding to control weeds (Milberg and Hallgren, 2004). Duary et al. (2005) recorded in dry-DSR that the loss in grain yield of rice was 69.5% and 78.8% over the best treatment. Season-long weedy condition caused more than 45% reduction in the grain yield of rice. Mishra and Singh (2008) reported that at Jabalpur, infestation of weed reduced the grain yield of D-SR by 60% compared with two-hand weeding. Similarly, Khaliq et al. (2013) from Pakistan also reported reduction in seed yield by 75%. Singh et al. (2011) in their on-farm experiments reported that rice yield losses due to uncontrolled weed growth were least as 12% in transplanted rice but otherwise as large as 85%, where rice had been sown to dry cultivated fields or to puddled soil, rising to 98% in D-SR sown without soil tillage.

8.3 WEED MANAGEMENT PRACTICES

Weeds are the number one biological constraint to the adoption and production of DSR systems. Weeds in different direct-seeded systems can cause rice yield losses of up to 50% and these losses are after one-hand weeding (or partial weed-free conditions) in weed-infested fields (Chauhan, 2012; Hossain et al., 2016). Hand weeding is the most common practice adopted by farmers to remove weeds from the paddy field, but nonavailability of labors at peak period and escalating wages make it difficult and nonprofitable to adopt this practice. Manual weeding can be done only when weeds have reached a sufficient size to be pulled out easily by hand. By that time, irrevocable yield losses may have already occurred. Some weed species, namely, *Echinochloa colona* and *Echinochloa crus-galli*, are difficult to distinguish from rice at the early stage and they escape hand weeding, reduce rice yield, and produce seeds to infest crops in subsequent seasons. For these reasons, herbicides are being promoted to control weeds and they are easy to use. Besides hand weeding and chemical method, weed problem in rice can be managed by various cultural practices such as brown manuring, crop rotation, intercropping with green manuring crop, stale seed bed, method of sowing, fertilizer application, irrigation, etc. The research works conducted on different weed management practices in the past for managing weeds and enhancing yield in rice have been reviewed and described below.

8.3.1   HAND WEEDING

Hand weeding is being practiced by Indian farmers since they initiated agriculture. It is very effective measures for annual weeds but ineffective or less effective against perennial weeds due to their regenerative capability. Manual weeding by hand is an efficient method for weed control. However, this is a labor intensive and is not practical for large areas. It is the most widely used weed control method, with availability of labor being the main limitation to its effectiveness. In some areas, adoption of line planting in transplanted rice has allowed the introduction of rotary weeders for cultivation between rice rows, considerably reducing labor requirements for weed control. In India, hand weeding rice at 15 and 30 DAS gave a 60% increase in yield over a single-hand weeding at 30 DAS. In Asia, weed control in upland rice has been reported to require 32–198 man days ha^{-1}, representing 17–57% of the total labor requirement for the crop. Raising cost of labor and their nonavailability lead to the search for alternative methods such as use of herbicide either alone or in combination with hand weeding (Singh et al., 2001; Rao and Nagamani, 2007; Rao et al., 2007). Bhan et al. (1980) reported that the crop tolerated weed growth for the initial 30 days without much damage to grain yield provided weeds were removed at 30 DAS. They also found that hand weeding at 15 and 45 or at 30 and 45 days stages produced grain yield similar to weed free, whereas if weeds were not removed after 60 DAS, the grain yield was at par to weedy check. Ali et al. (1985) observed from Coimbatore that hand weeding twice gave maximum grain yield (4651 kg ha^{-1}) of low land direct sown rice followed by hand weeding once (3990 kg ha^{-1}). Kumar and Gautam (1986) reported form Pantnagar, India that in DSR under puddled soil, hand weeding at 30 and 50 DAS gave higher yield (4.18 ha^{-1}) than any of the herbicide treatments.

8.3.2   CHEMICAL IN WEED CONTROL

Some important herbicides are indicating in Table 8.3 used in rice field for weed control describes as fallow:

TABLE 8.3   Chemical Herbicides Used in Weed Control.

Chemicals names	IUPAC	Molecular formula	Structures
Pendimethalin	N-(1-ethylpropyl)-2,6-dinitro-3,4-xylidine	$C_{13}H_{19}N_3O_4$	
Pretilachlor	2-chloro-N-(2,6-diethylphenyl)-N-(2-propoxyethyl) acetamide	$C_{17}H_{26}ClNO_2$	
Bispyribac-sodium	Sodium;2,6-bis[(4,6-dimethoxypyrimidin-2-yl)oxy]benzoate	$C_{19}H_{17}N_4NaO_8$	

8.3.2.1   PENDIMETHALIN

Pendimethalin belongs to dinitroanilines group and is active against grasses and broadleaf weeds in various crops. Pendimethalin, used as preemergence herbicide, inhibits germination and seedling development of susceptible weeds. Wherever weeds are expected to be a problem, it should be applied after planting and before emergence of rice and weeds. Because soil and weeds must be completely exposed to spray coverage, no floodwater should be on the field at the time of application. The residual activity of pendimethalin is activated by moisture. It is most effective when adequate rainfall or irrigation is received within 7 days after application.

Pendimethalin was most effective in reducing weed density compared to other weed management treatments, and reported to be effective against most of grassy weeds (Jabran et al., 2012; Sinha et al., 2008). Also, sequential application of pendimethalin followed by bispyribac lowered the weed count. Pre- and postemergent herbicides in sequential herbicide application were more effective in controlling initial as well as later flushes of weeds, respectively. Similar finding was also reported by Mahajan et al. (2009). Singh et al. (2005) found the *Sesbsnia* coculture + cutting residue incorporation treatment was not effective in reducing the population of grasses, sedges, and broadleaf which were the dominant weed species in DSR, thus resulting in highest weed population among rest of the weed management practices. Walia et al. (2008) reported that sequential application of pendi-methalin followed by bispyribac improved yield attributes of DSR.

Singh et al. (2005) found that each increment of pendimethalin from 0.4 to 2.0 kg ha^{-1} reduced both density and dry matter of all the weeds. Singh et al. (2007) reported that pendimethalin (1000 g a.i. ha^{-1}) or pretilachlor with safener (500 g a.i. ha^{-1}) as preemergence applications followed by one-hand weeding were effective in controlling weeds, increasing grain yield of D-SR, and resulting in higher net returns than the weed-free treatment. Sinha et al. (2008) reported that pendimethalin 1.0 kg ha^{-1} was particularly effective in controlling *Echinochloa colona* in dry-DSR when applied at 2 DAS and where the soil was moist after the field had been irrigated.

Bandyopadhyay and Choudhury (2009) reported that rate of dissipation of pendimethalin was very rapid (40–44%) in all the doses caused by presowing irrigation and rainfall after application. But 42–48% of applied pendimethalin persisted up to 10 DAT. According to Jabran et al. (2012), total weeds density and total weed dry weight were reduced by 73.1% and 75.96%, respectively, by the application of pendimethalin (0.825 kg ha^{-1}), as compared with the control. Parameters such as plant height, branches per panicle, productive tillers, 1000-grain weight, grain yield, biological yield, harvest index, and water productivity were statistically higher when pendimethalin (0.825 kg ha^{-1}) was applied than the control. The application of pendimethalin followed by bispyribac reported to give significantly higher weed control efficiency over other weed management practices due to lower dry matter accumulation of weeds at 60 DAS of crop growth. The main reason behind this was high efficiency of pendimethalin against complex weed flora reported in the findings of Khaliq et al. (2011).

8.3.2.2   PRETILACHLOR

Herbicides have been intensively used for weed control in many crops, especially in rice. Pretilachlor 50% emulsifiable concentrate is a selective systemic herbicide used as preemergence herbicide in transplanted rice fields for controlling grasses, broadleaf weeds, and sedges. It absorbed primarily by the germinating shoots, and secondarily by the roots, with translocation throughout the plant. Recently, increasing environmental contamination, economic pressure, and the development of herbicide resistance have led to a reduction in herbicide use in conventional farming (Lemerle et al., 2001). Among the various IWM treatments, application of pretilachlor followed by bispyribac significantly reported for higher weed control efficiency over other weed management practices due to lower dry matter accumulation of weeds at 60 DAS of crop growth. The main reason behind this was high efficiency of bispyribac against complex weed flora was seen in the findings of Khaliq et al. (2011).

Rao et al. (2007) observed that among the integrated treatments, preemergence application of pendimethalin at 1.0 kg ha^{-1} integrated with one-hand weeding at 30 DAS recorded the lowest weed growth and highest grain yield and was at par with all other treatments with hand weeding and herbicide integration. Pretilachlor 0.75 kg ha^{-1} followed by one-hand weeding at 30 DAS and pendimethalin 1.0 kg ha^{-1} followed by one-hand weeding were the cheapest IWM treatments that recorded highest grain yield of DSR. Singh et al. (2007) observed that among the pretilachlor with safener (500 g a.i. ha^{-1}) as preemergence applications followed by one-hand weeding were effective in controlling weeds, increasing grain yield of D-SR, and resulting in higher net returns than the weed-free treatment. According to Parthipan et al.'s (2013) observation among the herbicides, the preemergence application of pretilachlor + safener 0.45 kg/ha followed by one-hand weeding at 45 DAS was effective in controlling all weeds and registered higher yield attributes and yield in DSR which was at par with two-hand weeding.

8.3.2.3 BISPYRIBAC-SODIUM

Among popular postemergence selective herbicides, bispyribac-sodium is reported to be effective against most of grassy weeds (Ranjit and Suwanketnikom, 2005; Hussain et al., 2008). Bispyribac-sodium in rice field, have the potentiality to keep the weed below the economic threshold level. Hence, Bispyribac-sodium is to be evaluated for their bioefficacy of controlling wide range of weed flora, better crop growth, and yield of rice. Bispyribac reported to be effective against most of grassy weeds (Ranjit and Suwanketnikom, 2005; Hussain et al., 2008). Damalas et al. (2008) reported that application of bispyribac-sodium mixed with the insecticides carbaryl or dichlorvos showed reduced efficacy on *Echinochloa oryzoides* and *Echinochloa phyllopogon*, whereas increased efficacy on both species was observed for mixtures of bispyribac-sodium with diazinon as compared with the sole application of bispyribac-sodium. Hussain et al. (2008) observed that bispyribac-sodium 0.025 kg ha^{-1} gave higher weed control efficiency (90.5%) compared to ethoxysulfuron + iodosulfuron. Mahajan et al. (2009) reported that bispyribac-sodium gave an excellent control of sedges and broadleaf weeds over the control, thus causing a reduction in dry matter of weeds to the tune of 81.3%, 61.7%, 22.1%, and 31.2% over weedy field, respectively. It gave highest grain yield and maximum net returns in aerobic DSR. Kumaran et al. (2015) revealed that early postemergence application of bispyribac-sodium 10% SC 40 g ha^{-1} recorded higher weed control efficiency and lesser weed density and nutrient uptake at reproductive stage of the crop. Different weed management practices imposed on rice crop did not affect the germination of succeeding green gram.

8.3.2.4 SEQUENTIAL HERBICIDE APPLICATION

DSR fields are characterized by diverse weed flora (Rao et al., 2007), so a single herbicide cannot produce satisfactory and cost-effective weed control (Khaliq et al., 2011). Contrary to other upland cereals, single application of a particular herbicide seldom furnishes adequate weed control in DSR. Postemergence application of Bispyribac-sodium 0.025 kg ha^{-1} at 20 DAS in combination with preemergence pendimethalin 0.75 kg ha^{-1} gave effective weed control in DSR and resulted in grain yield at par with hand-weeded plots (Anonymous, 2007).

A field experiment conducted by Walia et al. (2008) to identify effective herbicides for the control of complex weed flora of DSR. Results of the experiment revealed that integration of preemergence application of pendimethalin 0.75 kg ha^{-1} with postemergence application of bispyribac-sodium 0.025 kg ha^{-1} gave better weed control and higher rice yield attributes thus consequently resulting in 372% increase in rice grain yield as compared to weedy field. Further, Walia et al. (2009) also confirmed the discussed findings in other field investigations. Mahajan et al. (2009) observed that sequential spray of preemergence application of pendimethalin 1.0 kg ha^{-1} followed by bispyribac-sodium 0.03 kg ha^{-1} at 15 DAS was found best for the control of weeds in DSR. Results of experiment conducted by Khaliq et al. (2011) revealed that pendimethalin followed by postemergence application of bispyribac-sodium gave more than 80% reduction in weed density and weed dry weight over control and concluded that sequential application of herbicides was better than alone in dry-DSR.

8.3.3 Sesbania COCULTURE/BROWN MANURING OF Sesbania

Sesbania, being a submergence tolerant plant, can be grown together with rice to suppress weeds (Torres et al., 1995). In *Sesbania* coculture, *Sesbania* is sown at 25–30 kg ha^{-1} together with rice. After 25–30 days of growth, when *Sesbania* is about 30–40-cm tall, it is killed with 2,4-dichlorophen-oxyacetic acid (2,4-D) at 0.5 kg ha^{-1}. Intercropping of *Sesbania* with DSR suppresses weed infestation. Being a legume crop, it also enhances soil fertility. But, the contribution from N fixation was estimated to be small because of intercropping and short growth duration (Singh et al., 2007). Hence, promoting the use of cover crops enhances the sustainability of crop production and reduces weed density.

Singh et al. (2003) reported that growing of *Sesbania* as an intercrop with DSR up to 30 DAS reduced the weed infestation by 30%. Gupta et al. (2006) documented that coculture of *Sesbania* in rice and its subsequent knock-down by 2,4-D ester reduced the weed population by nearly half without any adverse effect on rice yield. Anitha and Mathew (2010) carried out a field experiment on wet-DSR at Thrissur (Kerala) and it was concluded that concurrent growing of dhaincha along with wet-DSR significantly reduced the total population and dry matter production of weeds when compared with rice sown alone. This may be attributed to shading effect exerted by canopy of dhaincha. Growth and yield parameters (leaf area index, LAI; panicle weight; and filled grain %) of rice involving concurrent growth of dhaincha was significantly higher than that of pure crop rice. Kumar and Ladha (2011) suggested that the application of pendimethalin along with *Sesbania* cocul-ture will increase the effectiveness against weeds as pendimethalin is effec-tive in controlling grass weed species, which otherwise become difficult to control after knockdown of *Sesbania* because of their large size. *Sesbania*, on the other hand, is effective against broadleaf weeds and sedges. Maity and Mukherjee (2011) reported that in wet-DSR, farmers' practice recorded lowest weed dry weight which was closely followed by brown manuring resulting in highest weed control index and lowest weed competition index. In wet-DSR, brown manuring registered highest net returns and benefit:cost ratio.

8.3.4   IWM IN DSR

Rice is cultivated in India in a very wide range of ecosystems from irrigated to shallow lowlands, mid-deep lowlands, and deep water to uplands. Trans-planting is the major method of rice cultivation in India. However, trans-planting is becoming increasingly difficult due to shortage and high cost of labor, scarcity of water, and reduced profit. Thus, direct-seeding is gaining popularity among farmers of India as in other Asian countries. Direct-seeding constitutes both wet- and dry-seeding and it requires the need for seedlings, nursery preparation, uprooting of seedlings, and transplanting. Upland rice, which is mostly dry-seeded, is found in parts of Assam, Bihar, Chhattisgarh; Gujarat, Jharkhand, Kerala, Karnataka, Madhya Pradesh, Orissa, Uttar Pradesh, and West Bengal. The upland rice area is around 5.5 million ha which accounts for 12.33% of the total rice area of the country. The WSR is increasing in area in parts of Andhra Pradesh, Punjab, and Haryana. In the rice agroecosystems, ideal environment conditions are provided for optimal rice productivity, which are being exploited by the associated weeds.

IWM is a science-based decision-making process that coordinates the use of environmental information, weed biology, and ecology, and all available technologies to control weeds by the most economical means, while posing the least possible risk to people and the environment (Sanyal, 2008). The concept of IWM is not new. For example, the traditional practice of puddling soil to kill existing weeds and aid water retention, transplanting rice seedlings into standing water to achieve an optimum stand density, and maintaining standing water to suppress weeds, followed by one or several periods of manual weeding are well-established examples of IWM (Rao et al., 2007).

In DSR, weed flora is diverse where various weed species emerge in different flushes and single method of weed control may not be successful for raising crop. Hence, different methods such as chemical, cultural, and manual need to be adopted in an integrated manner for effective weed management. Sole reliance on a single herbicide can bring about inter- and intraspecific shifts as well as evolution of herbicide-resistant weed biotypes due to herbicide selection pressure (Wrubel and Gressel, 1994; Shrestha et al., 2010). Weed management must strive at minimizing the weed population to a level at which weeds occurrence has no ill effect on farmer's economic and ecological interests. Farmers have more options for controlling weeds by using different appropriate management practices thereby reducing the possibility of escapes and weed adaptation to any single weed management tactic. Cultural practices also have a direct impact on weed infestation in DSR. Integrating weed control and cultural practices for weed management can have a strong impact on weeds and need to be coordinated with other crop production practices that affect agroecosystems (Alsaadawi et al., 2011).

Maity and Mukherjee (2011) also observed that IWM practices in DSR reduced the nutrient depletion by weeds. Weedy plot recorded highest nutrient depletion by weeds due to maximum weed growth and their dry weight. Among weed management treatment, the highest NPK depletion by weeds was recorded under *Sesbania* coculture + cutting residue incorporation treatment. It was due to maximum total weed dry matter under the treatment as nutrient depletion is positively correlated with weed dry matter accumulation. Kandasamy and Chinnusamy (2005) observed that application of pretilachlor under drum seeding + dhaincha gave lowest density and dry weight of weeds and maximum grain yield of rice as compared to broadcasting of rice. Mishra and Singh (2008) observed that integration of pendimethalin 1.0 kg ha^{-1} or pretilachlor 0.75 kg ha^{-1} with hand weeding at 30 DAS proved quite effective against weeds and gave significantly higher yield attributes (number of rice plants m^{-1}, row length, and number of panicles m^{-1}) and higher grain yields and benefits than weedy fields. The foregoing discussion indicated that integration of pre- and postemergence broad-spectrum herbicides or preemergence herbicides with *Sesbania* intercropping/brown manuring could provide effective control of weeds in DSR. It eliminates the early competition due to weeds causing lower weed population and weed dry matter production. The elimination of early competition due to weeds promotes better utilization of the various resources by crop which ultimately reflects in higher grain yield. The herbicide doses, however, needs to be standardized for different locations.

8.4 SUMMARY AND CONCLUSION

All the weed management practices markedly lowered weed density, weed dry weight, and NPK depletion. In this chapter, an attempt has been made to summarize the results presented in experimental findings, and also to draw valid conclusions based on the significant findings of the present investigation entitled "*IWM practices in zero-till DSR (Oryza sativa L.)*." The efficiency of treatments was measured in terms of weed density (species wise), weed dry matter production, nutrient depletion, and weed control efficiency. The weed density, weed dry matter production, and weed control efficiency were recorded at 60 DAS. Various crop characters such as growth attributes, namely, plant height, LAI, number of tillers m^{-1} of row length, and dry matter accumulation m^{-1} of row length at various growth stages and yield-attributing characters such as number of effective tillers m^{-1} row length, panicle length, weight of panicle, test weight, yields (i.e., grain yield, straw, and biological yields), and harvest index (%) after crop harvest were recorded and suggested that the application of pretilachlor followed by bispyribac-sodium was more remunerative for weed management practices in DSR.

KEYWORDS

- bispyribac-sodium
- crop–weed competition
- green manuring
- *Oryza sativa*

- weed control efficiency
- weed dry weight
- zero-till

REFERENCES

Ali, A. M.; Sankaran, S.; Rao, R. S.; Bhanumurthy, V. B. Time of Application of Herbicides on *Echinocloa crusgalli* (L.) Beauv and *Cyperus difformis* L. in Low Land Direct Sown Rice. *Indian J Weed Sci.* **1985**, *17* (3), 1–8.

Alsaadawi, I. S.; Khaliq, A.; Al-Temimi, A. A.; Matloob, A. Integration of Sunflower (*Helianthus annuus* L.) Residues with a Preplant Herbicide Enhances Weed Suppression in Broad Bean (*Vicia faba* L.). *Planta Daninha* **2011**, *29* (4), 849–859.

Anitha, S.; Mathew, J. Direct and Residual Effect of Concurrent Growing of Dhaincha (Sesbaniaaculeata) in Wet-seeded Rice (*Oryza sativa*) on the Productivity of Rice–Rice Cropping System. *Ind. J. Agric. Sci.* **2010**, *80*, 487–92.

Anonymous. *Zero Tillage Rice Establishment and Crop Weed Dynamics in Rice and Wheat Cropping Systems in India and Australia*; 2nd Annual 2007–2008 Technical Report, Department of Agronomy, Punjab Agriculture University, Ludhiana, India, 2007, 25–32.

Bandyopadhyay, S.; Choudhury, P. P. Leaching Behaviour of Pendimethalin Causes Toxicity Towards Different Cultivars of *Brassica juncea* and *Brassica campestris* in Sandy Loam Soil. *Interdiscip. Toxicol.* **2009**, *2* (4), 250–253.

Bastiaans, E. W.; Kropff, M. J.; Kempachetty, N.; Kaan. A.; Migo, T. R. Can Stimulation Models Help Design Rice Cultivars That More Competitive Against Weeds. *Field Crops Res.* **1997**, *51*, 101–111.

Bhan, V. M.; Maurya, R. A.; Nagi, R. K. Characterization of Critical Stages of Weed Competition in Drill Seeded Rice. *Indian J. Weed Sci.* **1980**, *21* (1), 75–79.

Chauhan, B. S. *Weed Management in Direct-seeded Rice Systems*; International Rice Research Institute: Los Baños, Philippines, 2012; pp 1–20.

Chauhan, B. S.; Johnson, D. E. Row Spacing and Weed Control Timing Affect Yield of Aerobic Rice. *Field Crops Res.* **2011**, *121* (2), 226–231.

Damalas, C. A.; Dhima, K. V.; Eleftherohorinos, I. G. Bispyribac-sodium Efficacy on Early Watergrass (*Echinochloa oryzoides*) and Late Watergrass (*Echinochloa phyllopogon*) as Affected by Co-application of Selected Rice Herbicides and Insecticides. *Weed Technol.* **2008**, *22* (4), 622–627.

Duary, B.; Hossain, A.; Mondal, D. C. Integrated Weed Management in Direct Seeded Dry Sown Rice in Lateritic Belt of West Bengal. *Indian J. Weed Sci.* **2005**, *37*, 101–102.

Gopal, R.; Jat, R. K.; Malik, R. K.; Kumar, V.; Alam, M. M.; Jat, M. L.; Mazid, M. A.; Saharawat, Y. S.; McDonald, A.; Gupta, R. *Direct Dry Seeded Rice Production Technology and Weed Management in Rice Based Systems*; Technical Bulletin, International Maize and Wheat Improvement Center: New Delhi, India, 2010; p 28.

Gupta, R.; Jat, M. L.; Singh, S.; Singh, V. P.; Sharma, R. K. Resource Conservation Technologies for Rice Production. *Indian Farm.* **2006,** *56* (7), 42–45.

Hossain, M. M.; Begum, M.; Rahman, M. M.; Akanda, M. M. Weed Management on Direct-seeded Rice System—A Review. *Progress. Agric.* **2016,** *27,* 1–8.

Hussain, S.; Ramzan, M.; Akhter, M.; Aslam, M. Weed Management in Direct Seeded Rice. *J. Anim. Plant Sci.* **2008,** *18* (2–3), 86–88.

Jabran, K.; Farooq, M.; Hussain, M.; Ehsanullah, K. M. B.; Shahid, M.; Lee, D. J. Efficient Weeds Control with Penoxsulam Application Ensures Higher Productivity and Economic Returns of Direct Seeded Rice. *Int. J. Agric. Biol.* **2012,** *14* (6), 901–907.

Jayadeva, H. M.; Bhairappanavar, S. T. Chemical Weed Control in Drum Seeded Rice. *Indian J. Weed Sci.* **2002,** *34* (3–4), 290–292.

Juraimi, A. S.; Begum, M.; Mohd, Y. M. N.; Man, A. Efficacy of Herbicides on the Control Weeds and Productivity of Direct Seeded Rice Under Minimal Water Conditions. *Plant Prot. Q.* **2010,** *25* (1), 19–25.

Kandasamy, O. S.; Chinnusamy, C. In *Integration of Seeding Methods and Weed Control Practices in Drum Seeded Rice.* Extended Summaries, National Biennial Conference, April 6–9, 2005; ISWS, Punjab Agriculture University: Ludhiana, India, 2005; pp 10–11.

Khaliq, A.; Matloob, A. Weed-crop Competition Period in Three Fine Rice Cultivars Under Direct-seeded Rice Culture. *Pak. J. Weed Sci. Res.* **2011,** *17* (3), 229–243.

Khaliq, A.; Matloob, A.; Shafiq, H. M.; Cheema, Z. A.; Wahid, A. Evaluating Sequential Application of Pre and Post Emergence Herbicides in Dry Seeded Fine Rice. *Pak. J. Weed Sci. Res.* **2011,** *17* (2), 111–123.

Khaliq, A.; Matloob, A.; Ihsan, M. Z.; Abbas, R. N.; Aslam, Z.; Rasool, F. Supplementing Herbicides with Manual Weeding Improves Weed Control Efficiency, Growth and Yield of Direct Seeded Rice. *Int. J. Agric. Biol.* **2013,** *15,* 191–199.

Kumar, J.; Gautam, R. C. Effect of Various Herbicides on Yield and Yield Attributes of Direct Seeded Rice on Puddled Soil. *Indian J. Weed Sci.* **1986,** *14* (1), 56–58.

Kumar, J.; Singh, D.; Puniya, R.; Pandey, P. C. Effect of Weed Management Practices on Nutrient Uptake by Direct Seeded Rice. *Oryza* **2010,** *47* (4), 291–294.

Kumar, V.; Ladha, J. K. Direct-seeding of Rice: Recent Developments and Future Research Needs. *Adv. Agron.* **2011,** *111,* 297–413.

Kumaran, S. T.; Kathiresan, G.; Murali Arthanari, P.; Chinnusamy, C.; Sanjivkumar, V. Efficacy of New Herbicide (Bispyribac Sodium 10% SC) Against Different Weed Flora, Nutrient Uptake in Rice and Their Residual Effects on Succeeding Crop of Green Gram Under Zero Tillage. *J. Appl. Nat. Sci.* **2015,** *7* (1), 279–285.

Lemerle, D.; Gill, G. S.; Murphy, C. E. Genetic Improvement and Agronomy for Enhanced Wheat Competitiveness and Weed Management. *Aust. J. Agric. Res.* **2001,** *52,* 527–548.

Mahajan, G.; Chauhan, B. S.; Johnson, D. E. Weed Management in Aerobic Rice in North Western Indo-Gangetic Plains. *J. Crop Improv.* **2009,** *23* (4), 366–382.

Maity, S. K.; Mukherjee, P. K. Effect of Brown Manuring on Grain Yield and Nutrient Use Efficiency in Dry Direct Seeded *Kharif* Rice (*Oryza sativa* L.). *Indian J. Weed Sci.* **2011,** *42* (1–2), 61–66.

Mann, R. A.; Ahmad, S.; Hassan, G.; Baloch, M. S. Weed Management in Direct Seeded Rice Crop. *Pak. J. Weed Sci. Res.* **2007,** *13* (3–4), 219–226.

Milberg, P.; Hallgren, E. Yield Loss Due to Weeds in Cereals and Its Large Scale Variability in Sweden. *Field Crop Res.* **2004,** *86,* 199–209.

Mishra, J. S.; Singh, V. P. Integrated Weed Management in Dry-seeded Irrigated Rice (*Oryza sativa*). *Indian J. Agron.* **2008,** *53* (4), 299–305.

Moorthy, B. T. S.; Saha, S. Relative Performance of Herbicide Alone and in Combination with Hand Weeding in Rainfed Lowland Direct-seeded Rice. *Indian J. Weed Sci.* **2003,** *35* (3–4), 268–270.

Naresh, R. K.; Gupta, R. K.; Singh, R. V.; Singh, D.; Singh, B.; Prakash, S.; Misra, A. K.; Rathi, R. C.; Suraj, B. Promotion of Integrated Weed Management for Direct Seeded Rice in the North West India. *Progress. Agric. Int. J.* **2011,** *11* (2) 215–232.

Oerke, E. C.; Dehne, H. W. Safeguarding Production Losses in Major Crops and the Role of Crop Production. *Crop Prot.* **2004,** *23,* 275–285.

Parthipan, T.; Ravi, V.; Subramanian, E. Integrated Weed Management Practices on Growth and Yield of Direct-seeded Lowland Rice. *Indian J. Weed Sci.* **2013,** *45* (1), 7–11.

Ranjit, J. D.; Suwanketnikom, R. Response of Weeds and Yield of Dry Direct Seeded Rice to Tillage and Weed Management. *Kasetsart J. (Nat. Sci.)* **2005,** *39,* 165–173.

Rao, A. N.; Nagamani, A. In *Available Technologies and Future Research Challenges for Managing Weeds in Dry-seeded Rice in India,* Proceeding of the 21st Asian Pacific Weed Science Society Conference, Colombo, Sri Lanka, Oct 2–6, 2007, 2007; pp 391–400.

Rao, A. N.; Johnson, D. E.; Sivaprasad, B.; Ladha, J. K.; Mortimer, A. M. Weed Management in Direct-seeded Rice. *Adv. Agron.* **2007,** *93,* 153–255.

Sanyal, D. Introduction to the Integrated Weed Management Revisited Symposium. *Weed Sci.* **2008,** *56* (1), 140.

Shrestha, A.; Hanson, B. D.; Fidelibus, M. W.; Alcorta, M. Growth, Phenology, and Intraspecific Competition Between Glyphosate resistant and Glyphosate susceptible Horseweeds (*Conyza canadensis*) in the San Joaquin Valley of California. *Weed Sci.* **2010,** *58,* 147–153.

Singh, G. Integrated Weed Management in Direct-seeded Rice. In *Direct Seeding of Rice and Weed Management in the Irrigated Rice-wheat Cropping System of the Indo-Gangetic Plains*; Singh, Y., Singh, V. P., Chauhan, B., Orr, A., Mortimer, A. M., Johnson, D. E., Hardy, B., Eds.; International Rice Research Institute: Los Baños, Philippines; Directorate of Experiment Station, G. B. Pant University of Agriculture and Technology: Pantnagar, India, 2008; pp 161–176.

Singh, V. P.; Singh, G.; Singh, R. K. Integrated Weed Management in Direct Seeded Spring Sown Rice Under Rainfed Low Valley Situation of Uttaranchal. *Ind. J. Weed Sci.* **2001,** *33,* 63–66.

Singh, S.; Bhushan, L.; Ladha, J. K.; Gupta, R.; Naresh, R. K.; Singh, P. P. Weed Management in Zero-till Direct Seeded Rice: Some Promising Developments. *Rice-Wheat Inf. Sheet* **2003,** *47,* 7–8.

Singh, V. P.; Singh, G.; Singh, R. K.; Kumar, A.; Dhyani, V. C.; Kumar, M.; Sharma, G. Effect of Herbicides Alone and in Combination on Direct Seeded Rice. *Indian J. Weed Sci.* **2005,** *37* (3–4), 197–201.

Singh, S.; Bhushan, L.; Ladha, J. K.; Gupta, R. K.; Rao, A. N.; Sivaprasad B. Weed Management in Dry-seeded Rice (*Oryza sativa*) Cultivated in the Furrow-irrigated Raised-bed Planting System. *Crop Prot.* **2006,** *25,* 487–495.

Singh, V. P.; Singh, G.; Singh, S.; Kumar, P.; Kumar, A.; Sharma, G.; Singh, S.; Ladha, J. K.; Gupta, R. K.; Bhushan, L.; Rao, A. N.; Sivaprasad, B.; Singh, R. P. Evaluation of Mulching, Intercropping with *Sesbania* and Herbicide Use for Weed Management in Direct Seeded Rice (*Oryza sativa* L.). *Crop prot.* **2007,** *26* (4), 518–524.

Singh, Y.; Singh, V. P.; Singh G.; Yadav, D. S.; Sinha, R. K. P.; Johnson, D. E.; Mortimer, A. M. The Implications of Land Preparation, Crop Establishment Method and Weed Management on Rice Yield Variation in the Rice–Wheat System in the Indo-Gangetic Plains. *Field Crops Res.* **2011,** *121,* 64–74.

Singh, M.; Sairam, C. V.; Hanji, M. B.; Prabhukumar, S.; Kishor, N. Crop-weed Competition and Weed Management Studies in Direct Seeded Rice (*Oryza sativa*). *Indian J. Agron.* **2012,** *57,* (1) 38–42.

Sinha, N. K.; Kumar, G.; Singh, S. J. Eco-biological Study of Weed Flora in Kharif Cereals at Bhagalpur District of Bihar. *J. Appl. Biol.* **2005,** *15* (2), 124–126.

Sinha, R. K. P.; Singh, B. K.; Kumar, M,; Mortimer, A. M.; Johnson, D. E. Effect of Seed Rate, Weed Management, and Establishment Methods on Irrigated Rice in Bihar. In *Direct Seeding of Rice and Weed Management in the Irrigated Rice–Wheat Cropping System of the Indo-Gangetic Plains*; Singh, Y., Singh, V. P., Chauhan, B., Orr, A., Mortimer, A. M., Johnson, D. E., Hardy, B., Eds.; International Rice Research Institute: Los Baños, Philippines; Directorate of Experiment Station, G. B. Pant University of Agriculture and Technology: Pantnagar, India, 2008; pp 151–158.

Solunke, P. S.; Giri, D. G.; Rathod, T. H. Effect of Integrated Nutrient Management on Growth Attributes, Yield Attributes and Yield of Basmati Rice. *Crop Res.* **2006,** *32* (3), 279–282.

Torres, R. O.; Pareek, R. P.; Ladha, J. K.; Garrity, D. P. Stem Nodulating Legumes as Relay-cropped or Intercropped Green Manures for Lowland Rice. *Field Crops Res.* **1995,** *42* (1), 39–47.

Walia, U. S.; Bhullar, M. S.; Nayyar, S.; Walia, S. S. Control of Complex Weed Flora of Dry-seeded Rice (*Oryza sativa* L.) with Pre- and Post-emergence Herbicides. *Indian J. Weed Sci.* **2008,** *40* (3–4), 161–164.

Walia, U. S.; Bhullar, M. S.; Nayyar, S.; Sidhu, A. S. Role of Seed Rate and Herbicides on Growth and Yield of Direct Dry-seeded Rice. *Ind. J. Weed Sci.* **2009,** *41* (1–2), 33–36.

Wrubel, R. P.; Gressel, Z. Are Herbicide Mixtures Useful for Delaying the Rapid Evaluation of Resistance? A Case Study. *Weed Technol.* **1994,** *8,* 635–648.

Yadav, D.; Sushant, S.; Mortimer, A. M.; Johnson, D. E. Studies on Direct Seeding of Rice, Weed Control, and Tillage Practices in the Rice–Wheat Cropping System in Eastern Uttar Pradesh. In *Direct Seeding of Rice and Weed Management in the Irrigated Rice–Wheat Cropping System of the Indo-Gangetic Plains*; Singh, Y., Singh, V. P., Chauhan, B., Orr, A., Mortimer, A. M., Johnson, D. E., Hardy, B., Eds.; International Rice Research Institute: Los Baños, Philippines; Directorate of Experiment Station, G. B. Pant University of Agriculture and Technology: Pantnagar, India, 2008; pp 131–137.

PART 4

Postharvest Processing for Rice Quality Improvement

CHAPTER 9

EFFECT OF PARBOILING ON DIFFERENT PHYSICOCHEMICAL AND COOKING PROPERTIES OF RICE

DEEPAK KUMAR VERMA[1]*, MAMTA THAKUR[2],
DIPENDRA KUMAR MAHATO[3], SUDHANSHI BILLORIA[1],
and PREM PRAKASH SRIVASTAV[1]

[1]Department of Agricultural and Food Engineering, Indian Institute of Technology, Kharagpur 721302, West Bengal, India

[2]Department of Food Engineering and Technology, Sant Longowal Institute of Engineering and Technology, Longowal 148106, Punjab, India

[3]Indian Agricultural Research Institute, Pusa Campus, New Delhi 110012, India

*Corresponding author.
E-mail: deepak.verma@agfe.iitkgp.ernet.in; rajadkv@rediffmail.com

ABSTRACT

In Asia, the most of energy needs are fulfilled by the consumption of rice which is taken in different forms in several parts of Asia. The nutritional value of rice is greatly affected by processing conditions, degree of milling, variety, and growing environment. The quality rice can be obtained by the process of parboiling which is an energy- and labour-intensive hydrothermal premilling treatment commonly employed in rice. This process involves the conditioning, cooking, and drying of rice so that the starch can be partially gelatinized to seal the cracks to increase the milling yield. The effect of parboiling varies as per the intrinsic properties of grain in addition to the processing conditions. The parboiling increases the hardness of rice thus decreasing the pasting properties and making it easy to digest. Further, the parboiling is effective in increasing the shelf life of paddy rice with minimal changes on its nutritional quality. This chapter therefore discusses the detailed process of parboiling and its effect on the physicochemical and cooking characteristics of rice.

9.1   INTRODUCTION

Rice (*Oryza sativa* L.) being consumed by two-thirds of the world's population contributes to 75% of the daily calorie intake in Asia (FAO, 2009; Wynn, 2009; Verma and Srivastav, 2017). Rice processing comprise several operations to convert it into a desired product (Roberts, 1979). The consumption of rice depends on the consumers from different regions. Rice is preferred variably in different parts of the world. For example, Japanese consume mostly well-milled sticky rice (Deshpande et al., 1982). Americans on the other hand choose semimilled long-grain rice, whereas people in the Indian subcontinent prefer parboiled rice (Lyon et al., 1999). Rice has nutritious value and the processing conditions greatly influence the composition. The degree of milling has significant effects on the nutritional composition (Juliano and Bechtel, 1985). The germ and outer layer of the starchy endosperm are the places in rice grains, whereas nutritional constituents', namely, proteins, fats, vitamins, and minerals are concentrated (Juliano and Bechtel, 1985; Itani et al., 2002). The variation of these components depends upon the growing environments, rice varieties, cultivation methods, and processing conditions. Rice is harvested as paddy rice, which consists of the hull, including the lemma, the palea, the larger lemma, and the caryopsis as shown in Figure 9.1. The composition of rice varies greatly depending on the variety and growing environment, milled rice is composed of approximately 77.6% starch, 6.3–7.1% protein, 0.3–0.5% crude fat, 0.3–0.8% ash, and 0.2–0.5% crude fiber at 14% moisture (Juliano and Bechtel, 1985). The rice bran contains 12–15% protein, 15–20% lipid, and 1% starch. It also constitutes 70% of crude fiber, 51% ash, 65% thiamin, 39% riboflavin, and 54% niacin in the brown rice (Champagne et al., 2004).

Rice is a rice source of starch. Starch consists of linear fraction of amylose and a highly branched fraction of amylopectin. Typically, the amylose contents in the rice are in the range of 0–25% such as waxy (0–2%), short grain (15–23%), medium grain (14–18%), and long grain (20–25%) (Mitchell, 2009; Verma et al., 2012, 2013, 2015). Starch is semicrystalline in nature with alternating crystalline and amorphous structures. The crystalline structure is composed of amylopectin short chains forming parallel double helices with interhelical water present (Gidley, 1987); the amorphous structure consists of the branching regions. When viewed in polarized light, starch granules reveal a birefringent cross known as the maltese cross as a result of the semicrystalline structure (Whistler and BeMiller, 1997).

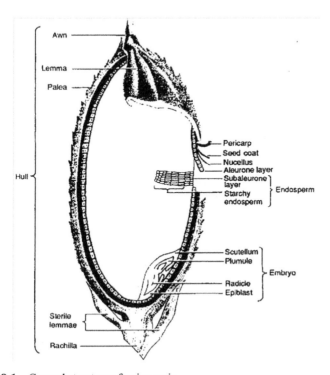

FIGURE 9.1   General structure of a rice grain.

Source: Juliano, B. O. Grain Structure, Composition and Consumers' Criteria for Quality. In *Rice in Human Nutrition*; International Rice Research Institute (IRRI): Philippines and Food and Agriculture Organization of the United Nations: Rome, Italy, 1993; pp 35–59. http://www.fao.org/docrep/t0567e/t0567e00.htm (accessed Jan 09, 2017).

9.2   PARBOILING AND PARBOILING PROCESS

Parboiling of rice consists of several steps such as soaking, steaming, drying, etc. The parboiled rice is mostly preferred in Asian countries such as India, Nepal, Bangladesh, and Sri Lanka. The parboiling can be done either by traditional or modern methods (Rahaman et al., 1996). The parboiling technique has become the vital part of food-processing industries in the developing countries of Indian subcontinent (Bhattacharya, 1985). The major change during parboiling is starch gelatinization (Bhattacharya, 1985; Itoh and Kawamura, 1985; Kimura, 1991; Kimura et al., 1993, 1995; Islam et al., 2001) along with changes in milling, cooking, and eating quality of rice. These qualities are significantly affected by the extent of parboiling treatment. Rice develops a dark color during parboiling at higher temperature, so a low temperature (80–100°C) is preferable to avoid this defect (Bhattacharya, 1985; Kimura et al., 1993). The other factors responsible for the quality of rice are processing equipment and various conditions of operation. Figure 9.2 depicts an overview of local parboiling processes and Figure 9.3 depicts all the major steps involved during the parboiling process such as soaking, steaming, and drying described by Roy et al. (2006).

9.2.1 ADVANTAGE OF PARBOILING PROCESS

In literature, there are various reports available that claim the advantages for parboiled rice over raw milled rice (Pillaiyar, 1990; Sujatha et al. 2004); they are as follows:

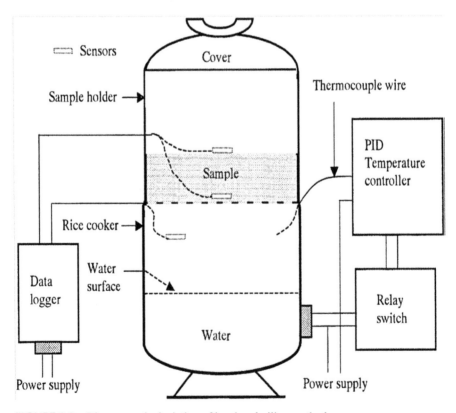

FIGURE 9.2 Diagrammatic depiction of local parboiling methods.

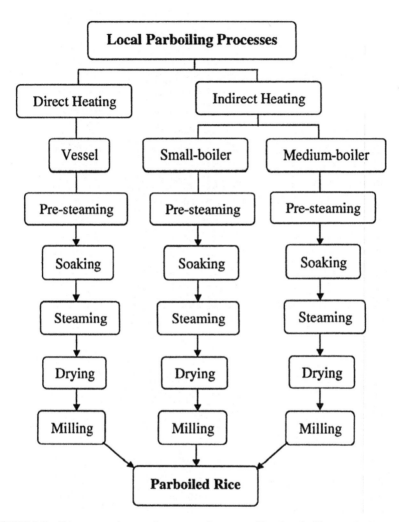

FIGURE 9.3 Diagrammatic overviews on major steps of local parboiling methods.

Source: Reprinted from Roy, P.; Shimizu, N.; Shiina, T.; Kimura, T. Energy Consumption and Cost Analysis of Local Parboiling Processes. *J. Food Eng.* **2006,** *76*, 646–655. © 2006, with permission from Elsevier.

- A better and higher recovery of head rice yield (HRY) as whole grains during milling and resistant to breakage resulting lower broken rice
- Easier shelling (removal of the hull) during milling
- Better retention and improved nutrient availability (particularly vitamins)

- Improved digestibility with high protein efficiency ratio
- Stabilization of the higher oil content in the bran resulting in lower glycemic index
- Change in taste, a translucent hard grain, and firmed cooked rice texture
- Decreased washing, less starch, and solid losses during cooking in the cooking water
- Better grain swelling during cooking and more swelling received when cooked to the desired softness
- Inactivation of enzymes and biological sanitation
- Decreased susceptibility and increase resentence against insect-pest attack during storage

9.3   PHYSICAL PROPERTIES OF PARBOILED RICE

Parboiling is a type of hydrothermal process. The crystalline starch is transformed into an amorphous because of irreversible swelling and fusion of starch. The parboiling process enhances different physicochemical properties of the rice. The indica varieties possess long, slender, and flat grains which facilitate the heat and water entry into the endosperm (Webb, 1975). The milled kernels of long-grain varieties are called "hard rice." There is little splitting with good cooked grains which remain separate. Besides this, there is high amylase content and high gelatinizing temperature. Therefore, long-grain rice is used for canned soups and quick cooked products. On the other hand, the short-grain varieties become soft, firm, and sticky after cooking. They possess low amylase content and low gelatinizing temperature. Therefore, they are used for making dry cereals and baby foods. It is a premilling treatment given to paddy prior to its milling to achieve maximum recovery of head rice and to minimize breakages. The breakage can be reduced by gelatinizing the starch. The major physical properties of parboiled rice are discussed briefly in the following sections.

9.3.1   KERNEL DIMENSIONS

The rice grains are categorized into long, medium, and short with respect to their length to width ratio. The values of length to width ratio for short, medium, and long brown rice are ≤ 2, 2.1–3, and ≥ 3.1, respectively, whereas for milled rice, the values are ≤ 1.9, 2–2.9, and ≥ 3 (Bergman et al., 2004; Shinde et al., 2014).

9.3.2 MILLING QUALITY

The milling quality is determined based on the total milled rice yield and the portion of whole kernels which is referred as HRY. These yields depend upon the variety of paddy as well as the milling process employed (Bergman et al., 2004).

9.3.3 CHALKINESS

The opacity nature of milled rice is referred to as chalkiness which usually disappears after cooking and negatively affects the quality of milled rice (IRRI, 2012).

9.4 PROGRESS AND DEVELOPMENT IN PARBOILING OF RICE

The parboiling of rice has been in practice since ancient time. Initially clay pots were employed for the purpose (Joachim, 2011). It is still widely practiced in Asian and European countries (Pillaiyar, 1981; Bhattacharya, 2004; Roy et al., 2007). It is adopted as a culture irrespective of caste and class (Joachim, 2011). In Asia, it is mainly employed for the purpose of better storage properties and special taste, whereas in North America and Europe, it is preferred for better cooking properties and fewer breakage of rice grains during the milling process. The scientific reasons and principles behind the parboiling are to attain sterile and nutritious rice by removal of microbes and some poisonous substance. The parboiling has been reported to reduce the breakages of grains during the milling. The nutritional composition including fat, protein, and amylose contents of the rice decreases while water uptake and thiamine contents show a sharp decrease upon parboiling (Ibukun, 2008). The initial washing of paddy rice removes various foreign materials such as clay and microbes attached to the surface of paddy rice. The soaking process assists the water absorption and dilution of toxins such as aflatoxins during storage of old stock. The soaking process also causes enzymes and pigments transformation as well as their movement into the endosperm (Ayamdoo et al., 2013). The steaming (at 80°C) facilitates killing of pathogens remained in the rice. The steaming under high pressure causes vitamins such as thiamine to enter into the endosperm. The process also helps deactivate the enzymes and make them easily available (Ayamdoo et al., 2013). With the increasing demand for parboiled rice in Africa and other parts of the world, the parboiling process has been modernized with scientific equipments to meet the consumption demand of the people. The different drying techniques such as fluidized bed drying (FBD) (Soponronnarit et al., 2006) and ohmic heating (Sivashanmugam and Arivazhagan, 2008) have replaced the traditional heating process. The parboiled rice contributes 15% of the world's milled rice and the demands are increasing day by day (Bhattacharya, 2004). The fabricated laboratory-scale parboiling setup has been designed to meet the increasing demands as well as for obtaining the better quality of parboiled rice (Islam et al., 2002a).

9.4.1 EFFECT OF PARBOILING ON DIFFERENT PROPERTIES OF RICE: CASE STUDY

Parboiling process significantly affects the different physicochemical properties of rice. They are discussed as following.

9.4.1.1 EFFECTS ON PHYSICAL PROPERTIES OF RICE

The parboiling process has prominent effects on the physical properties of rice. In order to study this, the effect of traditional, intermediate, and improved parboilers on certain physical parameters and cooking quality of rice, namely, NERICA 4 and Gambiaka grown and consumed in Benin were taken. The observation showed that the highest level of heat-damaged grains, that is, 90% was reduced to 17% by the use of improved equipment. The improved and intermediate parboiling technology produced hardness of 4 and 6 kg for Gambiaka and NERICA 4 variety, respectively, whereas traditional methods showed hardness of 4 and 3 kg, respectively, for Gambiaka and NERICA 4 (Fofana et al., 2011). The physical properties such as viscosity, hardness, volume expansion ratio, and solid content depend upon the conditions of parboiling treatments (Islam et al., 2001).

The appearance of grain is judged based on the opacity of the endosperm which depends on chalkiness. The chalky appearance is due to numerous air spaces present between loosely packed starch granules and its interaction with the light (Tashiro and Wardlaw, 1991). The prominent reason behind the chalkiness is the exposure of high temperatures during the ripening period (Tashiro and Wardlaw, 1991). Several studies have reported that the temperature and period of soaking and steaming influence whiteness of parboiled rice (Bhattacharya, 1996; Kimura et al., 1993); while color changes of brown rice is due to Maillard nonenzymatic browning reaction (Pillaiyar and Mohandas, 1981). The color of parboiled rice is due to leaching out of the nutrients from the bran (Framlingham and Anthony, 1996). The physical and functional properties of rice cultivars, namely, SR-1, SKAU-382, SKAU-345, Pusa-3, Koshar, K-332, and Jehlum were studied at soaking temperatures of 60°C, 70°C, and 80°C. They were then compared with the brown raw rice of the respective cultivars. The observation showed significant increase in hardness. The highest was reported in Jehlum but the data varied among the other cultivars. The pasting property decreased with the increase in temperature from 60°C to 80°C showing typical behavior such as high initial viscosity and lower peak viscosity with respect to raw rice. Besides this, water absorption index and water solubility indices increased upon increasing the soaking temperature (Mir and Bosco, 2013).

The discoloration is caused by Maillard reaction and the movement of hull and bran pigments into the endosperm during soaking step of parboiling process (Houston et al., 1956; Bhattacharya, 2004; Rordprapat et al., 2005; Lamberts et al., 2006; Parnsahkorn and Langkapin, 2013). It has been observed that Wells variety becomes more yellow than Jupiter upon parboiling and enhances yellowness of germinated Wells as compared to the germinated Jupiter. The yellowness of germinated Jasmine brown rice was due to the Maillard reaction as reported by Cheevitsopon and Noomhorm (2011, 2015) but Han (2015) contradicts and suggests that Maillard reaction negatively affected the whiteness of parboiled germinated rice with nonsignificant effects on its yellowness.

9.4.1.2   EFFECTS ON MILLING PROPERTIES OF RICE

Parboiling has significant effect on the milling properties of rice. It has been observed that parboiling reduces the number of broken rice in brown rice of Jupiter and Wells varieties (Cheevitsopon and Noomhorm, 2011). The broken rice declined by 12–17% in Wells as compared to 4% in Jupiter. According to the Federal Grain Inspection Service, Jupiter and Wells possess different kernel dimensions which are within the length-to-width ratio limitations prescribed for rough and brown rice cultivars. This reduction in number of broken rice is observed because of starch gelatinization (Bhattacharya, 2004). Therefore, parboiling greatly enhances the milling properties of rice.

9.4.1.3   EFFECTS ON CHEMICAL AND NUTRITIONAL PROPERTIES OF RICE

Parboiling of rice is effective in increasing the storage life of rice without much effect on its nutritional properties (Heinemann et al., 2006; Buggenhout et al., 2013; Min et al., 2014; Oli et al., 2014). Several chemical changes occur during parboiling which ultimately changes its nutritional properties. The major changes include movement of bran components into the rice caryopsis; inactivation of lipases, and starch gelatinization (Demont et al., 2012). Parboiling of unpolished black rice significantly reduced the fat content in comparison to nonparboiled unpolished black rice. On the other hand, parboiled unpolished red rice has similar fat content to that of nonparboiled unpolished red rice. The significant reduction in the fat content in the black rice is due to the detachment of oil bodies and their leaching out during hydration of rice at 60°C. Similarly, a decrease in ash content by 56.7% in the red rice was observed after parboiling treatment indicating huge amount of ash content in the bran layers of red rice caryopsis leaching out during the hydration at 60°C (Paiva et al., 2016). Contrary to this, Paiva et al. (2014) reported that the ash content in MPB-10 red rice was concentrated in the outer layers of the bran but homogeneously distributed in IAC-600 black rice cultivar. This is why there is higher reduction in ash content after parboiling. There is highest carbohydrate content in nonparboiled polished black rice and parboiled polished black rice as the protein and ash content in them are most affected as compared to those in red rice (Paiva et al., 2016). In another study conducted by Raghavendra Rao and Juliano (1970) with rough rice of seven varieties, it was observed that the least protein was extracted from parboiled rice. The starch of the parboiled rice has lower solubility as compared to that of raw rice. The variation in amylograph properties on parboiling is dependent on amylose content present.

9.4.1.4   EFFECTS ON COOKING PROPERTIES OF RICE

The appearance is vital part in cooking properties of rice. It gives first impression to the consumers. The appearance is based on the translucent character of the endoderm which is adverse to chalkiness. The chalkiness usually gets removed upon parboiling and cooking, so imparting no significant effects on cooking and eating qualities. However, higher levels of chalkiness decrease the physical quality and reduce milling recovery as well (Khush et al., 1979; Bhattacharya and Indudhara, 1982; Adu-Kwarten et al., 2003; Gayin et al., 2009). NERICA 4 and Gambiaka having chalkiness value of 5 have been reduced to 1 by parboiling methods. At the same time, the translucency of the endoderm is enhanced during parboiling because of pregelatinization of starch (Kondo, 2006). According to Adu-Kwarten et al. (2003) chalkiness greatly influences consumer preference for parboiled rice. The effect of parboiling on Gambiaka grain translucency showed prominent effects of traditional method over the other methods and also, similar result was observed for NERICA 4 by Fofana et al. (2011).

Gel consistency, on the other hand, enhances the tenderness property of the cooked rice (Cagampang et al., 1973). The softness increases with the increase in gel consistency and vice versa. Gel consistency was observed to be higher in nonparboiled samples as compared to parboiled samples of NERICA 1, 2, and 4. The gel consistency reduced after sixth week of storage in both parboiled and nonparboiled samples (Roseline et al., 2010). It has been observed that the water uptake ratio, volumetric expansion, and the amount of leached materials during cooking are decreased because of parboiling. The extent of decrease is dependent on the type of variety, parboiling condition, and their interactions (Patindol et al., 2008). Cooked rice hardness was seen to increase from 60.9 to 73.7 N for nonparboiled and from 62 to 81.8 N for parboiled samples. Similarly, stickiness increased from 2.9 to 3.7 N for nonparboiled and from 1.9 to 3.6 N for the parboiled samples (Rao and Juliano, 1970; Ali and Bhattacharya, 1980; Kato et al., 1983; Biswas and Juliano, 1988; Ong and Blanshard, 1995). The hardness of the cooked rice increased due to reassociation of gelatinized starch during parboiling (Ali and Bhattacharya, 1980; Biswas and Juliano, 1988; Ong and Blanshard, 1995). Besides this, Derycke et al. (2005) reported that a protein barrier may form during parboiling which restricts the leaching out of solids and increase in hardness during cooking of rice.

9.4.1.5   EFFECT ON THERMAL PROPERTIES OF RICE

The steaming operation of parboiling requires a lot of energy to produce steam. Gelatinization temperatures for Bisalayi, FARO 61, FARO 60, FARO 52, and FARO 44 range from 72.78–83.78°C, 73.90–82.55°C, 73.16–80.98°C, 64.24–77.70°C, and 64.37–75.26°C, respectively, whereas gelatinization enthalpy for the respective varieties are 5.76, 3.62, 3.15, 1.55, and 2.35 J/g. The rice varieties from West Africa have low gelatinization enthalpy (Traore et al., 2011; Odenigbo et al., 2013). During steaming of different varieties of paddies at 100°C for 5–20 min, there was no residual enthalpy of gelatinization of the parboiled samples at the different steaming times. The reason for this is low gelatinization enthalpy (1.55–5.76 J/g) of the raw parboiled rice (Ejebe et al., 2015). In a study conducted by Marshall et al. (1993) and Islam et al. (2002b) for the effect of parboiling conditions on gelatinization properties, a decrease in gelatinization enthalpy with severity of parboiling and its relation to the degree of starch gelatinization during heat processing was observed. Then, Marshall et al. (1993) proposed that rice starch only requires 40% gelatinization for maximum HRY to be achieved during parboiling. Besides this, Marshall et al. (1993) showed that the enthalpy of Lemont rice variety decreased from 14.6 to 2.2 J/g during conventional parboiling of rough rice in a pressure cooker for 8 min at 121°C and 85.6% of the rice were gelatinized during steaming process. The soaking temperature for rice has inverse relation with the enthalpy indicating that the degree of initial hydration of brown rice during soaking affects gelatinization properties of cooked rice (Han and Lim, 2009). It has been observed that the gelatinization temperature generally shifts to higher values and gelatinization enthalpy decreases as a result of parboiling. Onset gelatinization temperature ranges from 71°C to 75.8°C and 72.8°C to 81.6°C for the nonparboiled and parboiled head rice samples, respectively. Similarly, gelatinization enthalpy ranges from 9.8 to 10.6 J/g and 5.8 to 8.8 J/g for the nonparboiled and parboiled samples, respectively (Patindol et al., 2008). These results were also verified and are in agreements with the findings of Biliaderis et al. (1993); Ong and Blanshard (1995); Islam et al. (2002b); Lamberts et al. (2006); and Manful et al. (2008).

9.4.1.6   *REACTION KINETICS OF PARBOILING TREATMENT IN RICE*

Parboiling helps retain the nutritional properties and extends the storage life of paddy rice. Pressure parboiling employed in processing of rice can develop darkening of the finished products (Luh and Mickus, 1991). The color index is used to measure the extent of parboiling process (Johnson, 1965; Stermer, 1968). Bhattacharya and Rao (1966) developed the color of raw and parboiled rice by the variation in the soaking and steaming conditions during parboiling. On the other hand, pH of the soaking water has significant effect on the color development during parboiling (Jayanarayanan, 1964). In addition, kinetic parameters provide an understanding of the changes taking place during the parboiling (Arabshahi and Lund, 1985; Rao, 1986; Holdsworth, 1990). The information obtained can be employed for the new process design and optimization, scale-up of the fabricated equipment and for controlling the process and quality parameters of the finished product. The color development during parboiling has been reported to follow zero-order kinetics. Pressure, temperature, and time combination of steaming have a marked effect on color kinetics. Darkening of the sample accompanied by development of yellow color is prominent when paddy rice is parboiled for a longer time. Bhattacharya (1996) suggested that high pressures (up to 304 kPa) could be used but steaming time should be kept low in order to avoid sharp color development.

The effect of infrared (IR) and hot air (HA) drying conditions on drying kinetics of Leb Nok Pattani (LNP) rice and Suphanburi 1 (SP 1) parboiled rice and their qualities were studied by Tirawanichakul et al. (2012). It was observed that HA and IR parboiled rice drying could maintain high HRY and IR drying with 1.5 kW provided the highest HRY value. Despite this, the qualities analysis showed that whiteness, water absorption, cooking time, and pasting property were significantly different compared to reference samples. The drying techniques such as HA drying, IR drying (Delwiche et al., 1996; Das et al., 2003; Laohavanich and Wongpichet, 2008), and microwave (MW) drying (Therdthai and Zhou, 2009) are employed to reduce moisture content. IR technique is usually employed to reduce the drying time and enhance the bioproduct quality (Sandu, 1986; Ratti and Mujumdar, 1995; Paakkonen et al., 1999; Hebbar and Rastogi, 2001; Mongpraneet et al., 2002; Kian and Siaw, 2005). The use of IR along with FBD on paddy moisture reduction and milling quality could maintain the physical quality of rice grain kernels and reduce the specific energy consumption compared to HA drying (Meeso et al., 2004). With regards to cooking kinetics, it was observed that gelatinization of kernels of nonparboiled milled rice increased in a sigmoid pattern and the gelatinization process was completed in about 22 min (Billiris et al., 2012). The result was in agreement with the trends of Lucisano et al. (2009). Gelatinization kinetics of milled rice showed surface lipid content (SLC) from 0.15% to 0.55%. This suggested that milling nonparboiled rice to a lesser degree (up to 0.55% SLC) had no significant effect on gelatinization kinetics (Billiris et al., 2012).

9.5 SUMMARY AND CONCLUSION

The parboiling of rice has been in practice since ancient time. It has many advantages for health over the consumption of nonparboiled rice. The parboiling of rice increases its hardness while decreases the pasting properties, so it is easy to digest. The water absorption index and water solubility indices are higher in parboiled rice which even increases with the soaking temperature. This also decreases the number of broken rice as compared to the nonparboiled rice. This reduction in broken rice is due to starch gelatinization that seal fissures present in paddy rice during soaking and germination. Along with this, a slight decrease in the fat content of black rice is observed which is due to the detachment of oil bodies and also a decrease in ash content. Conclusively, parboiling is effective in increasing the shelf life of paddy rice with minimal changes on its nutritional quality.

ACKNOWLEDGMENTS

Authors are indebted to Department of Science and Technology, Ministry of Science and Technology, Government of India for an individual research fellowship (INSPIRE Fellowship Code No.: IF120725; Sanction Order No. DST/INSPIRE Fellowship/2012/686 and Date: Feb 25, 2013).

KEYWORDS

- parboiled rice
- cooking properties
- pasting property
- milling properties
- chalky appearance
- thermal properties
- soaking process

REFERENCES

Adu-Kwarten, E.; Ellis, W. O.; Oduro I.; Manful, J. T. Rice Grain Quality: A Comparison of Local Varieties with New Varieties Under Study in Ghana. *Food Control* **2003**, *14*, 507–514.

Ali, S. Z.; Bhattacharya, K. R. Pasting Behavior of Parboiled Rice. *J. Texture Stud.* **1980**, *11*, 239–245.

Arabshahi, A.; Lund, D. B. Consideration in Calculating Kinetic Parameters from Experimental Data. *J. Food Proc. Eng.* **1985**, *7, 239–251.*

Ayamdoo, J. A.: Demuyakor, B.: Dogbe, W.; Owusu, R. Parboiling of Paddy Rice, the Science and Perceptions of It as Practiced in Northern Ghana. *Int. J. Sci. Technol. Res.* **2013**, *2*(4). ISSN 2277-8616.

Bergman, C. J.; Bhattacharya, K. R.; Ohtsubo, K. Rice End-use Quality Analysis. In *Rice: Chemistry and Technology*, 3rd ed.; Champagne, E. T., Ed.; AACC: St. Paul, USA, 2004; pp 415–422.

Bhattacharya, K. R. Parboiling of Rice. In *Rice Chemistry and Technology*; Juliano, B. O., Ed.; American Association of Cereal Chemists, Inc.: St. Paul, Minnesota, 1985; pp 289–348.

Bhattacharya, S. Kinetics on Color Changes in Rice Due to Parboiling. *J. Food Eng.* **1996,** *29*, 99–106.

Bhattacharya, K. R. Parboiling of Rice. In *Rice Chemistry and Technology*; Champagne, N. E. T., Ed.; American Association of Cereal Chemists, Inc.: St. Paul, Minnesota, 2004; pp 329–404.

Bhattacharya, K. R.; Subba Rao, P. V. Effect of Processing Conditions on Quality of Parboiled Rice. *J. Agric. Food Chem.* **1966,** *14*, 476–479.

Bhattacharya, K. R.; Sowbhagya, C. M.; Indudhara Swamy, Y. M. Quality of Profiles of Rice: A Tentative Scheme for Classification. *J. Food Sci.***1982,** *47*, 564.

Biliaderis, C. G.; Tonogai, J. R.; Perez, C. M.; Juliano, B. O. Thermophysical Properties of Milled Starch as Influenced by Variety and Parboiling. *Cereal Chem.* **1993,** *70*, 512–516.

Billiris, M. A.; Siebenmorgen, T. J.; Meullenet, J. F.; Mauromoustakos, A. Rice Degree of Milling Effects on Hydration, Texture, Sensory and Energy Characteristics. Part 1. Cooking Using Excess Water. *J. Food Eng.* **2012,** *113*, 559–568.

Biswas, S. K.; Juliano, B. O. Laboratory Parboiling Procedures and Properties of Parboiled Rice from Varieties Differing in Starch Properties. *Cereal Chem.* **1988,** *65*, 417–23.

Buggenhout, J.; Brijs, K.; Celus, I.; Delcour, J. A. The Breakage Susceptibility of Raw and Parboiled Rice: A Review. *J. Food Eng.* **2013,** *117*, 304–315.

Cagampang, G. B.; Perez, C. M.; Juliano, B. O. A Gel Consistency Test for Eating Quality of Rice *J. Sci. Food Agric.* **1973,** *24*, 1589–1594.

Champagne, E. T.; Wood, D. F.; Juliano, B. O.; Bechtel, D. B. The Rice Grain and Its Gross Composition. *Rice Chem. Technol.* **2004,** *3*, 77–100.

Cheevitsopon, E.; Noomhorm, A. Effects of Parboiling and Fluidized Bed Drying on the Physicochemical Properties of Germinated Brown Rice. *Int. J. Food Sci. Technol.* **2011,** *46*, 2498–2504.

Cheevitsopon, E.; Noomhorm, A. Effects of Superheated Steam Fluidized Bed Drying on the Quality of Parboiled Germinated Brown Rice. *J. Food Process. Pres.* **2015,** *39*, 349–356.

Das, I.; Das, S. K.; Bal, S. Specific Energy and Quality Aspects of Infrared (IR) Dried Parboiled Rice. *J. Food Eng.* **2003,** *62*, 9–14.

Delwiche, S. R.; McKenzie, K. S.; Webb, B. D. Quality Characteristics in Rice by Near-infrared Reflectance Analysis of Whole-grain Milled Samples. *Cereal Chem.* **1996,** *73* (2), 257–263.

Demont, M.; Zossou, E.; Rutsaert, P.; Ndour, M.; Mele, P. V.; Verbeke, W. Consumer Valuation of Improved Rice Parboiling Techniques in Benin. *Food Qual. Prefer.* **2012,** *23*, 63–70.

Derycke, C.; Vandeputte, G. E.; Vermeylen, R.; De Man, W.; Goderis, B.; Koch, M. H. J.; Delcour, J. A. Starch Gelatinization and Amylose–Lipid Interactions During Rice Parboiling Investigated by Temperature Resolved Wide Angle X-ray Scattering and Differential Scanning Calorimetry. *J. Cereal Sci.* **2005,** *42*, 334–43.

Deshpande, S. S.; Bhattacharya, K. R. The Texture of Cooked Rice. *J. Texture Stud.* **1982,** *13*, 31–42.

Ejebe, C.; Danbaba, N.; Ngadi, M. Effect of Steaming on Physical and Thermal Properties of Parboiled Rice. *Eur. Int. J. Sci. Technol.* **2015,** *4*, 71–80.

FAO (Food and Agricultural Organization). *Rice in the World*, Report of the Fifth External Programme and Management Review of the International Plant Genetic Resources

Institute (IPGRI); FAO: Rome, Italy, 2001. http://www.fao.org/wairdocs/tac/x5801e/x5801e08.htm (accessed Dec 11, 2009).

Fofana, M.; Wanvoeke, J.; Manful, J.; Futakuchi, K.; Van Mele, P.; Zossou, E.; Bleoussi, T. M. R. Effect of Improved Parboiling Methods on the Physical and Cooked Grain Characteristics of Rice Varieties in Benin. *Int. Food Res. J.* **2011**, *18*, 715–721.

Framlingham, N.; Anthony, R. S. Studies on the Soak Water Characteristics in Various in Paddy Parboiling Methods. *Bioresour. Technol.* **1996**, *55*, 259–261.

Gayin, J.; Manful, J. T.; Johnson, P. N. T. Rheological and Sensory Properties of Rice Varieties from Improvement Programme in Ghana. *Int. Food Res. J.* **2009**, *16*, 167–174.

Gidley, M. J. Factors Affecting the Crystalline Type (A—C) of Native Starches and Model Compounds: A Rationalisation of Observed Effects in Terms of Polymorphic Structures. *Carbohydr. Res.* **1987**, *161* (2), 301–304.

Han, A. Effect of Germination and Parboiling on Milling, Physicochemical, and Textural Properties of Medium- and Long-grain Rough Rice. Theses and Dissertations, University of Arkansas, Fayetteville, 2015.

Han, J. A.; Lim, S. T. Effect of Presoaking on Textural, Thermal, and Digestive Properties of Cooked Brown Rice. *Cereal Chem.* **2009**, *86* (1), 100–105.

Hebbar, H. U.; Rastogi, N. K. Mass Transfer During IR Drying of Cashew Kernel. *J. Food Eng.* **2001**, *47*, 1–5.

Heinemann, R. J. B.; Behrens, J. H.; Lanfer-Marquez, U. M. A Study on the Acceptability and Consumer Attitude Towards Parboiled Rice. *Int. J. Food Sci. Technol.* **2006**, *41*, 627–634.

Holdsworth, S. D. Kinetic Data—What Is Available and What Is Necessary. In *Processing and Quality of Foods. High Temperature Short Time (HTST) Processing*; P. Zeuthen et al., Ed.; Elsevier Applied Science: London, 1990; Vol. 1, pp 74–90.

Houston, D. F.; Hunter, I. R.; Kester, E. B. Cereal Storage Effects, Storage Changes in Parboiled Rice. *J. Agric. Food Chem.* **1956**, *4*, 964–968.

Ibukun E. O. Effect of Prolonged Parboiling Duration on Proximate Composition of Rice. Department of Biochemistry, Federal University of Technology: P. M. B. 704, Akure, Nigeria. *Sci. Res. Essay* **2008**, *3* (7), 323–325.

IRRI, Rice Knowledge Bank-postproduction Course. International Rice Research Institute. http://www.knowledgebank.irri.org/postproductioncourse/ (accessed Aug 21, 2012).

Islam, M. R.; Shimizu, N.; Kimura, T. Quality Evaluation of Parboiled Rice with Physical Properties. *Food Sci. Technol. Res.* **2001**, *7*, 57–63.

Islam, R.; Roy, P.; Shimizu, N.; Kimura, T. Effect of Processing Conditions on Physical Properties of Parboiled Rice. *Food Sci. Technol. Res.* **2002a**, *8* (2), 106–112.

Islam, M. R.; Shimizu, N.; Kimura, T. Effect of Processing Conditions on Thermal Properties of Parboiled Rice. *J. Food Sci. Technol. Res.* **2002b**, *8* (2), 131–136.

Itani, T.; Tamaki, M.; Arai, E.; Horino, T. Distribution of Amylase, Nitrogen, and Minerals in Rice Kernels with Various Characters. *J. Agric. Food Chem.* 2002, *50*, 5326–5332.

Itoh, K.; Kawamura, S. Processing Conditions of Parboiled Rice and Its Qualities (Studies on Parboiled Rice, Part I). *Nippon Shokuhin Kogyo Gakkaishi* **1985**, *32*, 471–479 (in Japanese).

Itoh, K.; Kawamaura, S. Milling Characteristics of Parboiled Rice and Properties of the Milled Rice—Studies on Parboliled Rice. *J. Jpn. Soc. Food Sci. Technol.* **1991**, *38*, 776–783.

Jayanarayanan, E. K. Der Einfluss der Verarbeitungsbedingungen auf das Braunwerden von 'parboiled' Reis. *Nahrung* **1964**, *8, 129–137*.

Joachim, S. *Parboiling in Thailand and the World*. Kasetsart University, Bangkok, Thailand, 2011.

Johnson, R. M. Light Reflectance Meter Measures Degree of Milling and Parboiling of Parboiled Rice. *Cereal Chem.* **1965**, *42, 162–174*.

Juliano B. Criteria and Test for Rice Grain Quality. In *Rice Chemistry and Technology*; Juliano, B., Ed.; American Association of Cereal Chemists, Inc.: St. Paul, Minnesota, 1985; pp 443–513.

Juliano, B.; Bechtel, D. The Rice Grain and Its Gross Composition. In *Rice Chemistry and Technology*; Juliano, B., Ed.; American Association of Cereal Chemists, Inc.: St. Paul, Minnesota, 1985; pp 17–50.

Kato, H.; Ohta, T.; Tsugita, T.; Hosaka, Y. Effect of Parboiling on Texture and Flavor Components of Cooked Rice. *J. Agric. Food Chem.* **1983**, *31*, 818–23.

Khush, G. S.; Paule, C. M.; De la Cruz, N. M. In *Rice Grain Quality Evaluation and Improvement*, Proceedings of Workshop on Chemical Aspects of Rice Grain Quality, Manila, Philippines, 1979; International Rice Research Institute: Manila, Philippines, 1979; pp 21–31.

Kian, J. C.; Siaw, K. C. A Comparative Study Between Intermittent Microwave and Infrared Drying of Bioproducts. *Int. J. Food Sci. Technol.* **2005**, *40*, 23–39.

Kimura, T. Effects of Processing Conditions on the Hardening Characteristics of Parboiled Grain. *J. Jpn. Soc. Agric. Struct.* **1991**, *22*, 111–116.

Kimura, T.; Bhattacharya, K. R.; Ali, S. Z. Discoloration Characteristics of Rice During Parboiling (I): Effect of Processing Conditions on the Color Intensity of Parboiled Rice. *J. Soc. Agric. Struct. Jpn.* **1993**, *24*, 23–30.

Kimura, T.; Shimizu, N.; Shimohara, T.; Warashina, J. Trials of Quality Evaluation for Parboiled and Other Rice by Means of the near Infrared Spectroscopy and the Rapid Visco Analyser. *J. Jpn. Soc. Agric. Struct.* **1995**, *25*, 175–182.

Kondo, M. Analysis on the Effects of Air Temperature During Ripening and Grain Protein Contents on Grain Chalkiness in Rice. *Jpn. J. Crop Sci.* **2006**, *75,* 234–235.

Lamberts, L.; Brijs, K.; Mohamed, R.; Verhelst, N.; Delcour, J. A. Impact of Browning Reactions and Bran Pigments on Color of Parboiled Rice. *J. Agric. Food Chem.* **2006**, *54*, 9924–9929.

Laohavanich, J.; Wongpichet, S. Thin Layer Drying Model for Gas-fired Infrared Drying of Paddy. *Songklanakarind J. Sci. Technol.* **2008**, *30* (3), 343–348.

Lucisano, M.; Mariotti, M.; Pagani, M. A.; Bottega, G.; Fongaro, L. Cooking and Textural Properties of Some Traditional and Aromatic Rice Cultivars. *Cereal Chem.* **2009**, *86* (5), 542–548.

Luh, B. S.; Mickus, R. R. Parboiled Rice. In *Rice Utilization,* 2nd ed.; Luh, B. S., Ed.; Van Nostrand Reinhold: New York, NY, 1991; Vol. II, pp 51–88.

Lyon, B. G.; Champagne, E. T.; Vinyard, B. T.; Windham, W. R.; Barton, F. E. II.; Webb, B. D.; McClung, A. N.; Moldenhauer, K. A.; Linscombe, S.; McKenzie, K. S., et al. Effects of Degree of Milling, Drying Condition and Moisture Content on Sensory Texture of Cooked Rice. *Cereal Chem.* **1999**, *76*, 56–62.

Manful J. T; Grimm C. C.; Gayin, J.; Coker, R. D. Effect of Variable Parboiling on Crystallinity of Rice Samples. *Cereal Chem.* **2008**, *85*, 92–95.

Marshall, W. E.; Wadsworth, J. I.; Verma, L. R.; Velupillai, L. Determining the Degree of Gelatinization in Parboiled Rice: Comparison of Subjective and Objective Method. *Cereal Chem.* **1993**, *70* (2), 226–230.

Meeso, N.; Nathakaranakule, A.; Madhiyanon, T.; Soponronnarit, S. Influence of FIR Irradiation on Paddy Moisture Reduction and Milling Quality After Fluidized Bed Drying. *J. Food Eng.* **2004**, *65*, 293–301.

Min, B.; McClung, A.; Chen, M.-H. Effects of Hydrothermal Processes on Antioxidants in Brown, Purple and Red Bran Whole Grain Rice (*Oryza sativa L.*). *Food Chem.* **2014**, *159*, 106–115.

Mir, S. A.; Bosco, S. J. D. Effect of Soaking Temperature on Physical and Functional Properties of Parboiled Rice Cultivars Grown in Temperate Region of India. *Food Nutr. Sci.* **2013**, *4*, 282–288.

Mitchell, C. R. Rice Starches: Production and Properties. In *Starch: Chemistry and Technology*, 3rd ed.; BeMiller, J., Whistler, R., Eds.; Academic Press: Burlington, MA, 2009; pp 569–578.

Mongpraneet, S.; Abe, T.; Tsurusaki, T. Accelerated Drying of Welsh Onion by Far Infrared Radiation Under Vacuum Conditions. *J. Food Eng.* **2002**, *55*, 147–156.

Odenigbo, A. M.; Ngadi, M. O.; Ejebe, C.; Danbaba, N.; Ndindeng, S. A.; Manful, J. T. Study on the Gelatinization Properties and Amylose Content of Rice Varieties from Nigeria and Cameroun. *Int. J. Nutr. Food Sci.* **2013**, *2* (4), 181–186.

Oli, P.; Ward, D.; Adhikari, B.; Torley, P. Parboiled Rice: Understanding from a Material Science Approach. *J. Food Eng.* **2014**, *124*, 173–183.

Ong, M. H.; Blanshard, J. M. V. Texture Determinants of Cooked Parboiled Rice II: Rice Starch Amylose and the Fine Structure of Amylopectin. *J. Cereal Sci.* **1995**, *21*, 261–9.

Paakkonen, K.; Havento, J.; Galambosi, B.; Pyykkonen, M. Infrared Drying of Herb. *Agric. Food Sci. Finl.* **1999**, *8*, 19–27.

Paiva, F. F.; Vanier, N. L.; Berrios, J. J.; Pan, J.; Villanova, F. A.; Takeoka, G., et al. Physicochemical and Nutritional Properties of Pigmented Rice Subjected to Different Degrees of Milling. *J. Food Compos. Anal.* **2014**, *35* (1), 10–17.

Paiva, F. F.; Vanier, N. L.; Berrios, J. D. J.; Pinto, V. Z.; Wood, D.; Williams, T.; Pan, J.; Elias, M. C. Polishing and Parboiling Effect on the Nutritional and Technological Properties of Pigmented Rice. *Food Chem.* **2016**, *191*, 105–112.

Parnsahkorn, S.; Langkapin, J. Changes in Physicochemical Characteristics of Germinated Brown Rice and Brown Rice During Storage at Various Temperatures. *Agric. Eng. Int. CIGR J.* **2013**, *15*, 293–303.

Patindol J.; Newton J.; Wang Y. J. Functional Properties as Affected by Laboratory-scale Parboiling of Rough Rice and Brown Rice. *J. Food Sci.* **2008**, *73* (8). DOI: 10.1111/j.1750-3841.2008.00926.x.

Pillaiyar, P. Household Parboiling of Parboiled Rice. *Kishan World* **1981**, *8*, 20–21.

Pillaiyar, P. Rice Parboiling Research in India. *Cereal foods: World* **1990**, *35* (2), 225–227.

Pillaiyar, P.; Mohandas, R. Hardness and Color in Parboiled Rice Produced at Low and High Temperature. *J. Food Sci. Technol.* **1981**, *18*, 7–9.

Priestley, R. J. Studies on Parboiled Rice. Quantitative Study of the Effects of Steaming Various Properties of Parboiled. *Food Chem.* **1976,** *1,* 139–148.

Raghavendra Rao, S. N.; Juliano, B. O. Effect of Parboiling on Some Physico Chemical Properties of Rice. *J. Agric. Food Chem.* **1970,** *18,* 289–294.

Rahaman, M. A.; Miah, M. A. K.; Ahmed, A. Status of Rice Processing Technology in Bangladesh. *Agric. Mech. Asia Afr. Latin Am.* **1996,** *27,* 46–50.

Rao, M. A. Kinetics of Thermal Softening of Foods—A Review. *J. Food Proc. Pres.* **1986,** *10,* 311–329.

Ratti, C.; Mujumdar, A. S. Infrared Drying. In *Handbook of Industrial Drying*; Mujumdar, A. S. Ed.; Marcel Dekker, Inc.: New York, NY, 1995; pp 567–588.

Roberts, L. R. Composition and Taste Evaluation of Rice Milled to Different Degrees. *J. Food Sci.* **1979,** *44,* 127–129.

Rordprapat, W.; Nathakaranakule, A.; Tia, W.; Soponronnarit, S. Comparative Study of Fluidized Bed Paddy Drying Using Hot Air and Superheated Steam. *J. Food Eng.* **2005,** *71,* 28–36.

Roseline, B. T. M.; Fofana, M.; Innocent, B.; Futakuchi, K. *Effect of Parboiling and Storage on Grain Physical and Cooking Characteristics of the some NERICA Rice Varieties.* Second Africa Rice Congress, Bamako, Mali; Innovation and Partnerships to Realize Africa's Rice Potential: Bamako, Mali, 2010.

Roy, P.; Shimizu, N.; Shiina, T.; Kimura, T. Energy Consumption and Cost Analysis of Local Parboiling Processes. *J. Food Eng.* **2006,** *76,* 646–655.

Roy, P.; Shimizu, N.; Okadome, H.; Shiina, T.; Kimura, T. Life Cycle of Rice: Challenges and Choices for Bangladesh. *J. Food Eng.* **2007,** *79,* 1250–1255.

Sandu, C. Infrared Radiative Drying in Food Engineering a Process Analysis. *Biotechnol. Progress.* **1986,** *2,* 223–232.

Shinde, Y. H.; Vijayadwhaja, A.; Pandit, A. B.; Joshi, J. B. Kinetics of Cooking of Rice: A Review. *J. Food Eng.* **2014,** *123,* 113–129.

Sivashanmugam, P.; Arivazhagan, M. Experimental Studies on Parboiling of Paddy by Ohmic Heating. *Int. J. Food Eng.* **2008,** *4* (1) (Article 6). http:// www.bepress.com/ijfe/ vol4/iss1/art6. (accessed Feb 27, 2008).

Soponronnarit, S.; Nathakaranakule, A.; Jirajindalert, A.; Taechapairoj, C. Parboiling Brown Rice Using Superheated Steam Fluidization Technique. *J. Food Eng.* **2006,** *75,* 423–432.

Stermer, R. A. An Instrument for Objective Measurement of Degree of Milling and Color of Milled Rice. *Cereal Chem.* **1968,** *45,* 358–364.

Subrahmanyan, V.; Desikacliar, II. S. R.; Bhalia, D. S. Commercial Methods of Parboiling Paddy and Improvement of the Quality of Parboiled Rice. *J. Sci. Res.* **1955,** *14,* 110–114.

Sujatha, S. J.; Ahmed, R.; Bhat, P. R. Physico-chemical Properties and Cooking Qualities of Two Varieties of Raw and Parboiled Rice Cultivated in the Coastal Region of Dakshina Kannada, India. *J. Food Chem.* **2004,** *86* (2), 211–216.

Tashiro, T.; Wardlaw, I. F. The Effect of High Temperature on Kernel Dimensions and the Type and Occurrence of Kernel Damage in Rice. *Aust. J. Agric. Res.* **1991,** *42,* 485–496.

Therdthai, N.; Zhou, W. Characterization of Microwave Vacuum Drying and Hot Air Drying of Mint Leaves (Menthacordifoliaopiz ex fresen). *J. Food Eng.* **2009,** *91,* 482–489.

Tirawanichakul, S.; Bualuang, O.; Tirawanichakul, Y. Study of Drying Kinetics and Qualities of Two Parboiled Rice Varieties: Hot Air Convection and Infrared Irradiation. *Songklanakarin J. Sci. Technol.* **2012,** *34* (5), 557–568.

Traore, K.; McClung, A. M.; Fjellstrom, R.; Futakuchi, K. Diversity in Grain Physicochemical Characteristics of West African Rice, Including NERICA Genotypes, as Compared to Cultivars from the United States of America. *Int. Res. J. Agric. Sci. Soil Sci.* **2011,** *1* (10), 435–448.

Unnikrishnan, K. R.; Bhaltacharya, K. R. Influence of Varietal Difference on Properties of Parboiled Rice. *Cereal Chem.* **1987,** *64* (4), 315–321.

Verma, D. K.; Srivastav, P. P. Proximate Composition, Mineral Content and Fatty Acids Analyses of Aromatic and Non-aromatic Indian Rice. *Rice Sci.* **2017,** *24* (1), 21–31.

Verma, D. K.; Mohan, M.; Yadav, V. K.; Asthir, B.; Soni, S. K. Inquisition of some Physico-chemical Characteristics of Newly Evolved Basmati Rice. *Environ. Eco.* **2012,** *30* (1), 114–117.

Verma, D. K.; Mohan, M.; Asthir, B. Physicochemical and Cooking Characteristics of some Promising Basmati Genotypes. *Asian J. Food Agro-Indus.* **2013,** *6* (2), 94–99.

Verma, D. K.; Mohan, M.; Prabhakar, P. K.; Srivastav, P. P. Physico-chemical and Cooking Characteristics of Azad Basmati. *Int. Food Res J.* **2015,** *22* (4), 1380–1389.

Webb, B. D. Cooking, Processing and Milling Properties of Rice. In *Six Decades of Rice Research in Texas*; Texas Agricultural Experiment Station: College Station, TX, 1975; pp 97–106.

Whistler, R. L.; BeMiller, J. N. *Carbohydrate Chemistry for Food Scientists.* Eagan Press: St. Paul, Minnesota, 1997; pp 117–151.

Wynn, T. In *Rice Farming: How the Economic Crisis Affects the Rice Industry.* Presented at the Rice Producers Forum, USRPA, Houston, TX, USA, Aug 20, 2008. http://www. ricefarming.com/home/2009_JanProducersForum.html (accessed May 7, 2009).

CHAPTER 10

NUTRITIONAL QUALITY EVALUATION OF DIFFERENT RICE CULTIVARS

DEEPAK KUMAR VERMA[1]*, PREM PRAKASH SRIVASTAV[1], and MUKESH MOHAN[2]

[1]Department of Agricultural and Food Engineering, Indian Institute of Technology, Kharagpur 721302, West Bengal, India

[2]Department of Agricultural Biochemistry, College of Agriculture, Chandra Shekhar Azad University of Agriculture and Technology, Kanpur 208002, Uttar Pradesh, India

*Corresponding author.
E-mail: deepak.verma@agfe.iitkgp.ernet.in; rajadkv@rediffmail.com

ABSTRACT

Rice is one of the most nutritious staple foods, especially for people in Asia and its composition varies greatly from one region to another region. It is the rich source of carbohydrates and moderate source of proteins and fats. It also contains fair amount of B complex vitamins along with minerals such as Ca, P, Mg, Fe, Cu, etc. However, some cultivars of rice are low in nutrients while others are rich of valuable nutrients. Rice is considered as the queen among cereals due to its nutritional quality and higher digestibility. Several factors such as soil fertility, fertilizer application, environmental conditions, etc. affect the nutritional values of rice among which most important is the variation in cultivar. This chapter covers the recent research information on different rice varieties, their composition, and difference in their nutritional aspects due to the cultivar variation.

10.1 INTRODUCTION

Rice (*Oryza sativa* L.) is a major dietary component of people in most of the countries. It is appreciated as one of the most nutritious staple food among the cereals (Verma et al., 2012, 2013, 2015) and is highly consumed in Asia and Africa and less in the European Union (Vlachos and Arvanitoyannis,

2008). This staple food retains the main source of energy and protein which provide 700 calories/day-person for about 3000 million people of the world's population (Vlachos and Arvanitoyannis, 2008). Rice is a rich source of carbohydrate (CHO); it contains a moderate amount of protein and fat and also a source of vitamin B complex such as thiamine (vitamin B_1), riboflavin (vitamin B_2), and niacin (vitamin B_3) (Fresco, 2005). Rice CHO is mainly a starch which is composed of amylose and amylopectin. The grain of rice constitutes water 12%, starch 75–80%, and protein only 7% with a full complement of amino acids (AAs). Its protein is highly digestible (93%) with excellent biological value (74%) and protein efficiency ratio (2.02–2.04%) owing to the presence of higher concentration (~4%) of lysine (Eggum, 1969, 1973, 1977; Bressani et al., 1971; Juliano, 1985b, 1993a). Minerals like calcium (Ca), magnesium (Mg), and phosphorus (P) are present along with some traces of iron (Fe), copper (Cu), zinc (Zn), and manganese (Mn) (Oko et al., 2012). Although the nutritional values of rice varies with different cultivars, soil fertility, fertilizer application, and other environmental conditions, the following trend still exists by comparison with other cereals: low fat content after the removal of the bran, low protein content (about 7–10%), and higher digestibility of protein. Due to its nutritional quality and higher digestibility, rice is considered as the queen among cereals (Anjum et al., 2007). Freshly harvested rice grains contain about 80% CHO which includes starch, glucose, sucrose, dextrin, etc.

Researches are generally focused on such chemical and nutritional attributes of rice, and many researches have been conducted. Since rice is characterized by certain unique nutritional features, discussed above, the proposed chapter describes method of chemical and nutritional quality evolution of rice and explains that how various varieties compete with and differ from each other in their important nutritional quality attributes.

10.2   PROXIMATE COMPOSITION ANALYSIS

Composition of the grain makes it a palatable food of high energy value, which leads the nutritionist to have a major interest in the composition of the kernel. Analysis of the accumulated chemical composition in food is important, and for cereal in all cases, it shows that the highest percentage proximate composition of the staple food crops is CHO followed by crude protein (CP), moisture, crude fiber (CF), ash, and fat for rice, whereas in tubers, moisture is highest (Edeogu et al., 2007). The degree of milling and polishing determines the amount of nutrients removed. Rice grain quality affects the nutritional and commercial value of grains and is of immense importance to those involved in producing, processing, and consuming rice (Yi et al., 2009). Rice quality does not only imply with the physical look or qualities but also it includes both the characteristics, namely, chemical and cooking; therefore, it has become necessary to include the nutritional values derivable from rice into attention when choosing a particular rice variety (Oko and Onyekwere, 2010). Rice contains approximately 80% starch and 7% protein at 12% moisture (Wells, 1999). Polished rice is mainly made up of starch, protein, lipid, and moisture. The protein content of polished rice in 22 *japonica* rice varieties ranged from 5.9% to 7.9% and 6.0% to 13.6% in brown rice among 1518 Chinese *japonica* varieties (Lestari et al., 2009). Starch occurs in the endosperm as small many-sided granules (Kapri et al., 2017), whereas protein is present as particles that lie between the starch granules. Starch comprises 76.7–78.4% in polished rice with 14% moisture content (MC) (Huang et al., 1998). Rice grain also contains sugars, fat, dietary fiber, and minerals. The determination of food composition is fundamental to theoretical and applied investigations in food science and technology. This is often the basis for establishing the nutritional value and overall acceptance of the food from the consumers' stand point (Aganga and Tshwenyane, 2004). Rice can be distinguished among cereal grains on the basis of its important quality oriented attributes (moisture, CF, ash, CP, and crude fat contents). Proximate composition of rice cultivars reported from different geographic location of the world is depicted in Table 10.1.

10.2.1 MOISTURE CONTENT

Moisture content has a clear influence on all aspects of paddy and rice quality and it is essential that paddy be milled at the proper MC to obtain the highest head rice yield. Paddy is at its optimum milling potential at MC of 14% wet weight basis (Wilfred and Consultant, 2006). Grains with high MC are too soft to withstand hulling pressure which results in grain breakage and possibly grain pulverization. Grain that is too dry becomes brittle and has greater breakage (Pan et al., 2007). MC and temperature during drying process is also critical as it determines whether small fissures and/or full cracks are introduced into the grain structure (Park et al., 2001).

MC affects rice quality in several different ways. Of great significance is its effect on the keeping quality of all forms of rice. Sound dry rice can be maintained for year if properly stored, but only a few days are required for wet rice to spoil. Rough rice MC of 13% is commonly accepted as a safe level

TABLE 10.1A Proximate Composition of Rice Cultivars Reported from Different Geographic Location of India, Pakistan, and Bangladesh.

Rice varieties/cultivars	Proximate composition (%)						References
	CHO	MC	Fat	Protein	Fiber	Ash	
India							
Badshah Bhog	82.7 ± 0.24	8.90 ± 0.18	0.61 ± 0.01	7.23 ± 0.16	0.85 ± 0.00	0.59 ± 0.04	Verma and Srivastav (2017)
Bakul joha	76.6 ± 0.02	13.7 ± 0.12	1.6 ± 0.11	7.7 ± 0.03	0.2 ± 0.01	0.7 ± 0.02	Saikia et al. (2012)
Bas-370	–	11.85 ± 0.05	0.65 ± 0.08	7.02 ± 0.16	–	0.44 ± 0.06	Yadav et al. (2007)
Chak-haoamubi	78.0 ± 0.03	11.6 ± 0.04	1.0 ± 0.07	8.8 ± 0.05	0.3 ± 0.06	0.5 ± 0.07	Saikia et al. (2012)
CRM-839	90.4	9.75	1.23	7.79	–	–	Subudhi et al. (2013)
Dubraj	90.11	9.8	1.01	7.2	–	–	Subudhi et al. (2013)
Geetanjali	90.06	9.6	1.62	8.57	–	–	Subudhi et al. (2013)
Gopal Bhog	79.87 ± 0.94	10.11 ± 0.77	0.72 ± 0.06	8.78 ± 0.19	0.64 ± 0.01	0.52 ± 0.03	Verma and Srivastav (2017)
Govind Bhog	75.87 ± 0.53	13.57 ± 0.39	0.92 ± 0.01	9.27 ± 0.14	0.69 ± 0.00	0.38 ± 0.04	Verma and Srivastav (2017)
HBC-19	–	11.90 ± 0.03	0.54 ± 0.06	6.94 ± 0.14	–	0.67 ± 0.05	Yadav et al. (2007)
HKR120	–	12.15 ± 0.04	0.78 ± 0.08	5.59 ± 0.12	–	0.40 ± 0.05	Yadav et al. (2007)
IR 64	74.1 ± 17.6	13.60 ± 0.32	2.06 ± 0.14	7.95 ± 0.17	4.96 ± 0.01	1.27 ± 0.09	Deepa et al. (2008)
Jaya	–	12.55 ± 0.08	0.82 ± 0.07	5.46 ± 0.23	–	0.31 ± 0.04	Yadav et al. (2007)
Jyothi	72.8 ± 11.10	13.00 ± 0.24	2.60 ± 0.54	7.97 ± 0.50	5.82 ± 0.02	1.54 ± 0.09	Deepa et al. (2008)
Kalajira	90.2	9.9	0.25	7.58	–	–	Subudhi et al. (2013)
Kala namak	80.94 ± 0.91	10.59 ± 0.63	0.68 ± 0.01	7.23 ± 0.20	0.55 ± 0.00	0.56 ± 0.08	Verma and Srivastav (2017)
Kala namak	90.06	10.24	0.97	8.76	–	–	Subudhi et al. (2013)
Keteki joha	76.4 ± 0.06	12.1 ± 0.02	1.0 ± 0.07	9.9 ± 0.11	0.2 ± 0.06	0.8 ± 0.04	Saikia et al. (2012)
Ketekijoha	90.23	9.68	0.95	8.74	–	–	Subudhi et al. (2013)
Khushboo	78.87 ± 0.39	10.12 ± 0.12	0.77 ± 0.01	9.51 ± 0.25	0.48 ± 0.00	0.73 ± 0.05	Verma and Srivastav (2017)
Njavara	73.5 ± 13.21	13.10 ± 0.15	2.48 ± 0.50	9.52 ± 0.34	8.08 ± 0.03	1.42 ± 0.06	Deepa et al. (2008)
P-44	–	12.72 ± 0.05	0.72 ± 0.04	6.26 ± 0.27	–	0.67 ± 0.05	Yadav et al. (2007)
Poreiton chakhao	77.2 ± 0.01	13.5 ± 0.03	2.1 ± 0.08	6.6 ± 0.02	0.2 ± 0.12	0.5 ± 0.09	Saikia et al. (2012)
Pusa basmati-1	90.27	9.7	1.21	7.75	–	–	Subudhi et al. (2013)

TABLE 10.1A (Continued)

Rice varieties/cultivars	Proximate composition (%)							References
	CHO	MC	Fat	Protein	Fiber	Ash		
Pusa sungandh-2	91.0	9.37	1.37	8.2	–	–		Subudhi et al. (2013)
Salem samba	–	12.9	1.18	7.50	0.20	–		Ravi et al. (2012)
Sarbati	81.47 ± 0.36	11.25 ± 0.30	0.06 ± 0.01	6.87 ± 0.10	0.64 ± 0.00	0.35 ± 0.05		Verma and Srivastav (2017)
Sharbati	–	11.64 ± 0.06	0.66 ± 0.08	6.52 ± 0.12	–	0.41 ± 0.04		Yadav et al. (2007)
Swetganga	78.38 ± 0.12	12.01 ± 0.50	0.23 ± 0.06	8.73 ± 0.05	0.70 ± 0.00	0.66 ± 0.03		Verma and Srivastav (2017)
Todal	79.96 ± 0.18	12.20 ± 0.19	0.22 ± 0.01	7.09 ± 0.08	0.65 ± 0.16	0.54 ± 0.01		Verma and Srivastav (2017)
Toroari Basmati	90.65	9.35	1.38	8.96	–	–		Subudhi et al. (2013)
Tulsiphool	90.2	9.92	0.9	7.9	–	–		Subudhi et al. (2013)
Pakistan								
Basmati 198	78.56 ± 1.11	8.45 ± 0.21	2.36 ± 0.19	7.70 ± 0.34	0.85 ± 0.03	1.90 ± 0.05		Zubair et al. (2012)
Basmati 2000	78.92 ± 2.56	8.22 ± 0.16	2.14 ± 0.08	7.70 ± 0.38	0.89 ± 0.03	1.96 ± 0.05		Zubair et al. (2012)
Basmati 2000	74.34	11.06	3.64	6.84	1.68	2.44		Hayat et al. (2013)
Basmati 370	78.37 ± 1.73	7.62 ± 0.25	2.72 ± 0.09	8.75 ± 0.23	0.89 ± 0.03	1.48 ± 0.05		Zubair et al. (2012)
Basmati 385	76.98 ± 2.87	8.30 ± 0.20	2.70 ± 0.09	9.14 ± 0.31	0.71 ± 0.03	1.98 ± 0.06		Zubair et al. (2012)
Basmati 385	74.75	11.08	3.98	6.65	1.52	2.02		Hayat et al. (2013)
Basmati 515	78.18 ± 2.23	8.68 ± 0.32	2.26 ± 0.12	7.91 ± 0.22	0.82 ± 0.03	1.97 ± 0.05		Zubair et al. (2012)
Basmati Pak	78.92 ± 2.45	6.99 ± 0.20	1.92 ± 0.07	9.16 ± 0.26	0.92 ± 0.03	1.92 ± 0.04		Zubair et al. (2012)
Basmati Super	78.91 ± 1.21	6.84 ± 0.31	2.06 ± 0.09	9.10 ± 0.43	0.87 ± 0.02	1.98 ± 0.07		Zubair et al. (2012)
Basmati Supper	76.03	11.05	4.08	6.22	1.18	1.44		Islam et al. (2016)
Dr-83	71.21	9.19	5.45	8.12	2.31	3.16		Anjum et al. (2007)
IRRI 6	67.75	11.04	6.14	8.77	2.57	3.67		Anjum et al. (2007)
IRRI 6	79.43 ± 1.19	8.19 ± 0.18	2.06 ± 0.08	7.50 ± 0.24	0.81 ± 0.03	1.86 ± 0.07		Zubair et al. (2012)
IRRI 9	71.43	10.01	5.16	7.80	2.17	3.43		Anjum et al. (2007)
IRRI-6	72.88	11.12	3.64	8.24	1.68	2.44		Hayat et al. (2013)

TABLE 10.1A (Continued)

Rice varieties/cultivars	Proximate composition (%)						References
	CHO	MC	Fat	Protein	Fiber	Ash	
IRRI-9	73.16	11.24	3.75	7.65	1.74	2.46	Hayat et al. (2013)
KS 282	78.19 ± 2.57	9.02 ± 0.31	2.17 ± 0.11	7.75 ± 0.20	0.79 ± 0.02	1.95 ± 0.05	Zubair et al. (2012)
KSK 133	78.43 ± 1.32	9.08 ± 0.22	2.04 ± 0.09	7.52 ± 0.26	0.88 ± 0.02	1.94 ± 0.06	Zubair et al. (2012)
Sarshar	68.14	11.10	5.80	8.80	2.47	3.79	Anjum et al. (2007)
Bangladesh							
Kalijira	79.48	12.65	0.57	6.81	0.22	0.49	Ali et al. (2012),
Basmoti	78.49 ± 0.07	12.43 ± 0.04	0.72 ± 0.01	7.22 ± 0.02	0.63 ± 0.02	0.50 ± 0.01	Islam et al. (2016)
Begun Bichi	—	12.39 ± 0.02	0.680 ± 0.01	5.74 ± 0.035	—	1.070 ± 0.01	Asaduzzaman et al. (2013)
BRRI dhan-34	—	11.48 ± 0.01	1.331 ± 0.05	5.44 ± 0.548	—	1.210 ± 0.01	Asaduzzaman et al. (2013)
BRRI dhan-37	—	15.13 ± 0.01	0.782 ± 0.02	5.69 ± 0.075	—	1.050 ± 0.02	Asaduzzaman et al. (2013)
BRRI dhan-50	—	11.25 ± 0.01	0.873 ± 0.05	6.21 ± 0.015	—	1.460 ± 0.05	Asaduzzaman et al. (2013)
Chinigura	77.29 ± 0.08	12.60 ± 0.04	1.92 ± 0.02	7.61 ± 0.01	0.26 ± 0.01	0.30 ± 0.02	Islam et al. (2016)
Govindavogh	76.39 ± 0.04	12.93 ± 0.04	1.7 ± 0.01	7.93 ± 0.02	0.50 ± 0.02	0.48 ± 0.02	Islam et al. (2016)
Jeerashile	81.87 ± 0.05	10.21 ± 0.07	0.16 ± 0.01	7.08 ± 0.02	0.19 ± 0.02	0.48 ± 0.02	Islam et al. (2016)
Kalizira	—	11.55 ± 0.01	0.937 ± 0.01	5.18 ± 0.015	—	0.923 ± 0.00	Asaduzzaman et al. (2013)
Katarivogh	76.60 ± 0.11	12.44 ± 0.06	1.8 ± 0.02	8.31 ± 0.01	0.30 ± 0.02	0.47 ± 0.02	Islam et al. (2016)
Minikat	81.76 ± 0.06	11.04 ± 0.05	0.13 ± 0.01	6.31 ± 0.01	0.23 ± 0.01	0.52 ± 0.01	Islam et al. (2016)
Nagirshail	76.33 ± 0.13	12.15 ± 0.06	2.61 ± 0.01	8.11 ± 0.02	0.36 ± 0.02	0.43 ± 0.02	Islam et al. (2016)
Nurjahan	78.61 ± 0.03	12.48 ± 0.04	0.42 ± 0.02	7.35 ± 0.02	0.63 ± 0.02	0.49 ± 0.02	Islam et al. (2016)
Paijam	77.37 ± 0.08	13.22 ± 0.08	1.42 ± 0.02	7.50 ± 0.01	0.15 ± 0.02	0.34 ± 0.02	Islam et al. (2016)
Pari shiddho	77.69 ± 0.07	12.73 ± 0.05	1.32 ± 0.01	7.51 ± 0.01	0.24 ± 0.01	0.49 ± 0.02	Islam et al. (2016)
Parija	78.54 ± 0.04	10.45 ± 0.06	2.90 ± 0.02	7.37 ± 0.02	0.21 ± 0.01	0.52 ± 0.02	Islam et al. (2016)
Philipine Katari	—	12.11 ± 0.02	1.450 ± 0.02	3.23 ± 0.005	—	0.887 ± 0.06	Asaduzzaman et al. (2013)
Sawrna	80.36 ± 0.12	11.37 ± 0.07	0.09 ± 0.01	7.44 ± 0.02	0.17 ± 0.01	0.57 ± 0.02	Islam et al. (2016)

TABLE 10.1B Proximate Composition of Rice Cultivars Reported from Different Geographic Location of Thailand, Malaysia, Philippines, and Taiwan.

Rice varieties/cultivars	CHO	MC	Fat	Protein	Fiber	Ash	References
Thailand							
Aroso	81.1 ± 1.24	8.0 ± 0.02	1.9 ± 0.02	6.95 ± 0.07	1.5 ± 0.02	0.53 ± 0.06	Ebuehi and Oyewole (2008)
Bahng Gawk	75.04 ± 0.15	11.55 ± 0.04	2.86 ± 0.02	9.21 ± 0.13	3.63 ± 1.30	1.33 ± 0.01	Sompong et al. (2011)
Black waxy rice-96025	–	7.60 ± 0.21	1.93 ± 0.39	8.44 ± 0.03	0.26 ± 0.01	1.52 ± 0.12	Yodmanee et al. (2011)
Black waxy rice-96044	–	5.96 ± 0.06	1.67 ± 0.09	8.23 ± 0.17	0.29 ± 0.00	1.35 ± 0.07	Yodmanee et al. (2011)
Chormaiphai	–	6.08 ± 0.23	1.50 ± 0.04	8.46 ± 0.17	0.16 ± 0.05	1.58 ± 0.20	Yodmanee et al. (2011)
CN 1	–	10.43	0.81	7.59	–	0.40	Chaichaw et al. (2011)
Gam-Pai15	–	11.59 ± 0.12	–	6.62 ± 0.18	–	–	Maisont and Narkrugsa (2009)
Haek Yah	75.92 ± 0.11	12.38 ± 0.03	2.91 ± 0.05	7.40 ± 0.11	4.18 ± 0.14	1.40 ± 0.01	Sompong et al. (2011)
Hahng-Yi71	–	10.52 ± 0.08	–	8.34 ± 0.04	–	–	Maisont and Narkrugsa (2009)
Homkradunga	–	6.76 ± 0.26	1.47 ± 0.09	6.96 ± 0.10	0.28 ± 0.00	1.44 ± 0.10	Yodmanee et al. (2011)
Kamyan	–	8.65 ± 0.14	2.17 ± 0.04	6.63 ± 0.11	0.35 ± 0.05	1.64 ± 0.13	Yodmanee et al. (2011)
KDML 105	–	9.33	0.95	6.39	–	0.49	Chaichaw et al. (2011)
KDML105	–	11.81 ± 0.14	–	7.37 ± 0.16	–	–	Maisont and Narkrugsa (2009)
Kramrad	–	6.83 ± 0.26	1.44 ± 0.11	7.69 ± 0.09	0.28 ± 0.01	1.38 ± 0.10	Yodmanee et al. (2011)
Niaw Dam Pleuak Dam	71.99 ± 0.08	12.03 ± 0.13	3.65 ± 0.05	10.85 ± 0.09	3.41 ± 0.24	1.48 ± 0.02	Sompong et al. (2011)
Niaw Dam Pleuak Khao	74.09 ± 0.48	12.59 ± 0.16	3.72 ± 0.06	8.17 ± 0.41	4.01 ± 0.58	1.42 ± 0.01	Sompong et al. (2011)
Niaw Dawk Yong	73.73 ± 0.10	12.01 ± 0.01	3.19 ± 0.06	9.62 ± 0.16	3.75 ± 0.79	1.45 ± 0.06	Sompong et al. (2011)
Niaw Lan Tan	75.20 ± 0.16	13.12 ± 0.16	3.08 ± 0.08	7.35 ± 0.06	3.27 ± 1.20	1.26 ± 0.05	Sompong et al. (2011)
Niaw Look Pueng	77.53 ± 0.01	11.45 ± 0.03	2.37 ± 0.06	7.16 ± 0.01	3.17 ± 0.84	1.50 ± 0.07	Sompong et al. (2011)
Niaw-Phrae1	–	10.72 ± 0.14	–	6.32 ± 0.03	–	–	Maisont and Narkrugsa (2009)

TABLE 10.1B (Continued)

Rice varieties/cultivars	Proximate composition (%)						References
	CHO	MC	Fat	Protein	Fiber	Ash	
Niaw-San-Pah-Tawng	–	11.25 ± 0.06	–	6.81 ± 0.03	–	–	Maisont and Narkrugsa (2009)
RD 6	–	10.69	0.47	6.15	–	0.21	Chaichaw et al. (2011)
RD-10	–	11.66 ± 0.12	–	7.03 ± 0.05	–	–	Maisont and Narkrugsa (2009)
RD-6	–	11.73 ± 0.20	–	6.35 ± 0.04	–	–	Maisont and Narkrugsa (2009)
RD-6	79.2 ± 2.08	9.44 ± 0.76	1.20 ± 0.68	6.98 ± 0.07	1.13 ± 0.16	1.96 ± 0.11	Moongngarm and Saetung (2010)
Red waxy rice-96060	–	8.19 ± 0.30	1.58 ± 0.04	8.18 ± 0.12	0.35 ± 0.05	1.78 ± 0.35	Yodmanee et al. (2011)
Sakon-Nakhon	–	11.36 ± 0.13	–	7.58 ± 0.01	–	–	Maisont and Narkrugsa (2009)
Sangyod	–	7.18 ± 0.22	1.65 ± 0.56	8.06 ± 0.03	0.26 ± 0.01	2.15 ± 0.05	Yodmanee et al. (2011)
Sew-Mae-Jan	–	12.26 ± 0.18	–	7.48 ± 0.18	–	–	Maisont and Narkrugsa (2009)
Sung Yod Phatthalung	76.27 ± 0.13	9.28 ± 0.06	2.67 ± 0.06	10.36 ± 0.04	4.51 ± 1.60	1.42 ± 0.12	Sompong et al. (2011)
Suphan-Buri1	–	10.79 ± 0.14	–	7.17 ± 0.05	–	–	Maisont and Narkrugsa (2009)
Malaysia							
Bario	82.23 ± 0.5	10.55 ± 0.2	0.26 ± 0.2	6.35 ± 0.6	7.17 ± 0.1	0.63 ± 0.0	Thomas et al. (2013)
Basmati	79.34 ± 0.1	11.23 ± 0.0	1.02 ± 0.1	7.75 ± 0.1	8.09 ± 0.2	0.48 ± 0.2	Thomas et al. (2013)
Black (imported)	78.26 ± 0.6	11.07 ± 0.2	0.07 ± 0.2	8.16 ± 0.3	8.47 ± 0.2	0.90 ± 0.2	Thomas et al. (2013)
Brown	78.21 ± 0.9	12.88 ± 0.0	1.74 ± 0.1	6.48 ± 0.0	8.37 ± 0.1	0.55 ± 0.0	Thomas et al. (2013)
Glutinous	78.89 ± 0.6	10.04 ± 0.1	1.12 ± 0.2	8.14 ± 0.1	7.47 ± 0.4	0.82 ± 0.1	Thomas et al. (2013)
White (local)	80.14 ± 1.1	12.08 ± 0.1	1.24 ± 0.0	5.96 ± 0.2	7.07 ± 0.6	0.39 ± 0.4	Thomas et al. (2013)
Philippines							
IR 62	76.54	12.41	1.33	9.28	1.05	0.44	Panlasigui et al. (1991)
IR 36	78.85	11.24	1.15	8.11	2.00	0.65	Panlasigui et al. (1991)

TABLE 10.1B (Continued)

Rice varieties/cultivars	Proximate composition (%)						References
	CHO	MC	Fat	Protein	Fiber	Ash	
IR 42	78.09	11.24	1.51	8.30	1.73	0.86	Panlasigui et al. (1991)
Taiwan							
KS142	–	12.42 ± 0.08	0.95 ± 0.14	7.10 ± 0.37	–	0.59 ± 0.00	Chen et al. (2004),
KSS7	–	11.80 ± 0.05	0.38 ± 0.02	6.40 ± 0.04	–	0.38 ± 0.00	Panlasigui et al. (1991)
TC189	–	12.22 ± 0.17	0.64 ± 0.04	7.47 ± 0.05	–	0.47 ± 0.02	Panlasigui et al. (1991)
TCN1	–	11.77 ± 0.08	0.71 ± 0.03	7.61 ± 0.04	–	0.56 ± 0.03	Panlasigui et al. (1991)
TCS10	–	12.74 ± 0.05	0.45 ± 0.03	7.16 ± 0.02	–	0.51 ± 0.01	Panlasigui et al. (1991)
TCS17	–	11.97 ± 0.02	0.37 ± 0.02	6.47 ± 0.02	–	0.53 ± 0.01	Panlasigui et al. (1991)
TCSW1	–	12.50 ± 0.19	0.91 ± 0.04	7.77 ± 0.02	–	0.52 ± 0.00	Panlasigui et al. (1991)
TCW70	–	12.51 ± 0.16	0.72 ± 0.02	7.02 ± 0.04	–	0.38 ± 0.02	Panlasigui et al. (1991)
TG5	–	12.78 ± 0.10	0.69 ± 0.04	6.64 ± 0.02	–	0.44 ± 0.01	Panlasigui et al. (1991)
TG8	–	11.86 ± 0.06	0.70 ± 0.03	6.68 ± 0.04	–	0.49 ± 0.01	Panlasigui et al. (1991)
TG9	–	12.88 ± 0.06	1.14 ± 0.09	6.67 ± 0.08	–	0.61 ± 0.01	Panlasigui et al. (1991)
TGW1	–	12.02 ± 0.06	1.39 ± 0.06	7.08 ± 0.08	–	0.61 ± 0.02	Panlasigui et al. (1991)
TN9	–	12.10 ± 0.04	0.62 ± 0.01	6.30 ± 0.01	–	0.54 ± 0.01	Panlasigui et al. (1991)
TNuS19	–	12.55 ± 0.09	0.77 ± 0.01	6.16 ± 0.04	–	0.59 ± 0.01	Panlasigui et al. (1991)

TABLE 10.1C Proximate Composition of Rice Cultivars Reported from Different Geographic Location of Nigeria, Ghana, and Tanzania.

Rice varieties/cultivars	Proximate composition (%)						References
	CHO	MC	Fat	Protein	Fiber	Ash	
Nigeria							
Awilo	84.01 ± 0.01	6.67 ± 0.01	2.50 ± 0.01	4.82 ± 0.01	1.50 ± 0.01	0.50 ± 0.01	Oko and Ugwu (2011) and Oko et al. (2012)
Canada	85.09 ± 0.12	3.67 ± 0.01	3.50 ± 0.90	4.74 ± 0.01	2.00 ± 0.10	2.00 ± 0.10	Oko and Ugwu (2011) and Oko et al. (2012)
China	85.69	5.33	2.5	2.98	2.5	1.0	Oko et al. (2012)
Chinyereugo	86.12	5.33	1.0	4.55	1.5	1.5	Oko et al. (2012)
Cooperative	84.93	5.67	2.0	4.90	1.5	1.0	Oko et al. (2012)
E4077	85.33	5.67	3.0	4.00	1.5	0.5	Oko et al. (2012)
E4197	51.53	9.67	3.5	6.30	2.0	1.0	Oko et al. (2012)
E4212	82.56	5.00	1.5	7.94	2.0	1.0	Oko et al. (2012)
E4314	85.05	5.67	2.0	5.25	1.5	0.5	Oko et al. (2012)
E4334	85.57	5.30	1.5	5.60	1.0	1.0	Oko et al. (2012)
Ezichi	84.88	4.47	1.5	5.95	1.5	1.5	Oko et al. (2012)
Faro 14	83.45 ± 0.1	7.33 ± 0.01	0.50 ± 0.01	6.22 ± 0.01	1.50 ± 0.10	1.00 ± 0.00	Oko and Ugwu (2011)
Faro 14(I)	83.45	7.33	0.5	6.22	1.5	1.0	Oko et al. (2012)
Faro 14(II)	83.45	7.33	0.5	6.22	1.5	1.0	Oko et al. (2012)
Faro 15	86.03 ± 0.01	6.33 ± 0.10	1.00 ± 0.09	4.64 ± 0.10	1.50 ± 0.10	0.50 ± 0.01	Oko and Ugwu (2011)
Faro 15(I)	86.03	6.33	1.0	4.64	1.5	0.5	Oko et al. (2012)
Faro 15(II)	84.30	5.33	2.5	5.87	1.0	1.0	Oko et al. (2012)
IR68	51.50	35.67	3.5	6.30	2.0	1.0	Oko and Onyekwere (2010)
IR75395-2B-B-1-1-1-2-4	85.57	5.30	1.5	5.60	1.0	1.0	Oko and Onyekwere (2010)

TABLE 10.1C (Continued)

Rice varieties/cultivars	Proximate composition (%)						References
	CHO	MC	Fat	Protein	Fiber	Ash	
IR77384-12-17-3-18-2-B	85.33	5.67	3.0	4.0	1.5	0.5	Oko and Onyekwere (2010)
IR7764-3B-8-2-2-14-4	85.05	5.67	2.0	5.25	1.5	0.5	Oko and Onyekwere (2010)
Jamila	—	—	2.33 ± 0.19	7.78 ± 0.77	—	1.21 ± 0.06	Chinma et al. (2015)
Jeep	—	—	2.17 ± 0.50	10.29 ± 0.80	—	1.10 ± 0.03	Chinma et al. (2015)
Kwandala	—	—	2.15 ± 0.12	10.08 ± 0.39	—	1.39 ± 0.05	Chinma et al. (2015)
Mass (I)	85.90	5.67	1.0	5.43	1.0	1.0	Oko et al. (2012)
Mass (II)	86.82	4.33	2.0	3.85	2.5	0.5	Oko et al. (2012)
Mass(III)	85.36	4.67	2.5	4.47	1.5	1.5	Oko et al. (2012)
MR 219	—	—	2.41 ± 0.05	7.70 ± 0.68	—	1.06 ± 0.00	Chinma et al. (2015)
Ofada	78.3 ± 1.64	7.5 ± 0.08	2.6 ± 0.07	7.30 ± 0.14	3.5 ± 0.04	0.80 ± 0.02	Ebuehi and Oyewole (2008)
Onuogwu	84.25	6.00	2.0	5.25	1.5	1.0	Oko et al. (2012)
PSBRc50	82.56	5.0	1.5	7.9	2.0	1.0	Oko and Onyekwere (2010)
Sipi	76.92 ± 0.0	18.00 ± 0.10	0.50 ± 0.01	1.58 ± 0.01	2.00 ± 0.10	1.00 ± 0.01	Oko and Ugwu (2011) and Oko et al. (2012)
Ghana							
ANDY-11	76.90	13.00	1.20	7.42	0.11	1.37	Mbatchou and Dawda (2013)
Ex-Baika	—	11.8 ± 0.05	0.7 ± 0.06	5.8 ± 0.16	0.05 ± 0.02	0.5 ± 0.03	Diako et al. (2011)
Ex-Hohoe	—	11.6 ± 0.04	0.7 ± 0.07	5.6 ± 0.14	0.22 ± 0.01	0.7 ± 0.13	Diako et al. (2011)
IR12979-24-1	83.27	8.50	1.20	6.01	0.16	0.86	Mbatchou and Dawda (2013)
Jasmine 85	—	11.6 ± 0.04	0.5 ± 0.03	5.8 ± 0.10	0.11 ± 0.01	0.6 ± 0.01	Diako et al. (2011)
JASMINE-85	76.89	13.50	1.10	6.82	0.76	0.93	Mbatchou and Dawda (2013)
Marshall	—	12.2 ± 0.04	0.7 ± 0.07	5.9 ± 0.01	0.02 ± 0.01	0.6 ± 0.02	Diako et al. (2011)

TABLE 10.1C (Continued)

Rice varieties/cultivars	Proximate composition (%)						References
	CHO	MC	Fat	Protein	Fiber	Ash	
Royal Feast	–	11.7 ± 0.04	0.1 ± 0.01	5.3 ± 0.15	0.13 ± 0.02	0.4 ± 0.16	Diako et al. (2011)
Sultana	–	11.2 ± 0.15	0.1 ± 0.03	5.6 ± 0.14	0.04 ± 0.01	0.3 ± 0.02	Diako et al. (2011)
WITA-9	67.11	22.00	0.80	7.08	0.53	2.48	Mbatchou and Dawda (2013)
Tanzania							
Kaling'anaula	87.56 ± 0.330	–	0.85 ± 0.21	9.07 ± 0.021	0.29 ± 0.028	0.66 ± 0.035	Shayo et al. (2006)
Kihogo red	87.84 ± 0.056	–	0.70 ± 0.14	8.84 ± 0.028	0.73 ± 0.040	0.97 ± 0.028	Shayo et al. (2006)
Salama	87.24 ± 0.700	–	0.57 ± 0.98	9.30 ± 0.028	0.39 ± 0.028	0.18 ± 0.028	Shayo et al. (2006)
Salama M17	87.66 ± 0.091	–	0.70 ± 0.19	9.46 ± 0.021	0.36 ± 0.028	0.55 ± 0.014	Shayo et al. (2006)
Supa	87.37 ± 0.028	–	0.83 ± 0.11	7.94 ± 0.021	0.40 ± 0.014	0.83 ± 0.622	Shayo et al. (2006)

TABLE 10.1D Proximate Composition of Rice Cultivars Reported from Different Geographic Location of Brazil, the United States, Sri Lanka, and China.

Rice varieties/cultivars	Proximate composition (%)							References
	CHO	MC	Fat	Protein	Fiber	Ash		
Brazil								
MNACE0501	79.3 ± 0.6	10.4 ± 0.2	2.0 ± 0.1	7.2 ± 0.4	3.2 ± 0.1	1.2 ± 0.0		Ascheri et al. (2012)
MNACH0501	77.5 ± 1.2	11.6 ± 0.7	1.8 ± 0.1	7.8 ± 0.6	3.0 ± 0.3	1.3 ± 0.0		Ascheri et al. (2012)
MNAPB0405	78.5 ± 0.7	11.2 ± 0.4	2.3 ± 0.1	6.9 ± 0.5	3.3 ± 0.1	1.1 ± 0.0		Ascheri et al. (2012)
Traditional	78.3 ± 0.6	11.2 ± 0.5	1.6 ± 0.0	7.7 ± 0.1	2.5 ± 0.2	1.1 ± 0.1		Ascheri et al. (2012)
The United States								
Ahrent	–	9.0	0.48	9.3	–	–		Cameron and Wang (2005)
Cocodrie	–	11.0	0.50	8.3	–	–		Cameron and Wang (2005)
Cypress	–	10.5	0.38	7.2	–	–		Cameron and Wang (2005)
Drew	–	10.5	0.51	6.6	–	–		Cameron and Wang (2005)
Francis	–	10.7	0.33	7.2	–	–		Cameron and Wang (2005)
Wells	–	10.3	0.46	6.8	–	–		Cameron and Wang (2005)
XL7	–	10.3	0.18	7.0	–	–		Cameron and Wang (2005)
XL8	–	9.8	0.24	6.8	–	–		Cameron and Wang (2005)
Sri Lanka								
Sri Lanka Red Rice 1	75.45 ± 0.09	12.94 ± 0.03	1.15 ± 0.03	9.63 ± 0.04	2.82 ± 0.55	0.82 ± 0.14		Sompong et al. (2011)
Sri Lanka Red Rice 2	76.05 ± 0.27	11.12 ± 0.06	2.19 ± 0.10	9.52 ± 0.12	2.87 ± 0.59	1.12 ± 0.44		Sompong et al. (2011)
Sri Lanka Red Rice 3	79.27 ± 0.26	9.85 ± 0.15	1.17 ± 0.05	8.72 ± 0.06	2.88 ± 0.55	0.98 ± 0.19		Sompong et al. (2011)
China								
Chinese Black Rice	75.71 ± 0.37	11.26 ± 0.28	2.85 ± 0.09	8.44 ± 0.02	4.08 ± 0.54	1.74 ± 0.02		Sompong et al. (2011)
Chinese Red Rice	74.66 ± 0.17	11.90 ± 0.20	2.35 ± 0.03	9.72 ± 0.04	2.52 ± 0.24	1.37 ± 0.02		Sompong et al. (2011)

for storage for less than 6 months, whereas about 12% or less than this value of MC is recommended for long-term storage (Wasserman and Caklerwood, 1972; Johnston and Miller, 1973; Boiling et al, 1977). MCs of rough rice in access of 14% are designated as sample grade under the US standards for rice.

10.2.1.1   DETERMINATION OF MC

The determination of the total MC of the grinded powdered rice sample is done according to the procedure described in AACC (2000) Method No. 44-15 02. The MC of the sample is determined by weighing 5 g of sample into a preweighed moisture box and drying it in an air-forced draft oven (Fig. 10.1A) at a temperature of $105 \pm 5°C$ till the constant weight of dry matter is obtained (AACC, 2000). The percentage of total MC of the sample is calculated from the difference in weight of samples as follows:

$$\text{Moisture } (\%) = \frac{\text{Wt. of dried sample}}{\text{Wt. of original sample}} \times 100$$

10.2.1.2   VARIATION IN MC OF RICE VARIETIES/CULTIVARS

MC invariably affects the quality and palatability of rice grains (Oko and Onyekwere, 2010), which plays a significant role in determining the shelf life (Webb, 1985). Generally, maximum 14% of the MC is considered safe for the storage of processed rice (Ministry of Agriculture, 1988), although acceptable value around 12% is recommended for long-term storage and avoiding insect infestation and microbial growth (Adair et al., 1973; Cogburn, 1985), but high moisture higher than considered and recommended MC is not good for shelf life of stored rice because it boosts up the microbial growth (Weinberg et al., 2008). Adair et al. (1973) suggested that if rice cultivars contain MC within its acceptable limit (12%), they may be kept for long-term storage. On average, however, milled rice grain contains about 12% moisture (Juliano, 1993a,b).

The standard inspection for brown rice designated the value of 15% as the maximum MC; it is allowed, however, to be raised by 0.5% (Afsar et al., 2001). Although the equilibrium MC of the rice grains varies depending on the preceding conditions that they have undergone, it is determined by the compactness of the tissues filled with endosperm starch. Under high

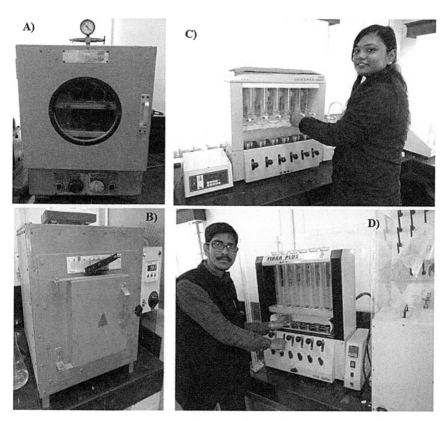

FIGURE 10.1   Apparatus (A) air-forced draft oven for determination of moisture content, (B) muffle furnace for determination of ash content, (C) Soxhlet apparatus (model: Pelican, Socplus, SCS06 RAS DLS, Pelican Equipments, India) for determination of fat content, and (D) apparatus (model: Fibra Plus, FES 6, Pelican Equipments, India) for determination of fiber content. (Picture credit: FST Lab, AgFE Department, IIT Kharagpur).

humidity conditions, it is higher with soft cultivars than with the hard type ones and also higher with opaque rice than with perfect rice. The equilibrium MC is greatly influenced by hysteresis. The value resulting from hydration as well as dehydration under an ambient relative humidity ranging from 11.1% to 86.7% was examined by Tsutsumi et al. (2003).

Gooding and Davies (1997) estimated the MC of rice grain during storage. The MC is also considered as an important feature because all grains are stored for a certain period before their ultimate utility. Grains with high MCs are difficult to store safely because these are more vulnerable to attack by pests and diseases as well as hydrolytic changes (Gooding and Davies,

1997). Generally, the MC in the rice grain can be determined by oven drying method. It has been observed that MC often varies from 7% to 11% in rice grain (Awan, 1996).

In Indian rice cultivars, Verma and Srivastav (2017) found high MC (13.57%) in Govind Bhog when compared with the other cultivars, whereas only Badshah Bhog (8.90%) found to have the lowest value. In 92 rice varieties, Devi et al. (2015) stipulated highest (11.6%) moisture in Pusa basmati-1 and Swarnamukhi (NLR 145) varieties, whereas the lowest MC with 7.13% showed by Bhuban variety. Subudhi et al. (2013) reported Kala namak possessed highest MC 10.24% followed by Tulsiphool (9.92%) and Kalajira (9.9%), whereas Geetanjali had the lowest value (9.6%). In short- and long-grain aromatic rice varieties, Ravi et al. (2012) estimated 9.3% grain moisture for indigenous organic Asian Indian rice variety *Salem samba*. The MC in aromatic pigmented rice varieties of Manipur (India), namely, Bakuljoha (13.7%), Ketekijoha (12.1%), Poreitonchakhao (13.5%), and Chak-haoamubi (11.6%) were reported by Saikia et al. (2012). Yadav et al. (2007) reported P-44 had higher MC (12.72%) in different Indian rice cultivars followed by Jaya (12.55%) and HKR120 (12.15%), whereas lowest MC was noted in Sharbati cultivars.

In Bangladeshi rice cultivars, Ali et al. (2012) reported 12.65% MC in scented rice *var. Kalijira* of Bangladesh. In six aromatic rice cultivars, Asaduzzaman et al. (2013) found BRRI dhan-37 had highest value (15.13%) followed by Begun Bichi (12.39%) and Philipine Katari (12.11%), whereas lowest value 11.25% noted BRRI dhan-50 cultivar. Twelve rice varieties investigated by Islam et al. (2016) and found variety Paijam (13.22%) had highest moisture and lowest 10.21% in Jeerashile, whereas rest almost all varieties in the study, namely, Basmoti, Chinigura, Govindavogh, Kata-rivogh, Nagirshail, Nurjahan, and Pari shiddho were noted to have more than 12% MC except Parija (10.45%), Minikat (11.04%), and Sawrna (11.37%).

In Brazil rice cultivars, Ascheri et al. (2012) observed in Brazilian geno-types MNACH0501 had highest MC (11.6%) followed by traditional variety (11.2%) and MNAPB0405 (11.2%), whereas MNACE0501 possesses low MC (10.4%). In 20 commercial samples of long-grain rice, Heinemann et al. (2005) reported the MC on the range of 9.39–13.50% in brown, parboiled brown, and parboiled milled rice.

In Ghana rice cultivars, Diako et al. (2011) reported in scented rice varieties, high MC found in *local varieties* Marshall followed by Ex-Baika (11.8%), Ex-Hohoe (11.6%), Jasmine 85 (11.6%), and Sultana, an *imported brands* had low MC. Mbatchou and Dawda (2013) obtained highest MC

22% possess by WITA-9 and lowest by IR12979-24-1 (8.50%) among four rice varieties cultivated in Ghana.

In Nigerian varieties, Fagbohun and Oluwaniyi (2015) reported increased MC 7.84% to 11.82% with 9.89 and 2.09 mean value and standard deviation, respectively, in Nigerian varieties during storage. Oko et al. (2012) observed E4197 variety noted for high values (9.67%) among the 20 Nigerian varieties followed by Sipi (8.0%), whereas least amount of MC 3.67% showed by Canada variety. In five major rice varieties of Nigeria, Oko and Ugwu (2011) found Sipi variety had 18.00% highest MC followed by Faro 14 (7.33%), Awilo (6.67%), and Faro 15 (6.33%), whereas lowest MC 3.67% confined by Canada variety. In lowland rice varieties, Oko and Onyekwere (2010) recorded highest moisture (35.67%) in IR68 and lowest (5.0%) in PSBRc50 variety. Ebuehi and Oyewole (2008) reported 7.5% MC in raw rice varieties Ofada, a local and indigenous rice cultivated in the southwest of Nigeria, whereas Aroso, a foreign and imported rice produced in Thailand, was noticed 8.0% MC by Ebuehi and Oyewole (2008).

In Thai rice variety, Wichamanee and Teerarat (2012) reported 12.91% moisture in red aromatic brown. Sompong et al. (2011) detected lowest MC (9.28%) in Sung Yog Phatthalung and highest (13.12%) in Niaw Lan Tan of red variety. Maisont and Narkrugsa (2009) reported Khao Dawk Mali105 (KDML105) had high MC 11.81% among 10 varieties, whereas highest and lowest MC values were found in Sew-Mae-Jan (SMJ) and Hahng-Yi71 (HY71) were 10.52% and 12.26%, respectively.

In locally grown and imported rice varieties of Malaysia, Thomas et al. (2013) recorded highest and lowest values of MC in Brown (12.88%) and Glutinous (10.04%) varieties, respectively. Anjum et al. (2007) found Sarshar (9.19%) had highest, whereas Dr-83 noted 11.10% as lowest MC in different Pakistani coarse rice varieties. Cameron and Wang (2005) reported 9.0% (Ahrent) to 11.0% (Cocodrie) in eight long-grain rice cultivars of the United States.

10.2.2 FAT CONTENT

Lipids are defined as natural substances which dissolve in an organic solvent. The lipid or fat content of rice is mainly in the bran fraction (20%, on dry matter basis), specifically as lipid bodies or spherosomes in the aleurone layer and bran. However, about 1.5–1.7% is present in milled rice, mainly as nonstarch lipids extracted by ether, and chloroform–methanol and cold water saturated butanol (Juliano and Goddard, 1986). Lipids in plant seeds are mainly studied for their physiological metabolism and food nutrition.

The fat content of rice is low and most of it is removed in the process of milling and is contained in the bran (Grist, 1975). Lipids are also known to influence viscoelastic properties by forming inclusion complexes with the helical structure of amylose (Hamakcr and Griffin, 1990). Maningat and Juliano (1980) found that defatting rice starch reduced both gelatinization temperature and gel viscosity of starch; Morrison and Azuclin (1987), however, could not find any predictive relationship between starch lipid content and viscoelastic properties.

10.2.2.1   DETERMINATION OF FAT CONTENT

The determination of fat in rice sample is done by using Soxhlet extraction as the method described by (AACC, 2000) Method No. 30-25 01. Approximately, 5 g of moisture free samples is weighed accurately into labeled thimbles. The dried boiling flasks (250 mL) are weighed correspondingly and filled with about 150 mL of petroleum ether (boiling point 60–68°C). After that, the Soxhlet apparatus (Fig. 10.1C) is assembled and allowed to reflux for 45 min. After boiling with petrol, the boiling flask is heated in a hot air oven until it is almost free of petroleum ether. After drying, it is cooled in a desiccator and weighed (AACC, 2000). The percentage fat in the sample is calculated using the following formula:

$$\text{Fat } (\%) = \frac{\text{Wt. of fat}}{\text{Wt. of original sample}} \times 100$$

10.2.2.2   VARIATION IN FAT CONTENT OF RICE VARIETIES/ CULTIVARS

The range of fat contents varied from 0.50% to 2.23% in different rice varieties (Taira and Itani, 1988; Sotelo et al., 1990; Tufail, 1997). In six aromatic and two nonaromatic rice cultivars of India, Verma and Srivastav (2017) found Govind Bhog had the highest fat content of 0.92%, followed by Khushboo (0.77%) and Gopal Bhog (0.72%), whereas Sarbati exhibited the lowest fat content (0.06%). Devi et al. (2015) recorded 0.5–3.77% fat content brown rice, which were noted significantly highest (3.7%) in rice variety MTU 1001 in followed by MSS 5, Nalini, and Sahyadri. Among the rice varieties, lowest content of fat is found in PR 115, SGT 1, and Barah Aavarodhi (Devi et al., 2015). Short- and long-grain aromatic rice varieties of India were studied by Subudhi et al. (2013); the highest fat percentage was found in Geetanjali (1.62%) followed by Toroari Basmati (1.38%) and Pusa Sungandh-2 (1.37%), whereas Kalajira had the lowest value (0.25%). Ravi et al. (2012) reported 0.5% and 1.18% fat for normal raw rice and *Salem samba* rice variety, respectively. Pigmented and nonpigmented aromatic rice cultivars of India were studied by Saikia et al. (2012) who found lowest (1%) fat in Keteki joha and Chak-haoamubi while highest (2.10%) in pigmented cultivars Poreiton chakhao followed by nonpigmented rice cultivar Bakul joha (1.6%). In different Indian basmati and nonbasmati cultivars, Yadav et al. (2007) found significant difference and reported Jaya, a coarse variety of nonbasmati had more fat content (0.82%) in comparison with basmati variety Bas-370 (0.65%) and nonbasmati fine variety Sharbati (0.66%), whereas rest cultivars, namely, HKR-120 and P-44 were followed to the Jaya with 0.78% and 0.72% fat content respectively. Basmati cultivar HBC-19 observed with low fat content in the study of Yadav et al. (2007).

Islam et al. (2016) reported fat in 12 rice varieties of Bangladesh, among which Parija (2.90%) possessed highest fat content followed by Nagirshail (2.61%) and Chinigura (1.92%), whereas Sawrna variety (0.09%) recorded for lowest fat value. Bangladeshi aromatic rice cultivars were significantly different for fat values reported by Asaduzzaman et al. (2013) who noticed Phillipine Katari (1.45%) variety had high fat content followed by BRRI dhan-34 (1.33%) and Kalizira (0.93%), while Begun Bichi (0.68%) had low fat value. Ali et al. (2012) also reported fat content (0.57%) in Bangladeshi scented milled rice varieties *Kalijira*.

The work of Hayat et al. (2013) on Pakistani rice varieties reported 3.64–4.08% fat content in IRRI-6 and Basmati Supper, respectively. Muhammad (2012) found 1.92–2.70% fat in different Pakistani varieties. Anjum et al. (2007) studied Pakistani rice varieties and found lowest fat in IRRI-9 (5.16%) while highest in IRRI-6 (6.14%) followed by Sarshar (5.80%) and Dr-83 (5.45%).

In kernels of long-grain rice cultivars of the United States, Cameron and Wang (2005) reported highest fat content on dry weight basis (Db) indicated by the cultivar Drew (0.51%) followed by Cocodrie, Ahrent, Wells, and Cypress with the fat content of 0.50, 0.48, 0.46, and 0.38, respectively, whereas lowest shown by hybrid cultivars XL7 (0.18%). In Brazilian commercial rice, Heinemann et al. (2005) reported statistically similar (*p*

> 0.05) mean value of crude fat in brown and parboiled brown rice, which were 2.69% and 2.65%, respectively, whereas parboiled milled and milled rice comparatively lower with 0.38% and 0.50% fat. In Japanese rice, Seki et al. (2005) reported highest fat content in pregerminated brown rice, that is, 24.8%, while lowest (1.5%) was found in white rice. According to Storck et al. (2005), the polished white rice contains 0.36% lipids (Db). The differences between five rice cultivars grown in Morogoro (Tanzania) for fat content were observed by Shayo et al. (2006); lowest (0.57%) in Salama, whereas highest (0.85%) in Kaling'anaula.

A comparative study of Chinma et al. (2015) on Nigerian rice varieties found higher fat content (2.41%) in the improved variety (MR 219) compared to the three Nigerian rice varieties, namely, Jamila (2.33%), Jeep (2.17%), and Kwandala (2.15%). Fagbohun and Oluwaniyi (2015) reported reduce fat contents of the rice from 2.04% to 1.57% after it was sundried and stored. In 20 local and newly introduced rice varieties cultivated in Nigeria, Oko et al. (2012) reported Canada and E4197 contained highest fat content (3.50%) followed by E4077 (3.0%), whereas Sipi, Faro14 (I), and Faro14 (II) varieties were noticed for lowest fat vales (0.5%). The work of Oko and Ugwu (2011) in Southeastern Nigerian rice variety showed 0.50–3.50% fat with higher and lower values in Canada and the varieties are Faro 14 and Sipi, respectively. In five new lowland hybrid rice varieties of Nigeria, Oko and Onyekwere (2010) found highest fat value in IR68 (3.5%), whereas lowest fat content (1.5%) found in PSBRc50 and IR75395-2B-B-1-1-1-2-4 varieties. Ibukun (2008) found 0.8–1.22% in parboiled paddy rice obtained from rice farmers of Nigeria. Ebuehi and Oyewole (2008) found 1.9–2.6% fat in raw rice varieties of Nigeria and Thailand, namely, Ofada and Aroso, respectively.

In 11 early and 5 late Indica rice varieties of China, Yong-Liang et al. (2008) reported early Indica rice variety M103s/Zhongzu1 (2.87%) had higher fat content followed by early variety M102s/Zhongzu1 (2.78%) and late variety Yuchi (2.66%), whereas lowest fat content found in early variety Liangyou301 (1.27%). The fat content in the Indica rice flour was found to be 1.65%, while in Japonica rice flour, it was 1.60% reported by Qin-lu et al. (2011).

Diako et al. (2011) found highest fat content (0.7%) in *local* scented rice varieties of Ghana, namely, Ex-Baika, Ex-Hohoe, and Marshall followed by Jasmine 85 (0.5%) and lowest (0.1%) reported in Royal Feast and Sultana *imported brands* scented rice varieties. Another study of four rice varieties cultivated in Kassena-Nankana District, Ghana was by Mbatchou and Dawda

(2013) who reported varieties, IR12979-24-1 and ANDY-11 were high fat (1.20%) containing varieties in comparison to JASMINE-85 (1.10%) and WITA-9 (0.80%).

Fari et al. (2011) reported fat content in four different Sri Lankan rice varieties used for rice noodles preparation which were ranged from lowest 0.56 ± 0.09% for Bathalagoda 300 to highest 1.36 ± 0.03% in Bathala-goda 94-1 followed by Ambalantota 306 (1.28%) and Bombuwala 272-6b (1.17%).

In Thai rice variety, Sompong et al. (2011) reported highest in Thai black glutinous rice varieties Niaw Dam Pleuak Khao (3.72%) and Niaw Dam Pleuak Dam (3.65%) followed by Thai red glutinous rice varieties Niaw Dawk Yong (3.19%) and Niaw Lan Tan (3.08%), whereas lowest in Sri Lanka Red Rice 1 (1.15%) and Sri Lanka Red Rice 3 (1.17%). Wichamanee and Teerarat (2012) reported Thai red aromatic brown rice in which fat content found 2.82%.

The fat content in the flour of six long-grain rice varieties of Ecuador, namely, INIAP 14, INIAP 15, INIAP 16, INIAP 17, F09, and F50 were studied by Cornejo and Rosella (2015) and found to have 0.63%, 0.59%, 0.60%, 0.47%, 0.79%, and 0.95%, respectively. In imported and local culti-vated rice varieties of Malaysia, Thomas et al. (2013) reported that locally cultivated rice varieties, namely, "Black rice" (0.07%) and "Brown rice" (1.74%) were noticed as lowest and highest fat content, respectively. Ascheri et al. (2012) reported MNAPB0405 (2.3 dag/kg) highest among different Brazilian red rice genotypes followed by MNACE0501 (2.0 dag/kg) and MNACH0501 (1.8 dag/kg) while lowest in traditional variety (1.6 dag/kg).

10.2.3 CP CONTENT

The protein content of milled rice is low in comparison with other cereals, although the whole rice grain content N × 5.95% rice ranged from 7.0% to 10.8% of which 70–80% is in the glulelin. The in vitro protein digestibility was relatively good (87.6–91.8%) as shown by Yousif (2000). Protein, as the other major constituent of rice, has not been thought to strongly influence cooking and eating qualities. When differences in gross protein content were examined in relation to texture of cooked rice, only a weak relationship was found, the higher protein rices being somewhat less tender than low protein rices (Onate et al, 1964; Juliano et al, 1965). Because commonly eaten rices generally contain about 7% protein and do not fluctuate widely form this level, protein content is not considered an important indicator of quality (Hamaker and Griffin, 1990).

Cultivars with early maturity generally exhibit a higher level of protein content compared with those with late maturity (Honjo, 1971). Among grains of different properties in appearance with an identical sample, white–belly rice also exhibits a lower content than the normal rice (Nagato et al., 1972). High ambient temperature (Honjo, 1971; Kataoka, 1975) and high water temperature (Honjo, 1971) increase protein content in rice. Brown rice from upland rice crops exhibited higher protein content than that of flooded rice by 30% (Taira, 1970).

The protein quality of rice depends on the composition of essential amino acids (EAAs) (FAO, 1970), but the balance of AA in rice protein is exceptionally good because of the averages content lysine is about 3.8–4.0% of the protein that's reason lysine called limiting AA (Jenning et al., 1979; Frei et al., 2003) but higher than any other cereal protein, whereas glutamic and aspartic acid are found in high amount (FAO, 2006). In composition of the rice, the EAAs, that is, histidine, proline, and threonine are varied widely. Abdul-Hamid et al. (2007) reported that the major AAs found (in decreasing order) were arginine, glutamic acid, aspartic acid, and serine. Rice protein is superior and more nutritious because of its unique composition of AAs and has a special benefit because its higher lysine content among the eight of the EAAs is found in delicately balanced proportions in comparison to any other cereal protein (Ahmed et al., 1998; Lásztity, 1984).

10.2.3.1   DETERMINATION OF CP CONTENT

There are different methods used to determine the protein content in rice, but micro-Kjeldahl analysis is preferred. The powdered rice samples are measured for CP content by Kjeldahl method (Fig. 10.2) as described in AOAC (1990), which involved protein digestion and distillation.

10.2.3.1.1   Protein Digestion

About 2.0 g of the sample is weighed into an ash less filter paper and put into a 250 mL Kjeldahl flask. Then, 5 g of digestion mixture (as catalyst) and 10 mL of 98% conc. H_2SO_4 (sulfuric acid) are added. The whole mixture is subjected to heating in the digestion chamber until transparent residue contents are obtained. Then, it is allowed to cool. After cooling, the digest

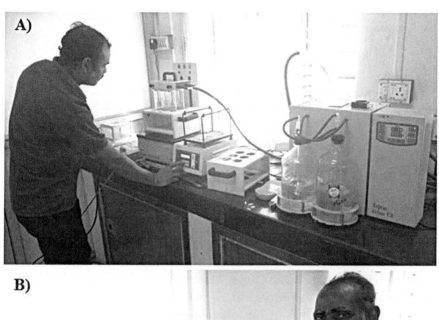

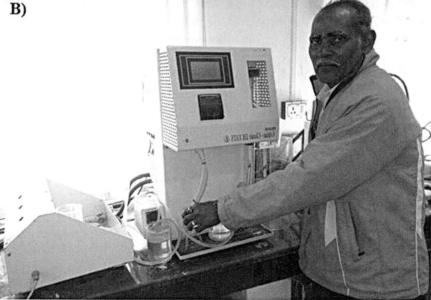

FIGURE 10.2   Kjeldahl analysis unit for protein determination. (A) Digestion unit and (B) distillation unit. (Picture credit: FST Lab, AgFE Department, IIT Kharagpur).

is transferred into a 100-mL volumetric flask and made up to the mark with distilled water and then distilled using distillation apparatus.

10.2.3.1.2 *Protein Distillation*

Before use, the distillation apparatus is steamed for 15 min after which a 100-mL conical flask containing 5 mL of 2% boric acid and one or two drops of mixed indicator is placed under the condenser such that the condenser tip is under the liquid. About 5.0 mL of the digest is pipetted into the body of the apparatus via a small funnel aperture. The digest is pushed down with distilled water followed by addition of three to four drops of phenolphthalein and 5 mL of 40% (W/V) NaOH solution. The digest in the condenser is steamed through until enough ammonium sulfate $[(NH_4)_2SO_4]$ is collected. The boric acid plus indicator solution changed color from red to green showing that all the ammonia liberated had been trapped. The solution in the receiving flask is titrated with 0.063 N hydrochloric acid up to a purple end point. Also, a blank is run through along with the sample. After titration, the % nitrogen is calculated using the following formula:

$$\% \text{ Nitrogen} = (V_s - V_B) \times M \text{ acid} \times 0.01401 \times 100W$$

where V_s is the volume (mL) of acid required to titrate sample, V_B is the volume (mL) of acid required to titrate the blank, M acid is the molarity of acid, and W is the weight of sample (g).

Then, percentage CP in the sample is calculated from the % nitrogen as % CP = % N $\times F$, where F (conversion factor) is equivalent to 6.25 (AOAC, 1990).

10.2.3.2 *VARIATION IN CP CONTENT OF RICE VARIETIES/ CULTIVARS*

Rice protein is a good source of EAAs for human nutrition. Many reports says that although rice is not rich in protein, but the protein quality of rice is far superior to other cereals (Janick and Whipkey, 2002), due to high content of lysine when compared to wheat, corn, and sorghum. Araullo et al. (1976) stated that brown rough milled rice consists of about 8% and 7% protein.

A study on randomized block design experiment in 2013 was conducted on Indian rice variety (including land races, improved varieties, aromatic varieties, and red rices) by Devi et al. (2015) at ICRISAT (International Crops Research Institute for the Semi-Arid Tropics) campus, Patancheru, Hyderabad, India. The study of Devi et al. (2015) showed high protein in Sahyadri (11.0%), a hybrid rice variety, whereas Swarna (5.9%) was noted as low protein rice. Rest varieties were possessed 10–12% protein content, that is, significantly highest. In aromatic long and short grain rice varieties of India, Subudhi et al. (2013) reported highest protein found in Toroari Basmati (8.96%) followed by Kala namak (8.76%) and Ketekijoha (8.74%) while lowest in Dubraj (7.2%). Ravi et al. (2012) estimated 6.8% and 7.50% protein for normal raw rice and *Salem samba* rice variety of India, respectively. Another experiment was conducted by Saikia et al. (2012) on aromatic rice and reported Keteki joha (9.9%) had highest protein on Db followed by Chak-haoamubi (8.8%) and Bakul joha (7.7%), whereas Poreiton chakhao (6.6%) was noted lowest. The findings of Saikia et al. (2012) on aromatic rice varieties stated to be good sources of protein. Yadav et al. (2007) reported protein contents in different Indian rice cultivars ranged from 5.46% to 7.02% which was noted highest in Bas-370 followed by HBC-19 (6.94%) and Sharbati (6.52%) while lowest in Jaya. Bhattacharjee et al. (2002) previously reported 7.1–8.9% protein for common Indian rice varieties of including scented ones.

In eight long-grain US rice cultivars, Cameron and Wang (2005) measured protein by micro-Kjeldahl method and observed differences were found in the protein content of rice cultivars; Ahrent (9.3%) was noted to have high protein content followed by Cocodrie (8.3%), and Cypress and Francis with 7.2%, whereas XL (6.8%) and Drew (6.6%) were ascribed to have low protein content. Champagne et al. (2005) noted 5.1–5.3% protein contents in rice variety grown in California, the United States.

In different Pakistani rice varieties, Awan (1996) showed protein 7.38–8.13%. Ansari et al. (2013) analyzed rice varieties cultivated in Sindh province of Pakistan and found 6.86% protein content which was highest in Jajai-77 followed by Shadab (9.57%) and IR-6 (8.59%) and lowest was noted in IR-8 (9.7%); the rest varieties were Khushboo-95 (8.21%), Shua-92 (7.76%), and Sarshar (7.74%). Hayat et al. (2013) studied polished and brown rice varieties of Pakistan, namely, Basmati Super, Basmati 385, Basmati 2000, IRRI-6, and IRRI-9; and reported 6.22–8.24% and 7.14–8.95% protein in polished and brown rice, respectively. However, IRRI-6 possessed highest (8.24%) protein among all the cultivars. Anjum et al. (2007) studied different Pakistani rice varieties and observed Sarshar (8.80%) had highest protein content and lowest protein content found in IRRI-9 (7.80%), whereas other cultivars, namely, IRRI-6 (8.77%) and Dr-83 (8.12%) also observed which passed marginal differences as compared to Sarshar.

Bangladeshi rice varieties were studied by Islam et al. (2016) and showed highest protein (8.31%) in Katarivogh rice variety followed by Nagirshail (8.11%) and Govindavogh (7.93%), whereas lowest values of protein were recorded in Minikat (6.31%). Hossain et al. (2015) investigated some selected rice cultivars produced from piedmont plain soils of Bangladesh for biochemical quality and recorded highest protein content in Paiza (7.8%) and lowest in Taipi (6.2%). Asaduzzaman et al. (2013) observed highest protein (6.21%) in BRRI dhan-50 variety in the comparison to other aromatic cultivars of Bangladesh, namely, Begun Bichi (5.74%), BRRI dhan-37 (5.69%), and BRRI dhan-34 (5.44%), whereas lowest protein values were noticed in Kalizira (5.18%) and Philipine Katari (3.23%). Ali et al. (2012) also reported 6.81% protein content in Bangladeshi scented milled rice varieties *Kalijira.*

The work of Ascheri et al. (2012) reported protein contents in red rice genotypes of Brazil; MNAPB0405 (6.9%) had highest protein contents followed by, namely, traditional variety (7.7%) and MNACE0501 (7.2%), whereas lowest contents were found in MNACH050 (17.8%). In the Brazilian Table of Food Composition, protein content of 4.3–18.2% in polished white rice were reported from Brazil and cited in the review of Walter et al. (2008), whereas the work of Storck et al. (2005) on Db shown 8.94% protein contains in polished white rice of Brazil. TACO (2006) tabulated protein content of 7.3% in long-grain brown and 7.2% in raw polished rice. Heinemann et al. (2005) recorded 5.71–7.42% protein contents in different 20 commercial rice samples of Brazil in which wide variation was noted as 6.34–7.42% for brown, 6.24–7.02% for parboiled brown, 5.71–6.71% for parboiled milled, and 6.10–7.20% for milled rice.

In rice sample of Ado-Ekiti, Nigeria, Fagbohun and Oluwaniyi (2015) observed increased protein content during storage from 11.45 to 12.56 with 12.10 and 0.94 mean value and standard deviation, respectively. A study conducted by Oko et al. (2012) on 20 local and newly introduced Nigerian rice varieties observed statistically significant variation and reported the newly introduced hybrid variety "E4212" with relatively highest protein content (7.94%), followed by E4197 (6.30%), Faro14 (I) (6.22%), and Faro 14 (II) (6.22%), whereas Sipi (1.58%) had the lowest protein content followed by China (2.98%). In the study of Oko and Ugwu (2011), Nigerian rice varieties were stipulated for protein content and found that the variety Faro 14 rice (1.58%) exhibited high content of protein and rest all the varieties, namely, Awilo, Canada, and Faro 15, found to have near to

5% protein except Sipi that possessed lowest protein content (6.22%). Oko and Onyekwere (2010) found that lowland rice varieties, namely, P5BRc50 had highest protein content (7.9%), whereas lowest (4.0%) was noted in IR77384-12-17-3-18-2-B. Ibukun (2008) recorded 6.55–8.40% protein in parboiled rice collected from rice farmers of Nigeria. The work of Ebuehi and Oyewole (2008) on Nigerian rice varieties recorded lowest protein content (6.95%) in Aroso and highest (7.30%) in Ofada.

The rice varieties of Ghana were studied by Mbatchou and Dawda (2013) for nutritional composition and found variation among the four rice varieties in their protein content. The variety IR12979-24-1 was noted as lower protein content (6.01%), whereas ANDY-11 (7.42%) was higher in protein content followed by WITA-9 (7.08%) and JASMINE-85 (6.82%). Diako et al. (2011) found lower protein content in scented rice varieties of Ghana. The ranges of protein were 5.3% (Royal Feast) to 5.9% (Marshall) followed by local varieties Ex-Baika and Jasmine 85 with 5.8% protein and rest varieties, Ex-Hohoe (local variety) and Sultana (imported brand), had 5.6% protein. The high levels of protein were reported in Ghana from local and breeding lines by Adu-Kwarteng et al. (2003) who stipulated higher levels in local varieties (6.78–10.5%) relative to the breeding lines (5.95–9.16%) which existed with significant differences ($p < 0.05$) in the protein contents.

In Sri Lankan eight popular rice varieties, Fari et al. (2011) found ranged of protein content on Db from $6.84 \pm 0.20\%$ (Bathalagoda 94-1) to $11.18 \pm 0.22\%$ (Labuduwa 366) comparatively low to high among the rice varieties which has red pericarp followed by Ambalantota 306 (10.02%) and Bombuwala 272-6b (9.76%). In the study by Sompong et al. (2011), different rice varieties were collected from China, Sri Lanka, and Thailand and found protein content appreciably highest (10.36%, that is, >7%) in all the tested varieties, especially the Thai varieties PD (10.4%) and SY (10.4%) which were noticed for highest protein values and were reported interesting for developing new food products (Sompong et al., 2011).

In Indonesian Indica Rice, highest protein was reported by Lestari et al. (2014) in Batang Gadis (10.33%) followed by Cimelati (10.23%) and Memberamo (10.22%), whereas the variety Jatiluhur (5.53%) exhibited lowest value. In locally grown and imported rice varieties of Malaysia, the protein content reported by Thomas et al. (2013) was found higher (8.16%) in black (imported) rice followed by Glutinous (8.14%) and Basmati (7.75%) varieties, whereas lowest protein showed by white local variety (5.96%). In red aromatic brown rice of Thailand, Wichamanee and Teerarat (2012) found 7.98% protein. A study of Maisont and Narkrugsa (2009) noted highest protein in HY71 (8.34%) followed by Sakon-Nakhon (7.58%) and SMJ (7.48%), whereas lowest in NPH1 (6.32%) among all varieties. In China, a study of Yong-Liang et al. (2008) on 11 early and 5 late Indica rice varieties reported high protein content in Liangyou103 (13.26%) followed by Zhefu802 (12.82%) and Jinyou402 (12.45%), whereas early Indica rice variety, M103s/20257 (9.96), noticed for lowest protein content. The in 12 selected local cultivars of rice grown in Morogoro Region of Tanzania were studied by Shayo et al. (2006); they found highest protein content in Salama M17 (9.46%), while lowest in Supa cultivars (7.94%). The rice cultivars Salama M17, Salama, and Kaling'anaula did not show the significant difference with each other, whereas Kihogo Red and Supa cultivars showed significant differences and the rest of the cultivars at $p < 0.05$. Liu et al. (2005) found 7.35–11.47% proteins in Japanese rice genotypes. From International Rice Research Institute (IRRI) of Philippines, Gomez (1979) reported the protein content of rice from 4.3% to 8.2%.

10.2.4   ASH CONTENT

Ash content is one of the important factors to assessing the purity of flour (Awan, 1996) which also plays an important role to reflect the mineral elements of a food sample (Bhat and Sridhar, 2008; Oko and Onyekwere, 2010; Mbatchou and Dawda, 2013) and gives an idea to determine the levels of essential minerals present in the food (Edeogu et al., 2007). Moreover, since greater amounts of rice bran are removed from the grain during milling and polishing, more minerals are lost (Wardlaw and Kessel, 2002). The high percentage of ash content may affect the sensory quality of the rice especially color and taste (Juliano, 1985d). The mineral with the highest concentration was nitrogen, which was used in the computation of CP.

10.2.4.1   DETERMINATION OF ASH CONTENT

The ash content in proximate analysis of a food sample gives an idea of the essential mineral elements present in the food and also plays an important role to determine their levels (Bhat and Sridhar, 2008). The determination of total ash content of rice samples is measured according to the method described in AOAC (1997) by keeping 5 g of sample in silica crucible at 550 ± 5°C for 4–6 h in a muffle furnace (Fig. 10.1B) (AOAC, 1997).

$$\text{Total ash content } (\%) = \frac{\text{Wt. of ash}}{\text{Wt. of original sample}} \times 100$$

10.2.4.2 VARIATION IN ASH CONTENT OF RICE VARIETIES/ CULTIVARS

Indian aromatic and nonaromatic rice cultivars were studied by Verma and Srivastav (2017) and found significant difference among all rice cultivars. The study of Verma and Srivastav (2017) noted highest amount of ash in aromatic rice cultivar Khushboo (0.73%) followed by Swetganga (0.66%) and Badshah Bhog (0.59%), whereas lowest ash content in Sarbati (0.35%). Devi et al. (2015) reported Pusa Basmati-1 (2.34%) had highest ash content among the 92 Indian rice varieties followed by Sabita (1.72%), MTU 1010 (1.67%), and Vasumati (1.66%), whereas the lowest ash content 0.43% was noticed in Shakti and Sashi varieties. Keteki joha, nonpigmented Indian aromatic rice reported by Saikia et al. (2012) possessed highest ash (0.8%) on Db followed by Bakul joha (0.7%), while the varieties, namely, Poreiton chakhao and Chak-haoamubi, had lowest ash content of 0.5%. Yadav et al. (2007) reported highest ash content (0.67%) in HBC-19 and P-44 rice followed by Bas-370 (0.44%) and Sharbati (0.41%). The rice variety Jaya (0.31) showed low ash content. The study of Yadav et al. (2007) stated that no significant difference in ash content of rice cultivars, namely, P-44 and HBC-19 with each other while Jaya was significantly different.

The report of Hayat et al. (2013) on Pakistani rice varieties showed that IRRI-9 variety had 2.46% of ash content and Basmati Supper noted with 1.44% ash content. Different rice varieties of Pakistan observed for ash content ranged from 3.16% to 3.79% by Anjum et al. (2007), whereas 0.54–6.04% was found in different milling fractions which were highest and lowest in bran and polished rice. Sarshar possessed highest ash content followed by IRRI-6 (3.67%) and IRRI-9 (3.43%) while lowest in Dr-83. Anjum et al. (2007) found the ash content in brown (1.42%) and white (0.66%) rice. In 12 locally produced polished rice of Bangladesh, Islam et al. (2016) found highest ash content in Sawrna (0.57%) followed by Parija and Minikat with 0.52%, whereas lowest value showed by Chinigura (0.30%). Asaduzzaman et al. (2013) studied Bangladeshi aromatic rice cultivars and reported ash content 1.460%, 1.210%, 1.070%, and 1.050% in BRRI dhan-50, BRRI dhan-34, Begun Bichi, and BRRI dhan-37, respectively, while the rice cultivars Kalizira (0.923%) and Philipine Katari (0.887%) were with low ash value. Bangladeshi scented rice *Kalijira* was reported with 0.49% ash content by Ali et al. (2012).

In Nigeria, sundried and stored rice in the study of Fagbohun and Oluwaniyi (2015) reported 4.51–4.25% reduced ash contents. Oko et al. (2012) found that Canada variety had high ash content (2.0%) followed by all those varieties such as Chinyereugo, Mass (I), and Ezichi noted with 1.5% protein content, whereas the varieties Awilo, E4077, E4314, Faro15 (I), and Mass (II) were low ash content with 0.50%. Another work of Oko and Ugwu (2011) was also on Nigerian rice variety and reported that Canada (2.00%) variety possesses highest ash content, whereas Faro 15 and Awilo were the varieties that had lowest ash content of 0.50%. Five lowland varieties of Nigerian rice were analyzed by Oko and Onyekwere (2010) and reported highest ash content of 1.0% in IR68, PSBRc50, and IR75395-2B-B-1-1-1-2-4 varieties, whereas lowest ash content of 0.5% in the varieties, namely, IR77384-12-17-3-18-2-B and IR77647-3B-8-2-2-14-4. The parboiled rice collected from Nigerian rice farmers reported by Ibukun (2008) to have 0.18–1.77% ash content. Ebuehi and Oyewole (2008) recorded highest ash content in Ofada (0.80%) and lowest in Aroso (0.53%) rice varieties of Nigeria.

The nutritional composition of rice varieties cultivated in Ghana were analyzed by Mbatchou and Dawda (2013). They found highest ash content in WITA-9 (2.48%) and lowest IR12979-24-1 (0.86%), whereas rest varieties, namely, ANDY-11 and JASMINE-85 were noted with 1.37% and 0.93% of ash content, respectively. Ex-Hohoe, a local variety of Ghana, was reported by Diako et al. (2011) with highest ash content of 0.7% followed by Jasmine 85 and Marshall had 0.6% ash with together while lowest was found in Sultana (0.3%), a scented rice variety of imported brand. Adu-Kwarteng et al. (2003) were studied local varieties and breeding lines of rice in Ghana and found that 0.56% of average values of ash content in local varieties and 0.52% in breeding lines with significant differences at $p < 0.05$.

In raw grains of four Brazilian red rice genotypes, Ascheri et al. (2012) reported ash content between 1.1 dag/kg (traditional variety and MNAPB0405) and 1.3 dag/kg (MNACH0501) followed by MNACE0501 (1.20 dag/kg). Heinemann et al. (2005) reported 1.15–1.29%, 0.91–1.46%, 0.49–0.60%, and 0.32–0.59% of ash content in Brazilian commercially rice cultivars, namely, brown, parboiled brown, parboiled milled, and milled rice, respectively. Storck et al. (2005) categorized rice cultivars on the basis of chemical composition and reported 0.38% ash (Db) in polished white rice.

In Malaysian rice varieties, Thomas et al. (2013) found highest ash content (0.90%) in black (imported) and lowest (0.39%) in white (local) rice among the studied varieties, whereas rest varieties, namely, Basmati, Brown, Bario, and Glutinous had 0.48%, 0.55%, 0.63%, and 0.82%, respectively. Thai red aromatic brown rice was reported by Wichamanee and Teerarat (2012) with 1.80% ash content. Ash content among the rice varieties collected from Thailand, China, and Sri Lanka was studied by Sompong et al. (2011) who reported highest ash contents of 1.7%, 1.5%, and 1.5% in black Chinese and two Thai (LP and PD) rice varieties, respectively. There are significant differences among all studied rice varieties, except black varieties because these tend to have high ash contents (Sompong et al., 2011). Shayo et al. (2006) studied rice cultivars grown in Morogoro region of Tanzania and observed rice cultivar *Kihogo red* had highest ash content (0.97%) followed by *Supa* (0.83%) and *Kaling'anaula* (0.66%), whereas lowest ash content was found in *Salama* (0.18%). Dikeman et al. (1981) reported 5.2–6.6% total ash in kernel of short grain rice cultivars of the United States.

10.2.5 CHO CONTENT

Rice is a rich source of CHO. Rice CHOs are considered as main source of energy for more than one-third of the world's population (Subudhi et al., 2013). They are mainly starch which is composed of amylose and amylo-pectin (Juliano, 1993a,b). Rice starch is constituted about 90% of the dry matter in milled rice reported by Juliano (1993a,b) and present only in the endosperm cells of mature rice for all intents and purposes (Bechtel and Pomeranz, 1977; Wang et al., 1995a,b; Krishnan et al., 2001; Fujita et al., 2003, 2009, 2011; Li et al., 2004; Wang et al, 2004; Xu-run et al., 2014). Mbatchou and Dawda (2013) reported that a high level of starch makes the individual grains stick to each other, while low starch content prevents well from sticking of the grains together after cooking. The low CHO content in rice sample means to say they may be attributed to its high MC which also affects the milling quality and other environmental factors (Katz and Weaver, 2003). The high percentage of CHO contents of the rice varieties shows that rice is a good source of energy (Oko and Ugwu, 2011). Freshly harvested rice grains contain about 80% CHOs which include starch, glucose, sucrose, dextrin, etc. (Yousaf, 1992). The rice varieties exhibiting fairly high CHO content lead to individual grains remaining separated from each other after cooking (Verma and Srivastav, 2017). However, the CHO content of Indian aromatic and nonaromatic rice cultivars were observed to have more than

75% (Verma and Srivastav, 2017), and have to meet somewhat similar and near about the desired range (80%) (Juliano, 1985c; Juliano and Bechtel, 1985).

10.2.5.1 DETERMINATION OF CHO CONTENT

The determination of total CHO content in rice sample is generally done by the difference of total content of ash, moisture, fat, and protein subtracted by 100 expressed as the following formula:

$$\% \text{ Carbohydrate} = 100 - \left(\% \text{ of ash} + \% \text{ of moisture} + \% \text{ of fat} + \% \text{ of protein}\right)$$

The report *Food Energy—Methods of Analysis and Conversion Factors* of a technical workshop held during December 3–6, 2002, conducted by Food and Agriculture Organization of the United Nations (FAOUN), Rome, Italy, suggested the formula to calculate total CHO and available CHO as given in the following formula (FAO , 2003):

For calculation of total CHO
> By difference:
>> 100 – [Wt. in g (protein + fat + water + ash + alcohol) in 100 g of food]
> By direct analysis:
>> Wt. in g (MS + DS + OS + PS, including fiber)

For calculation of available CHO
> By difference:
>> 100 – [Wt. in g (protein + fat + water + ash + alcohol + fiber) in 100 g food]
> By direct analysis:
>> Wt. in g (MS + DS + OS + PS, excluding fiber)*

where MS is the monosaccharides, DS is the disaccharides, OS is the oligsaccharides, PS is the polysaccharides. *May be expressed as weight (anhydrous form) or as the monosaccharide equivalents (hydrous form including water).

10.2.5.2 VARIATION IN CHO CONTENT OF RICE VARIETIES/CULTIVARS

In aromatic and nonaromatic rice cultivar of India, Verma and Srivastav (2017) reported fairly higher amount of CHO, that is, 75.87–82.70%, except

Govind Bhog (75.87%). In 92 rice varieties of India studied by Devi et al. (2015), it was reported that Bhuban and Prachi varieties had highest CHO. A study of Subudhi et al. (2013) on short- and long-grain varieties of aromatic rice found that the variety Pusa sugandh-2 (64.6%) had lowest CHO and highest in Kala namak (89.15%) aromatic variety followed by Tulsiphool (83.4%) and Ketekijoha (80.0%). The study suggested that short- and long-grain varieties of aromatic rice, namely, Kala namak, Ketekijoha, and Tulsiphool had more than 80.0% CHO content; thus, these varieties are considered as rich sources of CHO (Subudhi et al., 2013). In the study of Saikia et al. (2012), pigmented aromatic rice Chak-haoamubi (78.0%) was reported as high in CHO on Db followed by Poreiton chakhao (77.2%) and the nonpigmented aromatic rice,namely, Bakul joha (76.6%) and Keteki joha (76.4%) were low in CHO.

In Malaysian rice varieties, Thomas et al. (2013) reported lowest CHO in Brown (78.21%) and highest CHO in Boro (82.23%) rice varieties followed by white (local) (80.14%) and Basmati (79.34%). All the Malaysian rice varieties were found to have >79% CHO which was considered as good source of CHO (Thomas et al., 2013). Ascheri et al. (2012) observed the high contents of CHOs in red rice genotypes of Brazil. The rice genotype MNACH0501 had 77.5% CHO, whereas MNACE0501 genotype showed 79.3% CHO followed by MNAPB0405 (78.5%) and Traditional variety (78.3%) (Ascheri et al., 2012). Thai red aromatic brown rice was reported by Wichamanee and Teerarat (2012) to have 72.35% CHO. The procured rice varieties from Thailand, China, and Sri Lanka were studied by Sompong et al. (2011) and found 79.27% CHOs in red rice variety and 71.99% CHOs in black rice. The study of Sompong et al. (2011) reported all the rice varieties had higher CHO than 71%.

The brown and polished rice varieties of Pakistan studied by Hayat et al. (2013) showed that Basmati Supper (76.03%) had the highest CHOs while the CHO content in IRRI-6 was 72.88%. The study of Hayat et al. (2013) suggested that brown rice varieties lower CHOs content in comparison to polished rice varieties. Islam et al. (2016) investigated Bangladeshi rice varieties and reported highest CHO in Jeerashile (81.87%) followed by Minikat (81.76%) and Sawrna (80.36%), whereas Nagirshail (76.33%) noted as lowest CHO variety. *Kalijira*, a scented rice variety of Bangladesh, had 79.48% CHO, as studied by Ali et al. (2012).

CHOs were reported to reduce from 63.72% to 61.28% in sundried and stored rice, respectively (Fagbohun and Oluwaniyi, 2015). Fifteen indigenous and five newly introduced rice variety of Nigeria were used in the study of Oko et al. (2012) who observed highest CHOs (51.50–86.92%) in all the hybrid rice varieties, except E4197. The work of Oko and Ugwu (2011) on five major rice varieties of Nigeria reported high content of CHO. The rice variety Faro 15 (86.03%) showed higher CHO among the rice varieties followed by Canada (85.09%) and Awilo (84.01%), whereas Sipi variety exhibited lower CHO (76.92%). Oko and Onyekwere (2010) reported 51.53% as lowest CHO in IR68 variety and 85.57% as highest in IR75395-2B-B-1-1-1-2-4 followed by IR77384-12-17-3-18-2-B. Parboiled paddy rice samples obtained from farm of Nigeria were reported by Ibukun (2008) to have 76.22–78.20% CHO. Ebuehi and Oyewole (2008) found 81.1% CHO in Aroso and 78.3% in Ofada, both were the rice varieties of Nigeria, whereas 73.15–76.37% CHO were reported by Edeogu et al. (2007) in staple food crops in Ebonyi State of Nigeria. The rice varieties of Ghana cultivated in Kassena-Nankana District were studied by Mbatchou and Dawda (2013) and reported low CHO of 67.11% in WITA-9 variety, whereas the variety IR12979-24-1 exhibited high CHO with 83.27% followed by ANDY-11 (76.90%) and JASMINE-85 (76.89%).

10.2.6 CF CONTENT

Fibers are also one of the most important constituents of the food (Verma and Srivastav, 2017). They evolved in the functions of digestive system and also reduce the risk of intestinal disorders, if taken in form of fiber-rich foods in daily diet. There are a large number of reports on rice varieties showing study on the fiber content. Oko et al. (2012) reported that highest fiber contents are considered to be a good quality trait because of the good source of insoluble fiber in rice. The report of Food and Agriculture Organization (FAO) stated that the risks of bowel disorders are reduced by insoluble fiber and also it fights against the constipation (FAO, 1998). In milled rice grain, the bran layers (and the hull) are the places where CF is found in highest amount, whereas aleurone layers are known for lowest amount of CF (Noreen et al., 2009; Oko et al., 2012). Oko et al. (2012) also suggested that the valuable dietary fibers are provided by the bran layer of brown rice grain. However, another report in literature stated that a significant amount of dietary fiber and more nutrients are not contained by milled or polished white rice as compared to unmilled rice grain.

Most of the rices in the world are consumed in daily diet of human as white polished grain despite the valuable food content of brown rice; report of Oko et al. (2012) stated that half cup of cooked white rice is estimated to provide us about 0.3 g of dietary fiber, which is very lower than the cooked brown rice which provides 1.8 g dietary fiber. Thus, dietary fiber content of cooked brown rice is higher as compared to cooked white rice. Sotelo et al. (1990) observed 1.9% of fiber content in brown rice after milling, whereas 0.5–1.0% in well-milled rice as standard fiber content observed by Oko and Onyekwere (2010). One cup of cooked brown rice (i.e., 160 g) contains about 2.4 g of dietary fiber, which is equivalent to 8% and 9.6% of an average man's and woman's daily fiber needs, respectively, as stated in 1998 by FAO in the technical report of World Health Organization (WHO) (FAO, 1998). Yeager (1998) confirmed fiber as the ultimate healer, because of their ability which possess important role to decrease the blood cholesterol and sugar after meals in diabetics and also reduce the risk of intestinal disorders. Rice and rice brain with increase fiber content are also subject to human health consideration and reported by Abdul-Hamid et al. (2007) who described the health improvement by lowering the plasma cholesterol. The human diet of daily routine, especially of urban populations who consumed low-fiber-containing rice, resulted in a wide range of ailments and conditions (Oko et al., 2012). Rice rich in fiber content augment the functions of digestive system. According to the report of WHO, digestibility of rice may be affected by their fiber content. For instance, rice with high-fiber content results in lower digestibility, whereas low fiber content results in higher digestibility (WHO, 1985).

10.2.6.1   DETERMINATION OF CF CONTENT

For determination of CF, about 2 g of moisture and fat-free powdered rice sample is taken into a fiber flask and 150 mL of 1.25% (0.255 N) H_2SO_4 is added. Then, the mixture is heated under reflux with heating mantle for 1 h. The hot mixture is filtered and washed through a fiber sieve cloth. After filtration and washing, the difference obtained is thrown off and the residue is returned to the flask to which 150 mL of 1.25% (0.313 M) NaOH is added and heated under reflux for another 1 h (Fig. 10.1D). It is again filtered and washed with hot water before it is finally transferred in the preweighted crucible. The crucible with residue is oven dried at $105 \pm 5°C$ overnight to drive off moisture. The oven-dried crucible containing the residue is cooled in a desiccators and latter weighted (W_1) for ashing at 550°C for 4 h (AOAC, 1990). The crucible containing white and grey ash (free of carbonaceous material) is cooled in a desiccator and weighted to obtain W_2. The percentage of CF is calculated as follows:

$$\text{Fiber} \left(\% \right) = \frac{W_1 - W_2}{\text{Wt. of sample}} \times 100$$

10.2.6.2   VARIATION IN CF CONTENT OF RICE VARIETIES/ CULTIVARS

In six aromatic and two nonaromatic Indian rice cultivars, Verma and Srivastav (2017) reported below 1.0% fiber. Among all cultivars, aromatic cultivar Badshah Bhog had 0.85% fiber followed by Swetganga and Govind Bhog with 0.70% and 0.69% fiber, respectively, whereas lowest (0.48%) fiber found in Khushboo aromatic rice cultivar. Devi et al. (2015) studied 92 rice genotypes including land races, improved varieties, aromatic varieties, and red rice collected from the Southern part of India and found two rice varieties, namely, MTU 3626 and MTU 1010 which possessed highest fiber content (0.95%). Low fiber content was found in the majority of rice samples among the 92 rice genotypes. *Salem samba* is an Indian rice variety reported by Ravi et al. (2012) with 0.2% fiber. Saikia et al. (2012) studied pigmented (Poreiton Chakhao and Chak-hao-amubi) and nonpigmented rice (Keteki joha and Bakul joha) collected from Assam and Manipur states of India, respectively. The pigmented rice Chak-haoamubi and Poreiton Chakhao were reported for highest fiber (0.3%, Db) and lowest fiber (0.2%, Db), respectively, whereas the nonpigmented rice, namely, Bakul joha and Keteki joha with 0.2% fiber Db were followed to the Chak-haoamubi pigmented rice.

Pakistani rice varieties, namely, IRRI-6 for highest (2.57%) and IRRI-9 for lowest (2.17%) fiber content were noticed by Anjum et al. (2007), while rest varieties somewhat nearer to IRRI-6 were Sarshar and Dr-83 with 2.47% and 2.31% of fiber content, respectively. The study of Anjum et al. (2007) advised that polished rice and white rice were lowest in fiber on milling fractions than that in the rice bran. Sotelo et al. (1990) also observed decrease fiber contents (1.9% ± 0.6% mean value) due to the milling in their study on 12 Mexican rice varieties. Hayat et al. (2013) found 1.18% in Basmati Supper and 1.74% in IRRI-9 among differently studied all the rice varieties of Pakistan. Tufail (1997) reported varied amount of fiber content from 0.20% to 0.35% in different Pakistani rice varieties.

From rice varieties of Bangladesh, Nurjahan and Basmoti varieties reported by Islam et al. (2016) for high fiber (0.63%) followed by varieties

Govindavogh and Nagirshail with 0.50% and 0.36% fiber content, respectively, and the variety Paijam for low fiber. However, a study of Ali et al. (2012) on *Kalijira,* a scented rice variety of Bangladesh reported 0.22% CF.

A local white rice variety of Malaysia reported by Thomas et al. (2013) for high fiber (8.47%) and imported black rice variety for low fiber (7.07%), whereas the varieties, namely, Brown and Basmati followed to local white rice variety with 8.37% and 8.09% fiber content, respectively. The red rice variety of Brazil, MNAPB0405, was reported in the study of Ascheri et al. (2012) for high fiber (3.3 dag/kg) followed by MNACE0501 and MNACH0501 with 3.2 dag/kg and 3.0 dag/kg content of fiber, while traditional variety for low fiber (2.5 dag/kg). In Brazilian polished white rice, total dietary fiber of 2.87% on dry weight basis was reported by Storck et al. (2005). The rice cultivars in the Morogoro region of Tanzania were studied by Shayo et al. (2006), who found Kihogo red (0.73%) and Kaling'anaula (0.29%) for high and lower fiber content, respectively. Other cultivars, namely, Salama M 17 (0.36%) and Kaling'anaula (0.29%); Supa (0.40%) and Salama (0.39%) were reported to have significant difference with each other. Eggum et al. (1982) reported 0.7% in IR32 rice variety taken from the IRRI, Philippines.

From the nutritional composition study of rice varieties of Ghana, Mbatchou and Dawda (2013) observed 0.11–0.76% fiber as 0.11% in ANDY-11, 0.16% in IR12979-24-1, 0.53% in WITA-9, and 0.76% in JASMINE-85. Diako et al. (2011) studied scented rice varieties found 0.22%, 0.13%, and 0.11% fiber in Ex-Hohoe, Royal Feast (imported brand), and Jasmine 85 (local variety), whereas the varieties Ex-Baika (0.05%), Sultana (0.04%), and Marshall (0.02%) had noted to have below 1%.

In Ebonyi State of Nigeria, Oko et al. (2012) studied local and newly introduced rice varieties for their chemical composition and reported the varieties China and Mass with high content of fiber (2.5%) and Faro 25 (II), Mass (I), and E4334 with low content of fiber (1%). The other rice varieties, namely, Sipi, Canada, E4197, and E4212 had 2% fiber, which were following to the high-fiber contending rice varieties with significant difference to each other. The major varieties from the part of Southeastern Nigeria, namely, Awilo, Canada, Faro 14, Faro 15, and Sipi were studied by Oko and Ugwu (2011) who reported highest fiber content with 2.0% in Sipi and Canada varieties, whereas rest all were low (1.50%) fiber rice; Edeogu et al. (2007) reported 1.93–4.3% fiber in rice from their study on staple food crops from Ebonyi State of Southeastern Nigeria. In new lowland rice varieties of Nigeria studied by Oko and Onyekwere (2010), low fiber (1%) in

IR75395-2B-B-1-1-1-2-4 variety and high fiber (2%) in varieties IR68 and PSBRc50 was found. Raw paddy rice samples of Ondo State of Nigeria were used by Ibukun (2008) for study of prolonged parboiling effect on nutritional composition and reported 0.55–1.13% fiber in parboiled rice. The rice samples obtained from Ado Ekiti region of Southwest Nigeria were used in the study of Fagbohun and Oluwaniyi (2015) which showed that the changes in fiber contents of sundried and stored rice varied from 10.27% to 8.52%. Another study of Ebuehi and Oyewole (2008) on a local and indigenous rice Ofada produced in the Southwest of Nigeria reported 3.5% fiber, while a foreign and imported rice Aroso cultivated in Thailand reported for 1.5%. Wichamanee and Teerarat (2012) reported the red aromatic brown rice from the Sa Kaew province of Thailand with 2.07% fiber. Sompong et al. (2011) studied 13 colored rice varieties collected from of Thailand (Bahng Gawk, Haek Yah, Niaw Dam Pleuak Dam, Niaw Dam Pleuak Khao, Niaw Dawk Yong, Niaw Lan Tan, Niaw Look Pueng, and Sung Yod Phatthalung), China (Chinese Black Rice and Chinese Red Rice), and Sri Lanka (Sri Lanka Red Rice-1, Sri Lanka Red Rice-2, and Sri Lanka Red Rice-3). The red Thai variety, Sung Yod Phatthalung had highest (4.51%) fiber content, whereas the Chinese Red Rice had lowest (2.52%) fiber content in the study of Sompong et al. (2011). All the Sri Lankan varieties, namely, Sri Lanka Red Rice-1, Sri Lanka Red Rice-2, and Sri Lanka Red Rice-3 were found with 2.82%, 2.87%, 2.88% fiber content, respectively.

10.2.7 MEASUREMENT OF FOOD ENERGY VALUE

About 20% of the world's dietary energy is supplied by major food commodity of the world trade 'Rice," as report of FAO and IRRI said. The energy value of any food is an important parameter in the health point of view that generally is determined by the available energy received from food materials through cellular respiration (Thomas et al., 2013). The important property of this parameter indicates the useful energy content of foods in the term of calorific value and thereby its value as fuel. The caloric values of some important grains and its products are given in Table 10.2. The gross food energy (FE) values was determined and calculated by multiplying the levels of proteins, lipids, and CHOs in each sample by their respective caloric values: 4, 4, and 9.4 kcal (Osborn and Voogt, 1978; Schakel et al., 1997), according to the conversion factor using the following equation also referred by Verma and Srivastav (2017):

$$Food\ energy\ (FE) = (\%CP \times 4) + (\%F \times 9) + (\%CHO \times 4)$$

where FE = food energy (in kcal/g); CP = crude protein; F = fat; and CHO = carbohydrate.

TABLE 10.2 Caloric Values for Grain and Grain's Products.

Grain and grain's products	Caloric values (in kcal/g)		
	Protein	Fat	Total carbohydrate
Barley (*Hordeum vulgare* L.)			
Barley—whole grain	3.55	8.37	3.95
Corn (*Zea mays* L.)			
Cornmeal, whole ground	2.73	8.37	4.03
Macaroni, spaghetti	3.91	8.37	4.12
Oat (*Avena sativa* L.)			
Oatmeal—rolled oats	3.46	8.37	4.12
Pearl millet (*Pennisetum glaucum* L.)			
Pearl millet—whole grain	3.55	8.37	3.95
Rice (*Oryza sativa* L.)			
Rice, brown	3.41	8.37	4.12
Rice, white or polished	3.82	8.37	4.16
Rye (*Secale cereale* L.)			
Rye flour—light	3.41	8.37	4.07
Rye flour—whole grain	3.05	8.37	3.86
Sorghum (*Sorghum bicolor* L.)			
Sorghum—wholemeal	0.91	8.37	4.03
Wheat (*Triticum aestivum* L.)			
Wheat—97–100% extraction	3.59	8.37	3.78
Wheat—70–74% extraction	4.05	8.37	4.12

Source: FAO (2003).

10.2.7.1 VARIATION IN FE VALUES OF RICE VARIETIES/CULTIVARS

In the study of Verma and Srivastav (2017) on Indian rice cultivars, Badshah Bhog (365.23 kcal/100 g) an aromatic rice cultivar showed the highest energy among all the samples analyzed, followed by Gopal Bhog (361.07 kcal/100 g) and Khushboo (360.44 kcal/100 g), whereas the lowest value were recorded by Govind Bhog (348.79 kcal/100 g). Islam et al. (2016)

reported FE 2834.31 kcal/kg in Nurjahan and 3017.27 kcal/kg in Parija among the all 12 locally produced rice varieties of Bangladesh.

Ascheri et al. (2012) found that MNACE0501, red rice genotype of Brazil, had high FE (363.9 kcal/100 g) followed by MNAPB0405 and traditional variety with 362.4 kcal/100 g and 358.7 kcal/100 g grain of energy values, respectively, whereas MNACH0501 had lowest energy value with 357.5 kcal/100 g grain. TACO (2006) reported 357 kcal/100 g and 364 kcal/100 g grains energy value in brown and polished (raw) red rice genotypes, respectively. In the study of Oko et al. (2012), the Nigerian hybrid rice genotype Canada (398.82 J/kg) reported high FE value whereas E4197 (262.94 J/kg) for low FE value.

The red and black rice varieties of Thailand, China, and Sri Lanka were collected by Sompong et al. (2011) for physicochemical and antioxidative properties study. The majority of rice varieties were reported with high energy values, that is, more than 350.72 cal/100 g from Sompong et al. (2011) study. The white (local) and black (imported) rice from Malaysia were reported by Thomas et al. (2013) for higher lowest energy values with 1523.57 kJ/100 g and 1457.72 kJ/100 g grain, whereas the other rice varieties, namely, Glutinous and Basmati followed the white (local) rice variety with 1502.65 kJ/100 g and 1498.46 kJ/100 g grain energy value. From the study of Sompong et al. (2011) and Thomas et al. (2013), studied experimental rice varieties may be concluded and recommended for human consumption in their daily diet due to high FE value.

10.3   MINERAL CONTENT ANALYSIS

The mineral profile present in composition of rice grain is considerably affected by the available soil nutrients application (Juliano and Bechtel, 1985). The report of Juliano and Bechtel (1985) said number of papers available in literature which investigated the diverse sampling, preparation, and analytical methods during the period of crop growth. The minerals of rice grain are broadly classified into two main groups; they are as follows:

1. *Macrominerals:* These are required in larger amount for growth and development of the body (Nielsen, 2013a). For instance, calcium (Ca), phosphorus (P), potassium (K), sodium (Na), magnesium (Mg), chloride (Cl), and sulfur (S).
2. *Microminerals:* These are required in minute/trace amount for growth and development of the body as compared to macrominerals

(Nielsen, 2013b). For instance, cobalt (Co), copper (Cu), iron (Fe), manganese (Mn), molybdenum (Mo), iodine (I), zinc (Zn), and selenium (Se).

Mineral elements are very important and essential constituents for growth and development of different part of body such as bone, teeth, tissue, muscle, blood, nerve cells, etc. (Champe and Harvey, 1994; Wang et al., 2011). In addition, various physiological functions, namely, digestion and utilization of nutrients, muscle contraction, neurotransmission, and production of hormones are also performed within the human body (Champe and Harvey, 1994; Wang et al., 2011). Among the microelements, Fe and Zn are two more important minerals compared to others for normal physiological growth and development of the human body (Verma and Srivastav, 2017). Sperotto et al. (2012) published a review in 2012 on Fe biofortification in rice which described that Fe- and Zn-fortified rice may prevent the mineral deficiency associated diseases. Some important antinutritional factors (ANFs), namely, phytate, trypsin inhibitor, oryzacystatin, and hemagglutinin–lectin are contained in the bran of rice (Juliano, 1993b). Among the ANFs, phytate bind with minerals decreasing bioavailability of Ca, Fe, P, Zn, and other trace elements to human. According to Fox and Tao (1989), phytic acid is usually found as a complex with essential minerals and proteins and produce an adverse effect on the bioavailability and digestibility of these essential nutrients.

10.3.1 DETERMINATION OF MINERAL CONTENT

The mineral profile of the rice grain is determined as par the methods described in AOAC (2000). According to the prescribed methods of AOAC (2000), powdered samples of rice grain are used as the following:

1. *Preparation of standard solution:* Solution of Fe, Zn, Na, K, Mg, Ca, and Cu is prepared and then diluted to a total volume of 100 mL to get 100 ppm and then 2–10 ppm. Standard solutions of 5–20 ppm are prepared for Fe, 5–20 ppm for Zn, 0.5–10 ppm for Na, 2.5–10 ppm for K, 5–20 ppm for Mg, 2–8 ppm for Ca, and 4–16 ppm for Cu are also prepared.
2. *Preparation of sample solution:* Each of the powdered rice sample (2.0 g) is first digested using nitric acid and perchloric acid in a volume ratio of 2:1. The digested sample is transferred into a 100-mL

volumetric flask and made up to mark with distilled water, filtered, and analyzed.

3. *Procedure:* Absorbance of standard solutions as well as sample solutions is taken by atomic absorption spectrophotometer (Fig. 10.3) using their respective cathode lamp such as Fe-cathode lamp, etc. Concentration of sample is determined by the graph between concentration along *X*-axis and absorbance along *Y*-axis.

10.3.2   VARIATION IN MINERAL CONTENT OF RICE VARIETIES/CULTIVARS

In the composition of rice grains, macroelements include Ca, Mg, and P are present, while Cu, Fe, Mo, Mn, and Zn are available in rice microelements (Yousaf, 1992; Oko and Onyekwere, 2010). Mg and Ca are the most abundant elements found in mineral profile of rice next to the K (Oko and Onyekwere, 2010). The bran and aleurone layers of the rice grain are important place, where minerals are found in highest and lowest amount, respectively (Anjum et al., 2007; Noreen et al., 2009). The minerals, namely, Ca, Fe, Mg, Mn, P, K, Na, and Zn are richly found in rice bran (Sunders, 1990; Hu et al., 1996; Xu and Godber, 1999; Frei et al., 2003; Satter et al., 2014; Raghav et al., 2016). The mineral, P, is accounted with considerable portion in the ash of rice caryopsis, whereas silica is the major element in hull ash (Juliano and Bechtel, 1985; Lu and Luh, 1991). Juliano and Bechtel (1985) reported levels of minerals higher in brown compared to milled rice. Among the minerals, K, Mg, and silicon are present also in large amounts in brown and milled rice, as reported by Juliano and Bechtel (1985). In 2004, the FAOUN highlighted to Fe and Zn during the celebration of *International Year of Rice-2004* as main components of red rice (nonspecific genotype) at the level of 5.5 and 3.3 mg/100 g rice, respectively (FAO, 2017).

The data on minerals contents of different rice varieties reported in literature from different countries such as Australia, Bangladesh, Brazil, China, India, Malaysia, Mexico, Nigeria, Pakistan, Philippines, South Korea, Thailand, the United States, and others are presented in Table 10.3.

In 2017, six aromatic rice cultivars [Gopal Bhog (Bihar and Odisha), Govind Bhog (Assam), Badshah Bhog (West Bengal, Odisha, Bihar, and Assam), Kala namak (Uttar Pradesh and Bihar), Swetganga (Odisha), Khushboo (Uttar Pradesh)] and two nonaromatic rice cultivars, Sarbati and Todal (Uttar Pradesh), were procured from different regions of India and

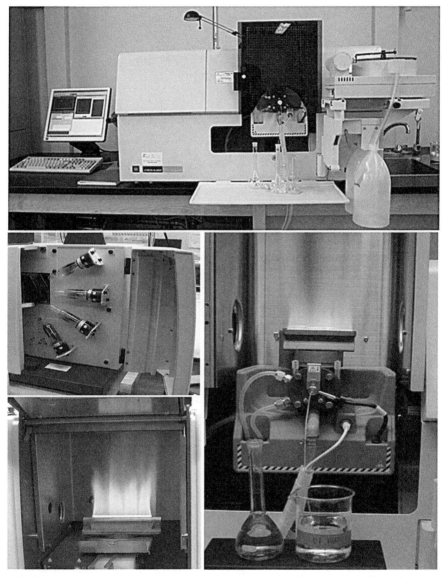

FIGURE 10.3 Atomic absorption spectrophotometer (AAS) apparatus. (A) Laboratory setup of AAS, (B) flame in AAS, and (C) sampling with a flame AAS.

analyzed by Verma and Srivastav (2017) for mineral profile in terms of overall mean expressed in mg/kg of rice on Db. Among the minerals, the higher Ca (98.75 mg/kg), Zn (17.00 mg/kg), and Fe (31.50 mg/kg) were in Gopal Bhog, whereas the highest Na (68.85 mg/kg) was in Badshah

TABLE 10.3 Mineral Content of Different Rice Cultivars vis-à-vis with Others' Study.

Author's work	Country	Mineral content composition (in mg/kg)						
		Ca	Na	Zn	Fe	Cu	K	Mg
Verma and Srivastav (2017)	India	62.95–98.75	41.44–68.85	9.3–13.25	0.8–31.5	4.1–15.95	265.85–500	83.5–182.45
Hashmi and Tianlin (2016)	Malaysia	–	52,530–137,610	93,500–20,530	7520–18,930	880–2740	185,730–1368,020	43,690–273,400
Islam et al. (2016)	Bangladesh	9.0–18.5	460.3–1188.7	–	0.002–0.930	11.0–18.5	03.8–34.1	01.3–06.1
Mohd Fairulnizal et al. (2015)	Thailand	45.2–58.7	25.8–38.5	10.6–11.7	3.8–26	0.2–1.3	–	92.2–138.6
Govarethinam (2014)	Malaysia	4.7–9.3	1.5–3.6	–	3.3–7.3	0.1–0.7	140–360	100–870
Da Silva et al. (2013)	Brazil	97–435	45–767	6.52–24.37	1.63–12.41	–	908–3472	106–1167
Mbatchou and Dawda (2013)	Ghana	46.96–123.64	29.03–115.68	10.00–12.50	13.67–21.35	14.07–38.13	756.82–1720.00	241.01–818.65
Tamanna et al. (2013)	Bangladesh	13–32	–	15–74	17–44	–	–	–
Diako et al. (2011)	Ghana	108–111	–	16.4–20.8	2.1–8.0	–	755–1170	0.1–288
Zeng et al. (2009)	China	144.9–188.2	–	22.0–34.4	28.70–50.60	10.6–17.1	2513.3–3483.5	1424.1–1913.3
Mohammed and Spyrou (2009)	Thailand	38–43	8–21	22–26	9–11	2–20	536–782	119–270
Deepa et al. (2008)	India	92.0–116	226–309	–	19.3–39.5	–	2480–3040	1500–2160
Ebuehi and Oyewole (2008)	Thailand and Nigeria	640–800	–	–	386–463	–	–	388–820
Ibukun (2008)	Nigeria	14.0–16.6	51.0–55.2	–	4.2–5.5	–	72.0–73.1	–
Anjum et al. (2007)	Pakistan	–	–	14.4–29.7	13.7–19.4	5.8–9.2	–	–
Heinemann et al. (2005)	Brazil	410–670	44–59	115–209	40–57	15–18	6546–18,171	1506–1688
Jung et al. (2005)	South Korea	74.3–117	2.94–23.7	14.8–19.7	5.41–18.0	–	0.148–0.279	680–1390
Scherz et al. (2000)	a	110–390	100	8–20	20–36	2.4–3	1500–2600	1100–1660
Marr et al. (1995)	Australia	30–110	0–190	13–21	5–57	1.4–13	2100–3000	1000–1300
Sotelo et al. (1990)	Mexico	26–79	98	16–31	8–25	–	1810–3680	–
Juliano and Bechtel (1985)	a	100–500	17–340	6–28	2–52	1–6	600–2800	200–1500
Wolnik et al. (1985)	The United States	26–79	–	7.2–21	1.4–10	0.53–5.1	125–1950	39.5–920
Dikeman et al. (1981)	The United States	219–246	–	12.8–15.8	21.2–26.4	2.10–2.72	3.9–4.3	1.2

Ca, calcium; Na, sodium; Zn, zinc; Fe, iron; Cu, copper; K, potassium; Mg, magnesium.
aNot known.
–Not reported.
Note: All data were converted in mg/kg.

Bhog and the highest K (500.00 mg/kg) was in Swetganga, Khushboo, and Sarbati. It is evident from the study that K was the most abundant mineral component in all Indian aromatic and nonaromatic rice cultivars and showed a range of 265.00–500 mg/kg (Verma and Srivastav, 2017). Another study on Indian rice variety conducted by Subudhi et al. (2013) who found Kala namak variety for high Fe (19.16 mg/kg), Pusa sugandh-2 for high Zn (32.04 mg/kg) and low Fe (7.09 mg/kg), and Ketekijoha for low Zn (14.33 mg/kg).

Different patterns of mineral composition in red rice genotypes were showed by Ascheri et al. (2012) with highest contents of K (198–268 mg/100 g), Mg (70–86 mg/100 g), and Ca (12–18 mg/100 g) and lowest contents of Zn (2–2.5 mg/100 g), Cu (0.1–0.4 mg/100 g), Mn (1.2–2.5 mg/100 g), and Fe (1.0–1.2 mg/100 g). A study on 20 Nigerian hybrid rice varieties, Oko et al. (2012) reported 0.13% Ca in Chinyereugo, 0.26% Mg in Faro1 (I), 0.55% P in E4197, and 0.23% K in E4212, while 0.17% Na in Canada, Faro15 (II), and Chinyereugo hybrid rice varieties. The content of Zn and Fe observed variations with 19–40 mg/kg and 14–18 mg/kg, respectively, in four different genotypes of red rice as studied by Pereira et al. (2009). Anjum et al. (2007) found significantly higher mineral contents in bran followed by polishing. The bran of rice was reported for highest Cu (1.69%), Fe (3.98%), Mn (5.12%), and Zn (4.69%), whereas polished rice for lowest Cu (0.28%), Fe (0.44%), Mn (0.51%), and Zn (1.12%) at different milling fractions. The varieties, namely, IRRI-6 reported for highest Fe (1.94%) and Zn (2.97%), IRRI-9 for Cu (0.92%), and Dr-83 for Mn (2.33%), while lowest Fe (1.37%), Zn (1.44%), Mn (1.57%) found in IRRI-9 and Cu in Dr-83 (0.58%). The amount of Fe, Zn, Mn, and Cu significantly vary in processes such as par boiling, milling, polishing, etc., as suggested by Anjum et al. (2007) study.

Ninety samples of Australian brown rice were analyzed by Marr et al. (1995) for K, P, S, Mg, Ca, Na, Al, Cu, Fe, Mn, and Zn. Highest Mn content reported in brown rice were compared to brown rice produced in other countries worldwide. In addition, the study of Marr et al. (1995) also reported highly significant positive correlations between the concentration levels of the mineral elements, namely, P, K, Mg, and S in studied rice varieties. Twelve Mexican varieties were studied by Sotelo et al. (1990) for mineral profile of rice and observed that milling process followed by polishing significantly reduces the K, Fe, and Zn contents of the rice, and also reported significant variation in mineral contents between varieties of brown and white rice. Similarly, Juliano (1985a) reported different concentrations of minerals in rice such as 11.6–34.9 mg/100 g Ca, 23.3–58.1 mg/100 g Mg, 81.4–151.2 mg/100 g K, 0.2–0.3 mg/100 g Zn, 0.7–2.7 mg/100 g Cu, 0.7–2.0 mg/100 g Mn, and 0.2–3.3 mg/100 g Fe. Watts and Dronzek (1981) reported 0.28% P,

0.30% K, 0.11% Mg, 17 ppm Fe, 14 ppm Mn, 51 ppm Zn, and 13 ppm Cu in 12 Canadian wild rice samples. Four Indica rice varieties, namely, IRRI-6, KS-282, Basmati 2000, and Basmati Super were investigated by Dikeman et al. (1981) for their mineral profile. The concentration level of minerals such as Na, K, Fe, and Zn were found significantly higher in brown rice tin as compared to white rice. The minerals, P and K were reported as the major mineral constituents of rice bran, with values up to 1633 mg/100 g. Twelve samples of Canadian wild rice were studied by Watts (1980) who reported P (0.28%), K (0.30%), Mg (0.11%), Fe (17 ppm), Mn (14 ppm), Zn (51 ppm), and Cu (13 ppm) in the mineral composition.

The presence of mineral elements in the composition of rice grain significantly differed due to various factors such as variation from variety to variety, milling fractions soil type, growing conditions, etc. (Shabbir et al., 2008).

10.4   FATTY ACIDS ANALYSIS

The fatty acids have drawn more attentions toward human health because of unique health benefits of rice fat besides dietary consumption of rice (Jenning and Akoh, 2009). They are the basic building block of all lipids as triglycerides comprises predominantly of fatty acids. The knowledge of the amount and type of fatty acids in the triglycerides is widely regarded as a useful criterion, because of the chemical tests for oil identity, purity can be related to the fatty acid composition (Rossell, 1991a). Both, the physical and chemical characteristics of the oil, are influenced by the kind and proportion of the fatty acids and the way in which they are positioned on the glycerol moiety.

Fatty acids are very important components of food because of their role in growth, development, and physiological functions of body (FAO/WHO, 1993). The abundance of unsaturated fatty acids (USFAs) in the oil is desirable from nutritional and health points of view as USFAs consumption will not lead to heart-related diseases. Besides the use of the oil as food in frying and baking, oils rich in USFAs have been reported to reduce the risk of heart diseases associated with cholesterol (Law, 2000). Saturated fatty acids (SFAs), namely, myristic, palmitic acid, and stearic acid reported by Verma and Srivastav (2017) are found in rice oil; and these can pose health risk such as atherosclerosis, a disease associated with heart attack, if have higher percentage of SFAs in the fatty acid profile of rice (Oluremi et al., 2013; Verma and Srivastav, 2017). The fatty acid profile of the rice in the study of

Verma and Srivastav (2017) showed that it is good for consumption if well refined because it contains good amounts of required fatty acids which could prevent heart disease.

Fatty acids play a key role in regulating the levels of cholesterol and there is devastating evidence to reduce the cholesterol; and other range of human disease such as coronary heart 66 disease, cancer, or inflammation (Kullenberg et al., 2012). Certik and Shimizu (1999) stated that for normal functioning of nervous, immune and inflammatory, cardiovascular, endocrine, respiratory, and reproductive systems, polyunsaturated fatty acids (PUSFAs) play a key role in the body. The function of the important fatty acids, namely, oleic, linoleic, and linolenic acid in a heart-healthy diet is known to reduce serum cholesterol. However, the type and quantity of fat used is very important. If the fat is unsaturated, low density lipids (LDL) value and total cholesterol decreases, whereas they increase when the fat is saturated. Individual fatty acids (stearic acid, oleic acid, and linoleic acid) are related to function of lowering total LDL cholesterol. Studies show the cholesterol-lowering effect of grain lipids or high-lipid bran products (Saikia and Deka, 2011). Fatty acids are also known for their functions in membrane fluidity and therefore modulate changes in function of receptors, transporters, and enzymes (Calder, 2003). Rice bran oil mainly contains linoleic acid and oleic acid (Sayre and Saunders, 1990; Bernal et al., 2011). In spite of low amount of PUSFAs, studies have revealed that rice bran oil has significant hypocholesterolemic effect in living biological systems as compared to other oils (Saikia and Deka, 2011). The effect has been attributed to components such as tocopherols, tocotrienols, oryzanol, and fatty acid (Jariwalla, 2001). Researchers evidently proved that unsaponifiable fractions in rice bran oil can compensate the high value of saturated fats and play a leading role in lowering the level of cholesterol (Most et al., 2005; Bernal et al., 2011).

Fatty acids occurring in the plants are classified according to their degree of unsaturation. Fatty acids can also be classified as short-, medium-, and long-chain fatty acids. Generally, short-chain fatty acids are those containing 4–10, medium chain containing 12–14, and long-chain fatty acids containing 16 or more than 16 carbon atoms. Fatty acids can also be classified as essential and nonessential fatty acids (Rossell, 1991b). The major fatty acids of these lipids are linoleic, oleic, and palmitic acids (Taira et al., 1988). Essential fatty acids in rice oil are linoleic acid and linolenic acid (Jaiswal, 1983). The content of essential fatty acids may be increased with temperature during grain development, but at the expense of reduction in total oil content (Taira et al., 1979). Glycolipids are mainly monoacyl lipids (fatty acids and lysophosphatides) complicated with amylose (Choudhury and Juliano, 1980). The glycolipid content is lowest for waxy starch granules (0.2%), is highest for intermediate amylose rice (1.0%), and may be slightly lower in high amylose rice (Juliano and Goddard, 1986). However, glycolipids contribute little to the energy content of the rice grain.

10.4.1   DETERMINATION OF FATTY ACIDS

Fatty acid methyl ester (FAME): FAME is environment friendly because of its complementary, nontoxic, biodegradable, and renewable properties (Cayh and Kusefoglu, 2008; Ngo et al., 2008; Lei et al., 2010). FAME has many uses, and nowadays, it is used as biodiesel (Hu et al., 2004; Salehpour and Dube, 2008). FAME is usually produced by esterifying or *trans*esterifying the vegetable oils or animal fats with methanol (Lotero et al., 2005; Ranganathan et al., 2008).

Preparation of FAME: The FAMEs are prepared according to the method of Appleqvist (1968) or other methods may also be considered for extraction of fatty acid from rice flour. In this line, the method suggested by Food Safety and Standards Authority of India (FSSAI) is used (FSSAI, 2012) as follows: Take 30–50 mg of melted fat (one drop) in a glass-stoppered test tube and add 1 mL of dichloromethane/benzene followed by 2 mL of 1% sodium methoxide solution (1 g sodium dissolved in 100 mL of anhydrous methanol). Hold the test tube at 60°C for 10 min. Cool and add 0.1 mL of glacial acetic acid followed by 5 mL of distilled water and 5 mL of petroleum ether (40–60°C). Mix the contents. Allow the layers to separate. Take out about 2 mL of the upper layer containing the methyl esters in a small tube and concentrate it by passing nitrogen gas before injecting to gas chromatograph

Fatty acid profiling by gas chromatography (GC): Extracted fatty acids from rice flour methylesterized using FAMEs and they are then subjected to GC. The instrument equipped with a flame ionization detector (FID) and an appropriate stationary phase column [column may be like fused silica capillary column (may be 50 m × 0.25 mm i.d. size or as appropriate, coated with CP-SIL 88 as the stationary phase), stainless steel column of 10 ft packed with 15% diethylene glycol succinate on chromosorb W (80–100 mesh) or any other]. In the study of Verma and Srivastav (2017), the oven temperature is programmed at 200°C for 13 min. The injector and FID are at 250°C, whereas in the method suggested by FSSAI (2012), column temperature is maintained at 185°C with 2.8 kg/cm² (25 mL/min) and 1 cm/min of flow rate of carrier gas nitrogen and chart speed, respectively. This may vary because of the composition of the instruments and their manufacturer. A reference standard of FAME mix is analyzed under the same operating conditions to determine the peak identity. The FAMEs are expressed as relative area percentage.

10.4.2 VARIATION IN FATTY ACIDS PROFILE OF RICE VARIETIES/CULTIVARS

Fatty acid composition of rice mainly comprises oleic acid (38.4%), linoleic acid (34.4%), and alpha-linolenic acid (2.2%) as USFAs, and palmitic (21.5%) and stearic (2.9%) acids as SFAs (Chotimarkorn et al., 2008). Rice contains linoleic acid (C18:2) as high as 30% of the total amount of fatty acids, whereas rice oil has an amount of 75% USFAs with 35% linoleic acid (CAC, 2003; Gunstone, 2004). Another report says that in rice oil, about 29–42% linoleic acid and 0.8–1.0% linolenic acid as essential fatty acids are found (Jaiswal, 1983). The saturated, monounsaturated, and PUSFAs are in the ratio of approximately 1:2.2:1.5 (Shin and Chung, 1998; Krishna, 2002). Usually, rice oil has oleic acid (38.4%), linoleic acid (34.4%), and linolenic acid (2.2%) and SFAs, namely, palmitic (21.5%) and stearic acid (2.9%) (Rukmini and Raghuram, 1991). Rice oil has an amount of 75% USFAs with 35% linoleic acid (CAC, 2003). Fatty acid compositions of rice oil investigated by many workers use different chromatographic techniques (Orthoefer, 1996; Kim et al., 1999; Zhou et al., 2002; Goffman et al., 2003; Rodrigues et al., 2004; Khatoon and Krishna, 2004; Anwar et al., 2005; Parrado et al., 2006; Chotimarkorn et al., 2008; Alim et al., 2008; Brenda et al., 2009; Ahmad et al., 2011; Verma and Srivastav, 2017).

Verma and Srivastav (2017) analyzed the fatty acid profile of six aromatic and two nonaromatic rice cultivars of India, accomplished by gas liquid chromatography. Oleic, linoleic, and palmitic acids were observed as major fatty acids, whereas linolenic, myristic, and stearic acids as minor in all the aromatic and nonaromatic rice cultivars. Myristic (0.27–4.10%), palmitic (2.60–31.91%), and stearic acids (2.28–6.47%) were found to be present in rice cultivars, which were altogether accounted as SFAs, whereas oleic (25.06–49.14%), linoleic (30.22–46.99%), and linolenic acids (0.89–1.27%) were the USFAs found in the samples. Oleic was absorbed as predominating USFAs followed by linoleic and linolenic acid. The nonaromatic rice cultivar Sarbati was found to have highest amount of oleic (49.14%), linoleic (46.99%), and linolenic acids (1.27%) while corre-sponding lowest amounts were observed in Govind Bhog (25.06%), Todal (30.22%), and Badshah Bhog (0.89%). The highest amount of myristic (4.60%) and palmitic acids (31.91%) were exhibited in Govind Bhog and stearic acid in Todal (6.47%), and corresponding lowest values were found in Badshah Bhog (0.27%), Sarbati (2.60%), and Gopal Bhog (2.28%) fatty acid profile. Myristic and stearic acids were not detected in aromatic and nonaromatic rice cultivars such as Swetganga, Khushboo, and Sarbati and also linolenic acid was not detected in Govind Bhog, Kala namak, Swet-ganga, Khushboo, and Todal.

Marco et al. (2006) also stated that rice is a natural source of USFAs, having more than 20% of oil. The high content of lipids in rice bran limits its use as a source of essential fatty acids, because the possibility of rapid rancidity; therefore, it is very necessary to stabilize, immediately after production. Amarasinghe and Gangodavilage (2004) investigated that major fatty acids palmitic, oleic, and linoleic comprised up to 90% of the total fatty acids composition of the rice oil. McPherson and Spiller (1995) studied whole grain rice for lipid contents (2–3%). Grains generally have about 75% of unsaturated lipids, , comprising of nearly equal amounts of oleic and linoleic acids and 1–2% of linolenic acid.

10.5   SUMMARY AND CONCLUSION

The chemical and nutritional composition in rice is very difficult to define with precision as preferences for quality evolution vary from country to country. Few from the scientists' and researchers' community realize its complexity and various quality parameters involved. The concept on quality evolution of rice varies according to the preparations methods for which sample rice grains are to be used. Some of the chemical and nutri-tional characteristics are desired by consumers which may place different emphasis on quality of rice cultivars containing varying amount of nutrients distinct to each other. A wide variation in chemical and nutritional composi-tion in different rice cultivars is found worldwide. Some rice cultivars are just a little with respect to chemical and nutritional quality, some moderate while others are highly. The variation between the chemical and nutritional quality of rice cultivars is due to several responsible factors such as envi-ronmental factors, agronomical conditions, genetic factors, and postharvest operations.

ACKNOWLEDGMENT

The authors are indebted to Department of Science and Technology, Ministry of Science and Technology, Government of India for an individual research fellowship (INSPIRE Fellowship Code No.: IF120725; Sanction Order No. DST/INSPIRE Fellowship/2012/686 and date: 25/02/2013).

KEYWORDS

- aromatic rice
- biofortification
- dietary component
- essential fatty acids
- fertilizer application
- grain moisture
- Indian rice cultivars

REFERENCES

AACC. *Approved Methods of the American Association of Cereal Chemists (Method No. 30-25 01, 44-15 02)*; American Association of Cereal Chemists: USA, 2000; Vol. 1.

Abdul-Hamid, A.; Sulaiman, R. R.; Osman, A.; Saari, N. Preliminary Study of the Chemical Composition of Rice Milling Fractions Stabilized by Microwave Heating. *J. Food Compos. Anal.* **2007,** *2*, 627–637. http://dx.doi.org/10.1016/j.jfca.2007.01.005 (accessedDec 28, 2017).

Adair, C. R.; Bollich, C. N.; Bowman, D. H.; Jodon, T. H.; Webb, B. D.; Atkins, J. G. Rice Breeding and Testing Method in the United States. In *Rice in the United States: Varieties and Production. US Dept. Agri. Handbook*; 1973, 289 (Revised); pp 22–27.

Adu-Kwarteng, E.; Ellis, W. O.; Oduro, J. T. I. Manful Rice Grain Quality: A Comparison of Local Varieties with New Varieties Under Study in Ghana. *Food Control* **2003,** *14* (7), 507–514.

Afsar, A. K. M. N.; Baqw, M.; Rawman, M.; Rouf, M. A. *Grades, Standards and Inspection Procedures of Fuce in Bangladesh. Bangladesh Food Management and Research Support Project*; Ministry of Food, Government of the People's Republic of Bangladesh, International Food Policy Research Institute. FMRSP Working Paper, 2001; 20. http://pdf.usaid.gov/pdf_docs/PNACN994.pdf. (accessed May 23, 2017).

Aganga, A. A.; Tshwenyane, S. Potentials of Guinea Grass (*Panicum maximum*) as Forage Crop in Livestock Production. *Pak. J. Nutr.* **2004,** *3* (1), 1–4.

Ahmad, M.; Samuel, S.; Zafar, M.; Khan, M. A.; Tariq, M.; Ali, S.; Sultana, S. Physicochemical Characterization of Eco-Friendly Rice Bran Oil Biodiesel. *Energy Sour. A Recover. Util. Environ. Eff.* **2011,** *33* (14), 1386–1397.

Ahmed, S. A.; Barua, L.; Das, D. Chemical Composition of Scented Rice. *Oryza* **1998,** *35* (2), 167–169.

Ali, M. A.; Hasan, S. M. K.; Mahomud, M. S.; Sayed, M. A. Processing and Storage of Instant Cooked Rice. *Bangladesh Res. Publ. J.* **2012,** *7* (3), 300–305.

Alim, M. A.; Lee, J. H.; Shin, J. A.; Lee, Y. J.; Choi, M. S.; Akoh, C. C.; Lee, K. T. Lipase-catalyzed Production of Solid Fat Stock from Fractionated Rice Bran Oil, Palm Stearin, and Conjugated Linoleic Acid by Response Surface Methodology. *Food Chem.* **2008,** *106* (2), 712–719.

Amarasinghe, B. M. W. P. K.; Gangodavilage, N. C. Rice Bran Oil Extraction in Sri Lanka: Data for Process Equipment Design. *Food Bioprod. Process.* **2004,** *82* (1), 54–59.

Anjum, F. M.; Pasha, I.; Bugti, M. A.; Butt, M. S. Mineral Composition of Different Rice Varieties and Their Milling Fractions. *Pak. J. Agric. Sci.* **2007,** *44* (2), 332–336.

Ansari, I. T.; Memon, A. N.; Ghanghro, A. B.; Sahito, M. A.; Narejo, N. T.; Umrani, J. H.; Khan, S.; Shah, A. M. Comparative Study of Physicochemical Properties of Mutant Rice Varieties Cultivated in Sindh. *Sindh Univ. Res. Jour. (Sci. Ser.)* **2013,** *45* (1), 17–20.

Anwar, F.; Anwer, T.; Mahmood, Z. Methodical Characterization of Rice (*Oryza sativa*) Bran Oil from Pakistan. *Grasas Aceites* **2005,** *56* (2), 125–134.

AOAC. *Official Methods of Analysis of Association of Official Analytical Chemists,* 15th ed.; AOAC: Arlington, VA, USA, 1990; pp 1–50.

AOAC. *Official Methods of Analysis of Association of Official Analytical Chemists,* 16th ed.; AOAC International: Washington, DC, 1997.

AOAC. *Official Methods of Analysis of Association of Official Analytical Chemists,* 17th ed.; AOAC: Maryland, USA, 2000; Vol. 1–2, pp 452–456.

Appleqvist, L. Å. Rapid Methods of Lipid Extraction and Fatty Acid Ester Preparation for Seed and Leaf Tissue with Special Remarks on Preventing the Accumulation of Lipid Contaminants. *Ark Kenci.* **1968,** *28,* 351–370.

Araullo, E. V.; Padua, D. B.; Graham, M. *Rice Postharvest Technology.* International Development Research Centre: Ottawa, Canada, 1976.

Asaduzzaman, M.; Haque, M. E.; Rahman, J.; Hasan, S. M. K.; Ali, M. A.; Akter, M. S.; Ahmed, M. Comparisons of Physiochemical, Total Phenol, Flavanoid Content and Functional Properties in Six Cultivars of Aromatic Rice in Bangladesh. *Afr. J. Food Sci.* **2013,** *7* (8), 198–203.

Ascheri, D. P. R.; Boêno, J. A.; Bassinello, P. Z.; Ascheri, J. L. R. Correlation between Grain Nutritional Content and Pasting Properties of Pre-gelatinized Red Rice Flour. *Rev. Ceres* **2012,** *59* (1), 16–24.

Awan, I. A. Physical and Biochemical Characterization of Some of the Pakistani Rice Varieties. M.Sc. (Hons) Dissertation, Department of Food Technology, University of Agriculture, Faisalabad, Pakistan, 1996.

Bechtel, D. B.; Pomeranz, Y. Ultrastructure of the Mature Ungerminated Rice (*Oryza saliva*) Caryopsis. The Starchy Endosperm. *Am. J. Bot.* **1977,** *64* (8), 966–973.

Bernal, J.; Mendiola, J. A.; Ibanez, E.; Cifuentes, A. Advanced Analysis of Nutraceuticals. *J. Pharm. Biomed. Anal.* **2011,** *55* (4), 758–774.

Bhat, R.; Sridhar, K. R. Nutritional Quality Evaluation of Electron Beam-irradiated Lotus (*Nelumbo nucifera*) Seeds. *Food Chem.* **2008,** *107,* 174–184.

Bhattacharjee, P.; Singhal, R. S.; Kulkarni, P. R. Basmati Rice: A Review. *Int. J. Food Sci. Technol.* **2002**, *37*, 1–12.

Boiling, H.; Hampel, II.; El Baya, A. W. Changes in Physical and Chemical Characteristics of Rice During Prolonged Storage. *II Rifo.* **1977**, *26* (1), 65.

Brenda, H.; Casimir, J.; Akoh, C. Characterization of Rice Bran Oil Structured Lipid. *J. Agric. Food Chem.* **2009**, *57*, 3346–3350.

Bressani, R.; Elias, L. G.; Juliano, B. O. Evaluation of Protein Quality. *J. Agric. Food Chem.* **1971**, *19*, 1028–1036.

CAC (Codex Alimentarius Commission). *The Need for Inclusion of Rice Bran Oil in the Standards for Named Vegetable Oils*; Joint FAO/WHO Food Standards Programme, Codex Committee on Fats and Oils: London, UK, Feb 3–7, 2003, 18th Session.

Calder, P. C. The Relationship between the Fatty Acid Composition of Immune Cells and Their Function. *Prostaglandins Leukot. Essent. Fatty Acids* **2003**, *79* (3–5), 101–108.

Cameron, D. K.; Wang, Y. A Better Understanding of Factors That Affect the Hardness and Stickiness of Long-grain Rice. *Cereal Chem.* **2005**, *82* (2), 113–119.

Cayh, G.; Kusefoglu S. Increased Yields in Biodiesel Production from Used Cooking Oils by a Two Step Process: Comparison with One Step Process by Using TGA. *Fuel Process Technol.* **2008**, *89*, 118–122.

Certik, M.; Shimizu, S. Biosynthesis and Regulation of Microbial Polyunsaturated Fatty Acid Production. *J. Biosci. Bioeng.* **1999**, *87*, 1–14.

Chaichaw, C.; Naivikul, O.; Thongngam, M. Effect of Heat-Moisture Treatment on Qualities of Gluten-Free Alkaline Rice Noodles from Various Rice Flour Varieties. *Kasetsart J. (Nat. Sci.)* **2011**, *45*, 490–499.

Champagne, E. T.; Bett-Garber, K. L.; Thompson, J.; Mutters, R.; Grimm, C. C.; McClung, A. M, Effects of Drain and Harvest Dates on Rice Sensory and Physicochemical Properties. *Cereal Chem.* **2005**, *82* (4), 369–374.

Champe, P. C.; Harvey, R. A. *Lippincott's Illustrated Reviews: Biochemistry,* 2nd ed.; Lippincott Raven Publishers: New Jersey, USA, 1994; pp 303–340.

Chen, J.; Lii, C.; Lu, S. Relationships Between Grain Physicochemical Characteristics and Flour Particle Size Distribution for Taiwan Rice Cultivars. *J. Food Drug Anal.* **2004**, *12* (1), 52–58.

Chinma, C. E.; Anuonye, J. C.; Simon, O. C.; Ohiare, R. O.; Danbaba, N. Effect of Germination on the Physicochemical and Antioxidant Characteristics of Rice Flour from Three Rice Varieties from Nigeria. *Food Chem.* **2015**, *185*, 454–458.

Chotimarkorn, C.; Benjakul, S.; Silalai, N. Antioxidant Components and Properties of Five Long-grained Rice Bran Extracts from Commercial Available Cultivars in Thailand. *Food Chem.* **2008**, *111*, 636–641.

Choudhury, N. H.; Juliano, B. O. Effect of Amylose Content on the Lipids of Mature Rice Grain. *Phytochemistry* **1980**, *19*, 1385–1389.

Cogburn, R. R. Rough Rice Storage. In *Rice Chemists and Technology, 2nd ed.*; Juliano, B. O., Ed.; The American Association of Cereal Chemists: St Paul, MI, USA, 1985; pp 265–287.

Cornejo, F.; Rosella, C. M. Physicochemical Properties of Long Rice Grain Varieties in Relation to Gluten Free Bread Quality. *LWT—J. Food Sci. Technol.* **2015**, *62* (2), 1203–1210.

Da Silva, D. G.; Scarminio, I. S.; Anuncia□ão, D. S.; Souza, A. S.; Da Silva, E. G. P.; Ferreira, S. L. C. Determination of the Mineral Composition of Brazilian Rice and Evaluation Using Chemometric Techniques. *Anal. Methods* **2013**, *5*, 998–1003.

Deepa, G.; Singh, V.; Naidu, K. A. Nutrient Composition and Physicochemical Properties of Indian Medicinal Rice—Njavara. *Food Chem.* **2008**, *106* (1), 165–171.

Devi, G. N.; Padmavathi, P.; Babu, V. R.; Waghray, K. Proximate Nutritional Evaluation of Rice (*Oryza sativa* L.). *J. Rice Res.* **2015**, *8* (1), 23–32.

Diako, C.; Sakyi-Dawson, E.; Bediako-Amoa, B.; Saalia, F. K.; Manful, J. T. Cooking Characteristics and Variations in Nutrient Content of Some New Scented Rice Varieties in Ghana. *Ann. Food Sci. Technol.* **2011**, *12* (1), 39–44.

Dikeman, E.; Bechtel, D. B.; Pomeranz Y. Distribution of Element in the Rice Kernel Determined by X-Ray Analysis and AAS. *Cereal Chem.* **1981**, *58*, 148–152.

Ebuehi, A. O. T.; Oyewole, A. C. Effect of Cooking and Soaking on Physical Characteristics, Nutrient Composition and Sensory Evaluation of Indigenous Rice and Foreign Rice Varieties in Nigeria. *Afr. J. Biotechnol.* **2008**, *6* (8), 1016–1020.

Edeogu, C. O.; Ezeonu, F. C.; Okaka, A. N. C.; Ekuma, C. E.; Elom, S. O. Proximate Compositions of Staple Food Crops in Ebonyi State, South Eastern Nigeria. *Int. J. Biotechnol. Biochem.* **2007**, *1*, 1–8.

Eggum, B. O.; Juliano, B. O.; Magnificat, C. C. Protein and Energy Utilization of Rice Milling Fractions. *J. Hum. Nutr.* **1982**, *31*, 371–376.

Eggum, B. O. *A Study of Certain Factors Influencing Protein Utilization in Rats and Pigs.* Publ. 406; Agricultural Research Laboratory: Copenhagen, 1973; p 173.

Eggum, B. O. Evaluation of Protein Quality and the Development of Screening Techniques. In *New Approaches to Breeding for Improved Plant Protein*; IAEA: Vienna, 1969; pp 125–135.

Eggum, B. O. Nutritional Aspects of Cereal Protein. In *Genetic Diversity in Plants*; Muhammad, A., Aksel, R., von Boustel, R. C., Eds.; Plenum Press, New York, 1977; pp 349–369.

Fagbohun, E. D.; Oluwaniyi, T. T. Mycoflora, Proximate Composition and Nutritional Changes During the Storage of *Oryza sativa*. *Food Sci. Qual. Manag.* **2015**, *40*, 108–116.

FAO (Food and Agriculture Organization). Calculation of the Energy Content of Foods—Energy Conversion Factors (Chapter-3). In *Food Energy—Methods of Analysis and Conversion Factors*. Report of a Technical Workshop, Rome, December 3–6, 2002, Food and Agriculture Organization of The United Nations: Rome, 2003. ftp://ftp.fao.org/docrep/fao/006/y5022e/y5022e00.pdf (accessed Jan 9, 2017).

FAO (Food and Agriculture Organization). Rice is Life, International Year of Rice 2004. The Rice and Human Nutrition. Food and Agriculture Organization of the United Nations: Rome, Italy, 2017. http://www.fao.org/rice2004/en/f-sheet/factsheet3.pdf. (accessed Jan 9, 2017).

FAO (Food and Agriculture Organization). Amino Acid Content of Foods and Biological Data on Proteins. Nutrition Division, FAO: Rome, 1970; Vol. 24, pp 122.

FAO (Food and Agriculture Organization). Obesity: Preventing and Managing Global Epidemic, World Health Organization (WHO) Technical Report, Geneva, Switzerland, 1998; pp 11–12.

FAO (Food and Agriculture Organization). Food and Agriculture Organization/International Rice Research Institute FAO Food and Nutrition Series, FAO, Rome, 2006, 2.

FAO/WHO. Fats and Oils in Human Nutrition. Report of a Joint Expert Consultation Organized by the Food and Agriculture Organization of the United Nations and the World Health Organization: Rome, October 19–26, 1993, Vol. 10, pp 19–26.

Fari, M. J. M.; Rajapaksa, D.; Ranaweera, K. K. D. S. Quality Characteristics of Noodles Made from Selected Varieties of Sri Lankan Rice with Different Physicochemical Characteristics. *J. Natl. Sci. Found. Sri Lanka* **2011**, *39* (1), 53–60.

Fox, M. R. S.; Tao, S. H. Antinutritive Effects of Phytate and Other Phosphorylated Derivatives. *Nutr. Toxicol.* **1989**, *3*, 59–62.

Frei, M.; Siddhuraju, P.; Becker, K. Studies on the In Vitro Starch Digestibility and Glycemic Index of Six Different Indigenous Rice Cultivars from the Philippines. *J. Food Chem.* **2003**, *83*, 395–400.

Fresco, L. Rice is Life. *J. Food Compos. Anal.* **2005**, *18*, 249–253.

FSSAI (Food Safety and Standards Authority of India). Oils and Fats. In *Manual of Methods of Analysis of Foods*; Lab. Manual-2, Food Safety and Standards Authority of India, Ministry Of Health and Family Welfare, Government of India: New Delhi, 2012.

Fujita, N.; Kubo, A.; Suh, S.-D.; Wong, K.-S.; Jane, J.-L.; Ozawa, K.; Takaiwa, F.; Inaba, Y.; Nakamura, Y. Antisense Inhibition of Isoamylase Alters the Structure of Amylopectin and the Physicochemical Properties of Starch in Rice Endosperm. *Plant Cell Physiol.* **2003**, *44*, 607–618.

Fujita, N.; Satoh, R.; Hayashi, A.; Kodama, M.; Itoh, R.; Aihara, S.; Nakamura, Y. Starch Biosynthesis in Rice Endosperm Requires the Presence of Either Starch Synthase I or IIIa. *J. Exp. Bot.* **2011**, *62* (14), 4819–4831.

Fujita, N.; Toyosawa, Y.; Utsumi, Y.; Higuchi, T.; Hanashiro, I.; Ikegami, A.; Akuzawa, S.; Yoshida, M.; Mori, A.; Inomata, K.; Itoh, R.; Miyao, A.; Satoh, H.; Nakamura, Y. Characterization of PUL-Deficient Mutants of Rice (*Oryza sativa* L.) and the Function of PUL on the Starch Biosynthesis in the Rice Endosperm. *J. Exp. Bot.* **2009**, *60*, 1009–1023.

Goffman, F. D.; Pinson, S.; Bergman, C. Genetic Diversity for Lipid Content and Fatty Acid Profile in Rice Bran. *J. Am. Oil Chem. Soc.* **2003**, *80* (5), 645–653.

Gomez, K. A. In *Effect of Environment on Protein and Amylose Content of Rice*. Proceedings of the Workshop on Chemical Aspects of Rice Grain Quality, International Rice Research Institute, Los Baños, Laguna, Philippines, 1979, pp 59–68.

Gooding, M. J.; Davies, W. P. *Wheat Production and Utilization: Systems, Quality and the Environment*; CAB International University Press: Cambridge, UK, 1997; p 355.

Govarethinam, B. A. P. A Comparative Study of Mineral Contents in Selected Malaysian Brown Rice and White Rice. M.Sc. Dissertation, Faculty of Science, University of Malaya, Kuala Lumpur, Malaysia, 2014.

Grist, D. H. *Rice*, 5th ed.; Longman: London, 1975.

Gunstone, F. D. Oils and Fats: Sources and Constituents. In *The Chemistry of Oils and Fats: Sources, Composition, Properties and Uses*; Gunstone, F. D., Ed.; Blackwell Publishing Ltd. and CRC Press: FL, USA, 2004; pp 1–35.

Hamakcr, B. R.; Griffin, V. K. Changing the Viscoelastic Properties of Cooked Rice Through Protein Disruption. *Cereal Chem.* **1990**, *67* (3), 261–264.

Hashmi, M. I.; Tianlin, J. S. Minerals Contents of Some Indigenous Rice Varieties of Sabah, Malaysia. *Int. J. Agric. For. Plant.* **2016**, *2*, 31–34.

Hayat, A.; Jahangir, T. M.; Alamgir, M. Effect of Germination Conditions on Proximate Chemical Composition of Some Pakistani Brown and Polished Rice Varieties. *Stud. J. Chem.* **2013**, *1* (3), 98–106.

Heinemann, R. J. B.; Fagundes, P. L.; Pinto, E. A.; Penteado, M. V. C.; Lanfer-Marquez, U. M. Comparative Study of Nutrient Composition of Commercial Brown, Parboiled and Milled Rice from Brazil. *J. Food Compos. Anal.* **2005**, *18*, 287–296.

Honjo, K. Studies in Protein Content in Rice Grains: Variations in Protein Content Among Rice Cultivars and Influences of Environmental Factors on the Protein Content. *Process. Crop Sci. Soc. Jpn.* **1971**, *40*, 183–189.

Hossain, M. A.; Bhattacharjee, S.; Armin, S. M.; Qian, P.; Xin, W.; Li, H. Y.; Burritt, D. J.; Fujita, M.; Tran, L. S. P. Hydrogen Peroxide Priming Modulates Abiotic Oxidative Stress Tolerance: Insights from ROS Detoxification and Scavenging. *Front. Plant Sci.* **2015**, *6*, 420.

Hu, W.; Wells, J. H.; Shin, T. S.; Godber, J. S. Comparison of Isopropanol and Hexane for Extraction of Vitamin E and Oryzanols from Stabilized Rice Bran. *J. Am. Oil Chem. Soc.* **1996**, *73*, 1653–1656.

Hu, J.; Du, Z.; Tang, Z.; Min, E. Study on the Solvent Power of a New Green Solvent: Biodiesel. *Ind. Eng. Chem. Res.* **2004**, *43*, 7928–7931.

Huang, Y. S.; Sun, Z. X.; Hu, P. S.; Tang, S. Q. Present Situation and Prospects for the Research on Rice Grain Quality Forming. *Chin. J. Rice Sci.* **1998**, *12* (3), 172–176.

Ibukun, E. O. Effect of Prolonged Parboiling Duration on Proximate Composition of Rice. *Sci. Res. Essay* **2008**, *3* (7), 323–325.

Islam, M. J.; Das, J.; Sentinu; Absar, N.; Hasanuzzaman, M. A Comparative Analysis in the Macro and Micro Nutrient Composition of Locally Available Polished Rice (*Oryza sativa* L.) in Bangladesh. *Int. J. Biol. Res.* **2016**, *4* (2), 190–194.

Jaiswal, P. K. Specification of Rice Bran Oil and Extractions. In *Rice Bran Oil, Status and Prospects*; Pradesh, A., Ed.; Southern Zone: Hyderabad, India, 1983; pp 64–77.

Janick, J.; Whipkey, A. *Trends in New Crops and New Uses*; ASHS Press: Alexandria, 2002; pp 100–103.

Jariwalla, R. J. Rice-bran Products: Phytonutrients with Potential Applications in Preventive and Clinical Medicine. *Drugs Exp. Clin. Res.* **2001**, *27* (1), 17–26.

Jenning, B. H.; Akoh, C. A. Effectiveness of Natural *Versus* Synthetic Antioxidants in a Rice Bran Oil-based Structured Lipid. *Food Chem.* **2009**, *114*, 1456–1461.

Jenning, P. R.; Coffman, W. R.; Kaufman, H. E. Grain Quality. In *Rice Improvement*; International Rice Research Institute: Los Baños, Manila, Philippines, 1979.

Johnston, T. H.; Miller, M. D. Rice in the United States: Varieties and Production. *Culture,* US Dept. Agri. Handbook; 1973; 289 (revised), pp 88–128.

Juliano, B. O. Polysaccharides, Proteins, and Lipids of Rice. In *Rice Chemistry and Technology*; Juliano, B. O., Ed.; American Association of Cereal Chemists: St Paul, Minnesota, 1985a; pp 59–174.

Juliano, B. O. *Factors Affecting Nutritional Properties of Rice Protein*; National Academy of Science and Technology (NAST): Philippines, 1985b; Vol. 7, pp 205–216.

Juliano, B. O. Production and Utilization of Rice. In *Rice Chemistry and Technology*; Juliano, B. O., Ed.; American Association of Cereal Chemists: St Paul, Minnesota, 1985c; pp 1–16.

Juliano, B. O. *Rice: Chemistry and Technology*; American Association of Cereal Chemists: St Paul, Minnesota, 1985d.

Juliano, B. O. Nutritional Value of Rice and Rice Diets. In *Rice in Human Nutrition*; Pub. International Rice Research Institute (IRRI), Philippines and Food and Agriculture Organization of the United Nations: Rome, Italy, 1993a; pp 61–84. http://www.fao.org/docrep/t0567e/t0567e00.htm (accessed Jan 9, 2017).

Juliano, B. O. Grain Structure, Composition and Consumers' Criteria for Quality. In *Rice in Human Nutrition*; Pub. International Rice Research Institute (IRRI), Philippines and Food and Agriculture Organization of the United Nations: Rome, Italy, 1993b; pp 35–59. http://www.fao.org/docrep/t0567e/t0567e00.htm (accessed Jan 9, 2017).

Juliano, B. O.; Onate, L. U.; DelMundo, A. M. Relation of Starch Composition, Protein Content and Gelatinization Temperature to Cooking and Eating Qualities of Milled Rice. *Food Technol.* **1965**, *19* (6), 116–119.

Juliano, B. O.; Bechtel, D. B. The Rice Grain and Its Gross Composition. In *Rice Chemistry and Technology, 2nd Ed.*; Juliano, B. O., Ed.; The American Association of Cereal Chemists: St Paul, MI, USA, 1985; pp 17–57.

Juliano, B. O.; Goddard, M. S. Cause of Varietal Difference in Insulin and Glucose Responses to Ingested Rice. *J. Qual. Plant Plant Foods Hum. Nutr.* **1986**, *36*, 35–41.

Jung, M. C.; Yun, S.-T.; Lee, J.-S.; Lee, J.-U. Baseline Study on Essential and Trace Elements in Polished Rice from South Korea. *Env. Geochem. Health* **2005**, *27* (5–6), 455–464.

Kapri, M.; Verma, D. K.; Ajesh, K. V.; Billoria, S.; Mahato, D. K.; Yadav, B. S.; Srivastav, P. P. Modified Pearl Millet Starch: A Review on Chemical Modification, Characterization and Function Properties. In *Engineering Interventions in Agricultural Processing*; Goyal, M. R., Verma, D. K., Eds.; Apple Academic Press, Inc.: NJ, USA, 2017; Vol. 13, pp 191–226.

Kataoka, K. Studies on Chemical Quality of the Rice Kernel: Effect of Temperature on Protein Content. *J. Tamagawa Univ.* **1975**, *15*, 96–100.

Katz, S. H.; Weaver, W. W. (Eds.). *Encyclopedia of Food and Culture*; Charles Scribner's Sons: New York, USA, 2003.

Khatoon, S.; Krishna, A. G. G. Fat Soluble Nutraceuticals and Fatty Acid Composition of Selected Indian Rice Varieties. *J. Am. Oil Chem. Soc.* **2004**, *81*, 939–943.

Kim, H. J.; Lee, S. B.; Park, K. A.; Hong, I. K. Characterization of Extraction and Separation of Rice Bran Oil Rich in EFA Using SFE Process. *Sep. Purif. Technol.* **1999**, *15*, 1–8.

Krishna, A. G. G. Nutritional Components of Rice Bran Oil in Relation to Processing. *Lipid Technol.* **2002**, *14*, 80–84.

Krishnan, S.; Ebenezer, G. A. I.; Dayanandan, P. Histochemical Localization of Storage Components in Caryopsis of Rice (*Oryza sativa* L.). *Curr. Sci.* **2001**, *80* (4), 567–571.

Kullenberg, D.; Taylor, L. A.; Schneider, M.; Massing, U. Health Effects of 813 Dietary Phospholipids. *Lipids Health Dis.* **2012**, *11*, 3.

Lásztity, R. *The Chemistry of Cereal Proteins*; CRC Press Inc.: Boca Raton, USA, 1984.

Law, M. Dietary Fat and Adult Diseases and the Implications for Childhood Nutrition: An Epidemiologic Approach. *Am. J. Clin. Nutr.* **2000**, *72*, 1291s–1296s.

Lei, H.; Ding, X.; Zhang, H.; Chen, X.; Li, Y.; Zhang, H.; Wang, Z. In Situ Production of Fatty Acid Methyl Ester from Low Quality Rice Bran: An Economical Route for Biodiesel Production. *Fuel* **2010,** *89,* 1475–1479.

Lestari, P.; Ham, T. H.; Lee, H. H.; Woo, M. O.; Jiang, W. PCR Marker-based Evaluation of the Eating Quality of Japonica Rice (*Oryza sativa* L.). *J. Agric. Food Chem.* **2009,** *57,* 2754–2762.

Lestari, P.; Reflinur; Koh, H. Prediction of Physicochemical Properties of Indonesian Indica Rice Using Molecular Markers. *HAYATI J. Biosci.* **2014,** *21* (2), 76–86.

Li, R.; Lan, S. Y.; Xu, Z. X. Studies on the Programmed Cell Death in Rice During Starchy Endosperm Development. *Agric. Sci. Chin.* **2004,** *3* (9), 663–670.

Liu, Z. H.; Cheng, F. M.; Cheng, W. D.; Zhang, G. P. Positional Variations in Phytic Acid and Protein Content Within a Panicle of Japonica Rice. *J. Cereal Sci.* **2005,** *41,* 297–303.

Lotero, E.; Liu, Y.; Lopez, D. E.; Suwannakarn, K.; Bruce, D. A.; Goodwin Jr, J. G. Synthesis of Biodiesel via Acid Catalysis. *Ind. Eng. Chem. Res.* **2005,** *44,* 5353–5363.

Lu, S.; Luh, B. S. Properties of the Rice Caryopsis (Chapter 11). In *Rice Production, 2nd ed.*; Luh, B. S., Ed.; Springer Science+Business Media: NY, USA, 1991, Vol. 1, pp 389–419.

Maisont, S.; Narkrugsa, W. Effects of Some Physicochemical Properties of Paddy Rice Varieties on Puffing Qualities by Microwave "Original." *Kasetsart J. (Nat. Sci.)* **2009,** *43,* 566–575.

Maningat, C. C; Juliano, B. O. Starch Lipids and Their Effect on Rice Starch Properties. *Starch/Staerke* **1980,** *32,* 76.

Marco, A. S.; Sanches, C.; Amante, E. R. Prevention of Hydrolytic Rancidity in Rice Bran. *J. Food Eng.* **2006,** *75,* 487–491.

Marr, K. M.; Batten, G. D.; Blakeney, A. B. Relationships Between Minerals in Australian Brown Rice. *J. Sci. Food Agric.* **1995,** *68* (3), 285–291.

Mbatchou, V. C.; Dawda, S. The Nutritional Composition of Four Rice Varieties Grown and Used in Different Food Preparations in Kassena-Nankana District, Ghana. *Int. J. Res. Chem. Environ.* **2013,** *3* (1), 308–315.

McPherson, R.; Spiller, G. A. Effects of Dietary Fatty Acids and Cholesterol on Cardiovascular Disease Risk Factors in Man. In *Handbook of Lipids in Human Nutrition*; Spiller, G. A., Ed.; CRC Press: Boca Raton, FL, 1995; pp 41–49.

Ministry of Agriculture. National Secretariat of Supply. Rules on Sorting, Packaging and Presentation of Rice. Portaria No. 269 of Nov 17, 1988, Brazil, 1988.

Mohammed, N. K.; Spyrou, N. M. Trace Elemental Analysis of Rice Grown in Two Regions of Tanzania. *J. Radioanal. Nucl. Chem.* **2009,** *281,* 79–82.

Mohd Fairulnizal, M. N.; Norhayati, M. K.; Zaiton, A.; Norliza, A. H.; Rusidah, S.; Aswir, A. R.; Suraiami, M.; Mohd Naeem, M. N.; Jo-Lyn, A.; Mohd Azerulazree, J.; Vimala, B.; Mohd Zainuldin, T. Nutrient Content in Selected Commercial Rice in Malaysia: An Update of Malaysian Food Composition Database. *Int. Food Res. J.* **2015,** *22* (2), 768–776.

Moongngarm, A.; Saetung, N. Comparison of Chemical Compositions and Bioactive Compounds of Germinated Rough Rice and Brown Rice. *Food Chem.* **2010,** *122,* 782–788.

Morrison, W. R.; Azuclin, M. N. Variation in (He Amylose and Lipid Contents and Some Physical Properties of Rice Starches. *J. Cereal Sci.* **1987,** *5,* 35.

Most, M. M.; Tulley, R.; Morales, S.; Lefevre, M. Rice Bran Oil, Not Fiber, Lowers Cholesterol in Humans1,2,3. *Am. J. Clin. Nutr.* **2005**, *81* (1), 64–68.

Muhammad, Z. Characterization of Selected Varieties of Rice (*Oryza sativa*) and Its By-Product (Rice Bran) For Valuable Nutrients and Antioxidants. Pakistan Research Repository. Ph.D. Thesis, Department of Chemistry and Biochemistry, Faculty of Sciences/University of Agriculture, Faisalabad, Pakistan, 2012.

Nagato, K.; Ebata, M.; Ishikawa, M. Protein Content of Developing and Mature Rice Grains. *J. Process. Crop Sci. Soc. Jpn.* **1972**, *41*, 472–479.

Ngo, H. L.; Zafiropoulos, N. A.; Foglia, T. A.; Samulski, E. T.; Lin, W. Efficient Two-step Synthesis of Biodiesel from Greases. *Energ. Fuel* **2008**, *22*, 626–634.

Nielsen, F. H. Macromineral Nutrition (Chapter 12). *Handbook of Nutrition and Food, 3rd ed.*; Berdanier, C. D., Dwyer, J. T., Heber, D., Eds.; CRC Press: New York, 2013a; pp 199–210. ISBN 978-1-4665-0572-8.

Nielsen, F. H. Trace Mineral Deficiencies (Chapter 13). In *Handbook of Nutrition and Food, 3rd Ed.*; Berdanier, C. D., Dwyer, J. T., Heber, D., Eds.; CRC Press: New York, 2013b; pp 211–226. ISBN 978-1-4665-0572-8.

Noreen, N.; Shah, H.; Anjum, F.; Masood, T.; Faisal, S. Variation in Mineral Composition and Phytic Acid Content in Different Rice Varieties During Home Traditional Cooking Processes. *Pak. J. Life Soc. Sci.* **2009**, *7* (1), 11–15.

Oko, A. O.; Onyekwere, S. C. Studies on the Proximate Chemical Composition and Mineral Element Contents of Five New Lowland Rice Varieties in Ebonyi State. *Int. J. Biotechnol. Biochem.* **2010**, *6* (6), 949–955.

Oko, A. O.; Ugwu, S. I. The Proximate and Mineral Compositions of Five Major Rice Varieties in Abakaliki, South-Eastern Nigeria. *Afr. J. Biotechnol.* **2011**, *6* (8), 1016–1020.

Oko, A. O.; Ubi, B. E.; Efisue, A. A.; Dambaba, N. Comparative Analysis of the Chemical Nutrient Composition of Selected Local and Newly Introduced Rice Varieties Grown in Ebonyi State of Nigeria. *Int. J. Agric. Sci.* **2012**, *2* (2), 16–23.

Oluremi, O. I.; Solomon, A. O.; Saheed, A. A. Fatty Acids, Metal Composition and Physico-Chemical Parameters of Igbemo Ekiti Rice Bran Oil. *J. Environ. Chem. Ecotoxicol.* **2013**, *5* (3), 39–46.

Onate, L. U.; DelMundo, A. M.; Juliano, B. O. Relationship Between Protein Content and Eating Quality of Milled Rice. *Philipp. Agric.* **1964**, *47*, 441.

Orthoefer, F. T. Rice Bran Oil: Healthy Lipid Source. *Food Technol.* **1996**, *50* (12), 62–64.

Osborn, D. R.; Voogt, P. Calculation of Calorific Value. In *The Analysis of Nutrients in Foods*; Academic Press: New York, USA, 1978, pp 239–240.

Pan, Z.; Thompson, J. F.; Amaratunga, K. S.; Anderson, P. T.; Zheng, X. Effect of Cooling Methods and Milling Procedures on The Appraisal of Rice Milling Quality. *J. Trans. ASAE* **2007**, *48* (5), 1865–1871.

Panlasigui, L. N.; Thompson, L. U.; Juliano, B. O.; Perez, C. M.; Yiu, S. H.; Greenberg, G. R. Rice Varieties with Similar Amylose Content Differ in Starch Digestibility and Glycemic Response in Humans. *Am. J. Clin. Nutr.* **1991**, *54*, 871–877.

Park, J. K.; Kim, S. S.; Kim, K. O. Effect of Milling Ratio on Sensory Properties of Cooked Rice and on Physicochemical Properties of Milled and Cooked Rice. *Cereal Chem.* **2001**, *78* (2), 1–156.

Parrado, J.; Miramontes, E.; Jover, M.; Gutierrez, J. F.; Teran, L. C.; Bautista, J. Preparation of a Rice Bran Enzymatic Extract with Potential Use as Functional Food. *Food Chem.* **2006,** *98* (4), 742–748.

Pereira, J. A.; Bassinello, P. Z.; Cutrim, V. A.; Ribeiro, V. Q. Comparison Between Agronomic, Culinary and Nutritional Characteristics in White and Red Rice Varieties. *Caatinga* **2009,** *22,* 243–248.

Qin-lu, L.; Hua-xi, X.; Xiang-jin, F.; Wei, T.; Li-hui, L.; Feng-xiang, Y. Physico-chemical Properties of Flour, Starch, and Modified Starch of Two Rice Varieties. *Agric. Sci. Chin.* **2011,** *10* (6), 960–968.

Raghav, P. K.; Agarwal, N.; Sharma, A. Emerging Health Benefits of Rice Bran—A Review. *Int. J. Multi. Res. Mod. Educ.* **2016,** *2* (1), 367–382.

Ranganathan, S. V.; Narasimhan, S. L.; Muthukumar, K. An Overview of Enzymatic Production of Biodiesel. *Bioresour. Technol.* **2008,** *99,* 3975–3981.

Ravi, U.; Menon, L.; Gomathy, G.; Parimala, C.; Rajeshwari, R. Quality Analysis of Indigenous Organic Asian Indian Rice Variety—*Salem samba. Indian J. Trad. Knowl.* **2012,** *11* (1), 114–122.

Rodrigues, C. E. C.; Filho, P. A. P.; Meirelles, A. J. A. Phase Equilibrium for the System Rice Bran Oil + Fatty Acids + Ethanol + Water + γ-Oryzanol + Tocols. *Fluid Phase Equilib.* **2004,** *216,* 271–283.

Rossell, J. B. Analysis and Properties of Oilseeds. In *Analysis of Oilseeds, Fats and Fatty Foods*; Rossell, J. B., Pritchard, J. L. R., Eds.; Elsevier Applied Science: New York, USA, 1991a; pp 80–98.

Rossell, J. B. Vegetable Oil and Fats. In *Analysis of Oilseeds, Fats and Fatty Foods*; Rossell, J. B., Pritchard, J. L. R., Eds.; Elsevier Applied Science: New York, USA, 1991b; pp 261–328.

Rukmini, C.; Raghuram, T. C. Nutritional and Biochemical Aspects of the Hypolipidemic Action of Rice Bran Oil. *J. Am. Coll. Nutr.* **1991,** *10,* 593–601.

Saikia, D.; Deka, S. C. Cereals: From Staple Food to Nutraceuticals. *Int. Food Res. J.* **2011,** *18,* 21–30.

Saikia, S.; Himjyoti, D.; Daizi, S.; Charu, L. M. Quality Characterization and Estimation of Phytochemical Content Capacity of Aromatic Pigmented and Non-pigmented Rice Varieties. *Food Res. Int.* **2012,** *46,* 334–340.

Salehpour, S.; Dube, M. A. Biodiesel: A Green Polymerization Solvent. *Green Chem.* **2008,** *10,* 321–326.

Satter, M. A.; Ara, H.; Jabin, S. A.; Abedin, N.; Azad, A. K.; Hossain, A.; Ara, U. Nutritional Composition and Stabilization of Local Variety Rice Bran BRRI-28. *Int. J. Sci. Technol.* **2014,** *3* (5), 306–313.

Sayre, B.; Saunders, R. Rice Bran and Rice Bran Oil. *Lipid Technol.* **1990,** *2,* 72–76.

Schakel, S. F.; Buzzard, I. M.; Gebhardt, S. E. Procedures for Estimating Nutrient Values for Food Composition Databases. *J. Food Compos. Anal.* **1997,** *10,* 102–114.

Scherz, H.; Senser, F.; Souci, S. W. *Food Composition and Nutrition Tables,* 6th ed.; CRC Press/Medpharm: Boca Raton, FL, USA, 2000; p 1182.

Seki, T.; Nagase, R.; Torimitsu, M.; Yanagi, M.; Ito, Y.; Kise, M.; Mizukuchi, A.; Fujimura, N.; Hayamizu, K.; Ariga, T. Insoluble Fiber is a Major Constituent Responsible for

Lowering the Post-Prandial Blood Glucose Concentration in the Pre-Germinated Brown Rice. *Biol. Pharm. Bull.* **2005,** *28* (8), 1539–1541.

Shabbir, M. A.; Anjum, F. M.; Zahoor, T.; Nawaz, H. Mineral and Pasting Characterization of Indica Rice Varieties with Different Milling Fractions. *Int. J. Agric. Biol.* **2008,** *10,* 556–560.

Shayo, N. B.; Mamiro, P.; Nyaruhucha, C. N. M.; Mamboleo, T. Physico-Chemical and Grain Cooking Characteristics of Selected Rice Cultivars Grown in Morogoro. *Tanzania J. Sci.* **2006,** *32* (1), 29–36.

Shin, D. H.; Chung, J. K. Changes During Storage of Rice Germ Oil and Its Fatty Acid Composition. *Korean J. Food Sci. Technol.* **1998,** *30,* 77–81.

Sompong, R.; Ehn, S.; Martin, L.; Berghofer, E. Physicochemical and Antioxidant Properties of Red and Black Rice Varieties from Thailand, China and Sri Lanka. *Food Chem.* **2011,** *124,* 132–140.

Sotelo, A.; Sousa, V.; Montalvo, I.; Hernandez, M.; Hernandez-Arago, I. Chemical Composition Fractions of 12 Mexican Varieties of Rice Obtained During Milling. *J. Cereal Chem.* **1990,** *67* (2), 209–212.

Sperotto, R. A.; Ricachenevsky, F. K.; de Abreu Waldow, V.; Fett, J. P. Iron Biofortification in Rice: It's a Long Way to the Top. *Plant Sci.* **2012,** *190,* 24–39.

Storck, C. R.; Silva, L. P.; Fagundes, C. A. A. Categorizing Rice Cultivars Based on Differences in Chemical Composition. *J. Food Compos. Anal.* **2005,** *18,* 333–341.

Subudhi, H.; Meher, J.; Singh, O. N.; Sharma, S. G.; Das, S. Grain and Food Quality Traits in Some Aromatic Long and Short Grain Rice Varieties of India. *J. Food Agric. Environ.* **2013,** *11* (3&4), 1434–1436.

Sunders, R. M. The Properties of Rice Bran as a Foodstuff. *Cereal Foods World* **1990,** *35* (7), 632–636.

TACO. *Brazilian Food Composition Table,* 2nd ed.; Campinas, NEPA-UNICAMP, 2006; p 113. http://www.unicamp.br/nepa/taco/contar/taco_versao2.pdf. (accessed Dec 22, 2016).

Taira, H. Protein Content in Hulled Grains of the Upland Rice Cultivars. *J. Nutr. Food* **1970,** *23,* 94–97.

Taira, H.; Itani, T. Lipid Content and Fatty Acid Composition of Brown Rice of Cultivars of the United States. *J. Agric. Food Chem.* **1988,** *36,* 460–462.

Taira, H.; Taira, H.; Maeshige, M. Influence of Variety and Crop Year on Lipid Content and Fatty Acids Composition of the Lowland Non-glutinous Brown Rice. *Jpn. J. Crop Sci.* **1979,** *48* (2), 220–228.

Taira, H.; Nakagahra, M.; Nagamine, T. Fatty Acid Composition of *Indica, Sinica, Japonica,* and *Japonica* Groups of Non-glutinous Brown Rice. *J. Agric. Food Chem.* **1988,** *50,* 3031–3035.

Tamanna, S.; Parvin, S.; Kumar, S.; Dutta, A. K.; Ferdoushi, A.; Siddiquee, M. A.; Biswas, S. K.; Howlader, M. J. H. Content of Some Minerals and Their Bioavailability in Selected Popular Rice Varieties from Bangladesh. *Int. J. Curr. Microbiol. Appl. Sci.* **2013,** *2* (7), 35–43.

Thomas, R.; Wan-Nadiah, W. A.; Bhat, R. Physiochemical Properties, Proximate Composition, and Cooking Qualities of Locally Grown and Imported Rice Varieties Marketed in Penang, Malaysia. *Int. Food Res. J.* **2013,** *20* (3), 1345–1351.

Tsutsumi, H.; Nishikawa, M.; Katagi, M.; Tsuchihashi, H. Adsorption and Stability of Suxamethonium and Its Major Hydrolysis Product Succinylmonochlorine Using Liquid Chromatograph Electrospray Ionization Mass Spectrometry. *J. Health Sci.* **2003,** *49* (4), 285–291.

Tufail, S. Cooking and Eating Quality of Some Cultivars of Rice. M.Sc. (Hons) Dissertation, Department of Food Technology, University of Agriculture, Faisalabad, Pakistan, 1997.

Verma, D. K.; Srivastav, P. P. Proximate Composition, Mineral Content and Fatty Acids Analyses of Aromatic and Non-aromatic Indian Rice. *Rice Sci.* **2017,** *24* (1), 21–31.

Verma, D. K.; Mohan, M.; Yadav, V. K.; Asthir, B.; Soni, S. K. Inquisition of Some Physico-chemical Characteristics of Newly Evolved Basmati Rice. *Envron. Ecol.* **2012,** *30* (1), 114–117.

Verma, D. K.; Mohan, M.; Asthir, B. Physicochemical and Cooking Characteristics of Some Promising Basmati Genotypes. *Asian J. Food Agro-Indus.* **2013,** *6* (2), 94–99.

Verma, D. K.; Mohan, M.; Prabhakar, P. K.; Srivastav, P. P. Physico-chemical and Cooking Characteristics of Azad Basmati. *Int. Food Res. J.* **2015,** *22* (4), 1380–1389.

Vlachos, A.; Arvanitoyannis, I. S. A Review of Rice Authenticity/Adulteration Methods and Results. *Crit. Rev. Food Sci. Nutr.* **2008,** *48,* 553–598.

Walter, M.; Marchezan, E.; Avila, L. A. Rice: Composition and Nutritional Characteristics. *Rural Sci.* **2008,** *38* (4), 1184–1192.

Wang, Z.; Gu, Y. J.; Hirasawa, T.; Ookawa, T.; Yanahara, S. Comparison of Caryopsis Development Between Two Rice Varieties with Remarkable Difference in Grain Weights. *Acta Bot. Sin.* **2004,** *46* (6), 698–710. (in Chinese with English Abstract).

Wang, Z.; Li, W. F.; Gu, Y. J.; Chen, G.; Shi, H. Y.; Gao, Y. Z. Development of Rice Endosperm and the Pathway of Nutrients Entering the Endosperm. *Acta Agron. Sin.* **1995a,** *21* (5), 520–527. (in Chinese with English Abstract).

Wang, Z.-Y.; Zheng, F.-Q.; Shen, G.-Z.; Gao, J.-P.; Snusted, D. P.; Li, M.-G.; Zhang, J.-L.; Hong, M.-M. The Amylose Content in Rice Endosperm Is Related to the Post-Transcriptional Regulation of the Waxy Gene. *Plant J.* **1995b,** *7* (4), 613–622.

Wang, K. M.; Wu, J. G.; Li, G.; Zhang, D. P.; Yang, Z. W.; Shi, C. H. Distribution of Phytic Acid and Mineral Elements in Three Indica Rice (*Oryza sativa* L.) Cultivars. *J. Cereal Sci.* **2011,** *54,* 116–121.

Wardlaw, G. M.; Kessel, M. *Prospective in Nutrition,* 5th ed.; McGraw-Hill: Boston, 2002; p 278.

Wasserman, T.; Caklerwood, D. L. Rough Rice Drying. In *Rice Chemistry and Technology;* Houslon, D. F., Ed.; The American Association of Cereal Chemists: St Paul, MI, USA, 1972.

Watts, B. M. Chemical and Physicochemical Studies of Wild Rice. Ph.D. Dissertation, Department of Plant Science, The University of Manitoba, Manitoba, 1980.

Watts, B. M.; Dronzek, B. L. Chemical Composition of Wild Rice Grain. *Can. J. Plant Sci.* **1981,** *61,* 437–446.

Webb, B. D. Criteria of Rice Quality in the US. In *Rice Chemistry and Technology, 2nd Ed.;* Juliano, B. O., Ed.; The American Association of Cereal Chemists: St Paul, MI, USA, 1985; pp 403–442.

Weinberg, Z.; Yan, Y.; Chen, Y.; Finkelman, S.; Ashbell, G.; Navarro, S. The Effect of Moisture Level on High-moisture Maize (*Zea mays* L.) under Hermetic Storage Conditions—*In Vitro* Studies. *J. Stored Prod. Res.* **2008,** 44, 136–144.

Wells, B. R. Rice Research Studies. In *Arkansas Agricultural Experiment Station*; Norman, R. J., Beyrolaty, C. A., Eds.; Fayetteville: Arkansas, USA, 1999; p 522.

WHO (World Health Organization). *Energy and Protein Requirements*. Report of Joint FAO/WHO/UNU Expert Consultation. WHO Tech. Rep. Ser. 724. WHO: Geneva, 1985.

Wichamanee, Y.; Teerarat, I. Production of Germinated Red Jasmine Brown Rice and Its Physicochemical Properties. *Int. Food Res. J.* **2012**, *19* (4), 1649–1654.

Wilfred, O. R.; Consultant, L. *Final Survey Report on the Status of Rice Production, Processing and Marketing in Uganda*. Japan International Cooperation Agency (JICA) in Collaboration with Sasakawa Africa Association Uganda, 2006. http://www.mofa.go.jp/mofaj/gaiko/oda/bunya/agriculture/pdf/uganda_report.pdf (accessed Jan 17, 2017).

Wolnik, K. A.; Fricke, F. L.; Capar, S. G.; Meyer, M. W.; Satzger, R. D.; Bonnin, E.; Gaston, C. Element in Major Raw Agricultural Crops in the United States. 3. Cadmium, Lead, and Eleven Other Elements in Carrots, Fields Corn, Onions, Rice, Spinach, and Tomatoes. *J. Agric. Food Chem.* **1985**, *33*, 807–811.

Xu, Z.; Godber, J. S. Purification and Identification of Components of γ-Oryzanol in Rice Bran Oil. *J. Agric. Food Chem.* **1999**, *47*, 2724–2728.

Xu-run, Y.; Liang, Z.; Fei, X.; Zhong, W. Structural and Histochemical Characterization of Developing Rice Caryopsis. *Rice Sci.* **2014**, *21* (3), 142–149.

Yadav, R. B.; Khatkar, B. S.; Yadav, B. S. Morphological, Physicochemical and Cooking Properties of Some Indian Rice (*Oryza sativa* L.) Cultivars. *J. Agric. Technol.* **2007**, *3* (2), 203–210.

Yeager, S. Fibre—The Ultimate Healer. In *The Doctors Book of Food Remedies*; Rodale Press, Inc.: Emmaus, Pennsylvania, USA, 1998, pp 184–185.

Yi, M.; New, K. T.; Vanavichit, A.; Chai-arree, W.; Toojinda, T. Marker Assisted Backcross Breeding to Improve Cooking Quality Traits in Myanmar Rice Cultivar Manawthukha. *Field Crops Res.* **2009**, *113*, 178–186.

Yodmanee, S.; Karrila, T. T.; Pakdeechanuan, P. Physical, Chemical and Antioxidant Properties of Pigmented Rice Grown in Southern Thailand. *Int. Food Res. J.* **2011**, *18* (3), 901–906.

Yong-Liang, X.; Shan-Bai, X.; Yun-Bo, L.; Si-Ming, Z. Study on Creep Properties of Indica Rice Gel. *J. Food Eng.* **2008**, *86*, 10–16.

Yousaf, M. Study on Some Physico-Chemical Characteristics Affecting Cooking and Eating Qualities of Some Pakistani Rice Varieties. M.Sc. Thesis, Department of Food Technology, University of Agriculture Faisalabad, Pakistan, 1992.

Yousif, N. E. Effect of Fermentation and Dry Cooking Following Fermentation on Protein Fractions and *In-vitro* Protein Digestibility of Sorghum, Corn and Rice. Ph.D. Thesis, University of Khartoum, Sudan, 2000.

Zeng, Y.; Wang, L.; Du, J.; Liu, J.; Yang, S.; Pul, X.; Xiao, F. Elemental Content in Brown Rice by Inductively Coupled Plasma Atomic Emission Spectroscopy Reveals the Evolution of Asian Cultivated Rice. *J. Integr. Plant Biol.* **2009**, *51* (5), 466–475.

Zhou, Z.; Robards, K.; Helliwell, S.; Blanchard, C. Ageing of Stored Rice: Changes in Chemical and Physical Attributes. *J. Cereal Sci.* **2002**, 35, 65–78.

Zubair, M.; Anwar, F.; Ali, S.; Iqbal, T. Proximate Composition and Minerals Profile of Selected Rice (*Oryza sativa* L.) Varieties of Pakistan. *Asian J. Chem.* **2012**, *24* (1), 417–421.

INDEX

Printed and bound by CPI Group (UK) Ltd, Croydon, CR0 4YY

23/10/2024

01777703-0010